Functional Foods

About the Authors

Professor Hari Niwas Mishra (www.fctliitkgp.in) A Professor of Food Technology in the Agricultural & Food Engineering Department Dr. Mishra is former Head and Chairman of the Post Harvest Technology Centre, IIT Kharagpur. He has done his Masters in Food Technology from G B Pant University of Agriculture and Food Technology and PhD in Post Harvest Technology from IIT Kharagpur.

A former President of Association of Food Scientists & Technologists (India), Professor Mishra has many laurels and awards to his credit. To name a few are All India Food Processors' Association (AIFPA) Presidents award for his outstanding contribution to growth and development of food processing industry, Dr JS Pruthi award for new product & process development, Best paper award of the Association of Food Scientists & Technologists (AFST) India, and other professional bodies. He has published more than 285 research papers and popular articles in refereed journals of national and international repute and has 6 Indian patents to his credit. Besides, he is on the editorial & referees boards of reputed journals and has guided more than 195 student research projects including 8 Post Doctoral and 30 Ph D research scholars.

Professor Mishra is an Expert Member, National Mission Steering Group (NMSG), Integrated Child Development Services (ICDS), Ministry of Women & Child Development, Government of India. He is also a member of the Board of Directors of Prasan Solutions, Cochin. He has served on several committees of the Ministry of Food Processing Industries (MoFPI), Department of Biotechnology (DBT), Food Safety and Standards Authority (FSSAI), Bureau of Indian Standards (BIS), Indian Council of Agricultural Research (ICAR), Council of Scientific & Industrial Research (CSIR).

Research interests of Professor Mishra include RTE convenience foods, health foods & nutraceuticals; extension of shelf life of high value perishable foods, processing of horticultural & plantation crop products, high pressure processing, algal food biotechnology, and TQM in food industry.

Dr. Mishra has worked in different capacities on various academic and administrative committees of IIT Kharagpur and many other institutions in the country and has visited several countries abroad. Dr Mishra is member of the Research Advisory Committee / Academic Council / Senate / Board of Studies and other academic committees of several Universities / Institutes.

Professional details including research publications of Professor Mishra can be seen at

http://www.iitkgp.ac.in/fac-profiles/showprofile. php? empcode = aSm XQ & depts_name=AG and also at www.fctliitkgp.in

Dr. Rajesh Kapur is a Doctorate from All India Institute of Medical Sciences; New Delhi. Currently, he is holding the position of Advisor, Govt. of India in the Department of Biotechnology (DBT) and is currently Head of Medical Biotechnology including public health, food and nutrition sciences.

Dr. Kapur has been steering Indian research in the area of food biotechnology for the past two decades and was instrumental in developing many functional and novel foods with health and nutrition claims which have been clinically verified.For promotion of biotech enterprise he was instrumental in setting up of an nutri- food cluster in the State of Punjab, India. He has been the "Managing Director" of Bharat Immunologicals & Biologicals Corporation Ltd (BIBCOL) a Public Sector Unit of the Govt. of India. He is at present the Trustee of the Wellcome Trust/DBT India Alliance set up to build excellence in the Indian biomedical scientific community by supporting future leaders in the field through fellowships by adopting international standards.

Dr. Navneet Singh Deora received his Bachelor's degree (B.E) in Food Processing Technology from AD Patel Institute of Technology (ADIT), Gujarat (Sardar Patel University). He did his Master's degree in Food Technology from Central Food Technological Research Institute (CFTRI), Mysore. He obtained his PhD from Agricultural & Food Engineering Department, Indian Institute of Technology (IIT), Kharagpur, India.

In Past, he was associated with Prasan Solutions Pvt. Limited; working as a research scientist and was instrumental in the development of Analytical Food Research laboratory. His research interest's focus on advancement in Cereal Science and Technology. He has been author or co-author of many research papers, books and patents. He has been in past recipient of scholarship from Department of Science and Technology (DST), Government of India, Department of Biotechnology (DBT), Government of India, International Travel Funding Fellowship, Indian Institute of Technology, Kharagpur.

Dr. Aastha Deswal received her Bachelor's degree from Delhi University in Food Technology. She did her Master's degree in Food Technology from Central Food Technological Research Institute (CFTRI), Mysore. She obtained her PhD from Agricultural & Food Engineering Department, Indian Institute of Technology, Kharagpur, India.

Dr. Aastha Deswal is currently associated with Bright Lifecare Pvt. limited as a Manager, Formulation Development. In Past, she was associated with Prasan Solutions Pvt. Limited as a research scientist and was instrumental in the New Product Formulation in the area of natural food preservatives. Her research interest is focused on functional food product development. She has been author or co-author of many research papers, books and patents.

Dr. Aastha Deswal has been in past recipient of many scholarship and awards. Few of her award includes Best oral presentation award at International Conference on Food Safety, Quality and Nutrition,(ICFSQN), Manchester Food Research Center, United Kingdom in 2012, International Travel Funding Fellowship, Department of Biotechnology (DBT), Govt. of India in 2012, NET (National Eligibility Test) Qualified in 2011, Best dissertation award for "Functional foods & beverages: Recent trends", MSc, CFTRI in 2009, University topper in B.A.Sc.(Hons) Food Tech., Delhi university for the year 2004-07, Best student award in graduation in 2007, All India 3rd rank in Indian Council of Agricultural Research–PG & JRF entrance exam in 2007.

FUNCTIONAL FOODS

Hari Niwas Mishra
Rajesh Kapur
Navneet Singh Deora
Aastha Deswal

NEW INDIA PUBLISHING AGENCY
New Delhi – 110 034

NEW INDIA PUBLISHING AGENCY
101, Vikas Surya Plaza, CU Block, LSC Market
Pitam Pura, New Delhi 110 034, India
Phone: + 91 (11)27 34 17 17 Fax: + 91(11) 27 34 16 16
Email: info@nipabooks.com
Web: www.nipabooks.com

Feedback at feedbacks@nipabooks.com

ISBN: 978-93-83305-98-8

Composed and Designed by NIPA

Preface

The past decade has witnessed a dramatic expansion in research activities worldwide to identify new foods and isolating functional ingredients which has resulted in accumulation of vast array of information and knowledge base on science and technology of functional foods. This book is a comprehensive resource and reference book that provides reliable and accurate information on different aspects of functional foods. It comprises of six sections including (1) market opportunities and recent trends, (2) ingredients and functionality, (3) formulation and processing technology, (4) packaging and storage stability, (5) prevention, disease control and bioavailability, and (6) future prospects of functional foods.

There is a large, growing market for unique functional food products that provide proven health benefits to consumers; this book will play an important role in the development of such products. A critical issue in the development of functional foods is health aspects and its role in disease prevention and control. Recent advances in functional food technology have allowed scientists to objectively investigate some of the most ancient concepts in the application of foods in the prevention and/or the treatment of diseases. In Section 5, a variety of examples are discussed indicating the role and action of functional ingredients in preventing disease. The book also addresses the key issues of processing and its effects on the bioavailability of bioactives.

The book presents an updated information on the regulatory status of functional foods in different countries, which is valuable in commercialization of these products. Also, core discussion on the reliable and economical scale up of laboratory-based extraction and purification techniques for different functional ingredients is presented in a detailed manner in the book.

The book is targeted for researchers and scientists in food technology and related fields. It is also intended for use by practitioners in functional food companies and food processing centres and will be of immense benefits to the health conscious individuals. We firmly believe the scientific community, producers and consumers of functional foods will be immensely benefited from this book, which may enlighten the public throughout the globe. There will be a greater appreciation of enjoyment of a safe and plentiful food supply for public health.

Information taken from the published literature has been duly cited; any missing is inadvertent. Comments and suggestions of the readers of the book are welcome for further improvements in future editions.

Authors

Contents

Section IV: Functional Food Products: Packaging and Storage Stability

Section V: Functional Food Products: Prevention, Disease Control and Bioavailability

Section VI: Future of Functional Foods

Section I: Functional Foods : Scope, Market and Opportunities

1

Functional Foods - Scope, Market Opportunities and Recent Trends

Introduction

The tenet **"Let food be thy medicine and medicine be thy food,"** said by Hippocrates nearly 2,500 years ago, is receiving renewed interest. In particular, there has been an explosion of consumer interest in the health enhancing role of specific foods or physiologically active food components called as functional foods [1]. Clearly, all foods are functional, as they provide taste, aroma, or nutritive value. Within the last decade, however, the term functional as it applies to food has adopted a different meaning - that of providing an additional physiological benefit beyond that of meeting basic nutritional needs.

The term functional foods was first introduced in Japan in the mid 1980s and refers to processed foods containing ingredients that aid specific bodily functions in addition to being nutritious. To date, Japan is the only country that has formulated a specific regulatory approval process for functional foods. Known as foods for specified health use (FOSHU), these foods are eligible to bear a seal of approval from the Japanese Ministry of Health and Welfare [2]. Currently, 100 products are licensed as FOSHU foods in Japan. In the United States, the functional foods category is not recognized legally. Irrespective of this, many organizations have proposed definitions for this new and emerging area of the food and nutrition sciences. The potential of functional foods to mitigate disease, promote health, and reduce health care costs is very significant. In the 21st century, there will be more dieticians in comparison to doctors who treat diseases. The responsibilities of government will be then mainly providing correct information on health and nutrition so as to help the people to differentiate between facts and fiction in matters of food and health.

High public awareness about the vital link between food and health has been increased during the last one decade or so.

New range of health foods under the names, viz. functional foods, medicinal foods and nutraceuticals are now becoming very popular in the developed countries. Indeed, we are witnessing a new paradigm shift in our approach to health and medicine. It is something like that we are moving from a crisis intervention based health care model to preventive health care model in which the food we eat will be given a special attention. Selection of such food articles particularly the vegetables and fruits in preventing diseases and discomforts or that promotes good health is termed as functional foods or nutraceuticals. In India, it was a tradition to have specific dietary regimes in different seasons that varied from region to region. The Indian tradition, particularly the traditional system of medicines like Ayurveda, gives a detailed account on the type of food to be taken by people in different seasons and in different agro-climatic conditions. According to a WHO estimate, about 80% of the world population relies on traditional systems of medicines for primary health care, where plants form the dominant component over other natural resources. The development of these traditional systems of medicines with the perspectives of safety, efficacy and quality will help not only to preserve this traditional heritage but also to rationalize the use of natural products in the health care. The plant species mentioned in the ancient texts of these Ayurveda and other Indian systems of medicines are being explored with the modern scientific approaches for better leads in the health care. Various medicines have been developed for treating diseases with the help of Ayurveda. Certain spices and herbs mentioned in Ayurveda such as garlic and turmeric have been shown to be helpful to people suffering from cardiovascular diseases [2, 3]. So, food can act as a preventive measure when it comes to diseases.

The main factors that led to the endorsement of the "functional food" concept are increasing life-expectancy in developed countries (often entailing longer hospital care), high health care costs, advances in food and ingredients technology, the need for publicly funded research institutions to publicize their findings, and the greater media coverage given to these findings and to health issues [4]. Many consumers opt for healthier versions of their favourite products, being unable or unwilling to adapt their diet to the suggested dietary guidelines. Developing economic powers (especially in the Asia-Pacific region) are seeking ways to ensure a vigorous population and a productive workforce. In addition, it is likely that current progress in technology for diagnosing genetic predisposition to disease, detecting the early stages of disease, and monitoring the effectiveness of therapy will urge people to orient toward functional foods [5]. The main health concerns influencing functional foods purchase are summarized in Table 1.1.

Table 1.1: Top health concerns influencing functional foods purchase [6]

Rank	USA	UK	Germany	France
1.	Weight loss	Gives energy system	Boosts immune	Gives energy
2.	Reduces cholesterol bones	Promotes healthy bones	Promotes healthy	Lowers cholesterol
3.	Promotes healthy bones	Promotes healthy teeth	Promotes healthy teeth	Increases disease resistance
4.	Gives energy	Reduces risk of breast cancer	Gives energy system	Boosts immune
5.	Prevents digestive problems	Lowers cholesterol	Promotes healthy gut	Prevents constipation

Food industry is expecting that consumers, after years of being bombarded with negative health messages to reduce the intake of fat, cholesterol, and sodium; are ready to welcome foods that emphasize the positive benefits of newly added ingredients, even if the scientific community has not yet clearly concluded their real benefits. Despite the continuously growing popularity of functional foods, scientists have identified few specific substances, or combinations of substances, that proved to reduce the risk of disease. Moreover, the precise definition of functional foods is not universal—it varies from country to country, or even from company to company, and is still evolving. In addition, the category of functional foods comprises nutraceuticals, designer foods, pharma-foods, and a plethora of variously named similar products. Therefore, it is doubtful whether the consumer will actually manage to benefit from this trend. Some products might lead to significant public health advances, and it makes sense to foster their development, marketing, and consumption. However, on the other hand, the absence of convincing scientific research, pent-up consumer demand and inadequate regulatory controls may create a situation in which the market place is flooded with products of dubious benefit and false, or in the best case, exaggerated claims. In brief, if governments do not require functional ingredients to be proven effective (and safe) before they become part of the food supply, if claims are not required to be adequately substantiated, and if functional ingredients are simply added to foods high in fat, cholesterol, sodium, or sugar, then dubious functional foods may merely amount to 21st century quackery.

Defining Functional Foods and Nutraceuticals

Functional foods

Although the term "functional food" has already been defined several times, so far there is no unitary accepted definition for this group of food. Each country has its own definition of functional foods. As stated by NoSeong and Jukes (2001), "Overlap and discrepancies exist for EU, USA, Japan, WHO, and Codex Alimentarius definitions and terms for functional, dietetic, medical, fortified or enriched foods, dietary supplements, health foods, nutraceuticals, and novel, designer, and pharma-foods" [7].

Japan is the birthplace of the term 'functional food'. Moreover, that country has been at the forefront of the development of functional foods since the early 1980s when systematic and large-scale research programmes were launched and funded by the Japanese government on systematic analysis and development of food functions, analysis of physiological regulation of function by food, and analysis of functional foods and molecular design. As a result of a long decision making process to establish a category of foods for potential enhancing benefits as part of a national effort to reduce the escalating cost of health care, the concept of foods for specified health use (FOSHU) was established in 1991 [8, 9]. These foods, which are intended to be used to improve people's health and for which specific health effects (claims) are allowed to be displayed, are included as one of the categories of foods described in the Nutrition Improvement Law as foods for special dietary use. According to the Japanese Ministry of Health and Welfare, FOSHU are:

- Foods that are expected to have a specific health effect due to relevant constituents, or foods from which allergens have been removed, and
- Foods where the effect of such an addition or removal has been scientifically evaluated and permission has been granted to make claims regarding the specific beneficial effects on health expected from their consumption.

Foods identified as FOSHU are required to provide evidence that the final food product, but not isolated individual component(s), is likely to exert a health or physiological effect when consumed as part of an ordinary diet.

The European Commission's Concerted Action on Functional Food Science in Europe (FuFoSE), coordinated by International Life Science Institute (ILSI) Europe defined functional food as follows: "a food product can only be considered functional if together with the basic nutritional impact it has beneficial effects on one or more functions of the human organism thus either improving the general and physical conditions or/and decreasing the risk of the evolution of diseases. The amount of intake and form of the functional food should be as it is normally expected for dietary purposes. Therefore, it could not be in the form of pill or capsule just as normal food form" [10]. On the contrary to this latter statement, since 2001 FOSHU products in Japan can also take the form of capsules and tablets, although a great majority of products are still in more conventional forms [11]. Several organizations have given different definitions of functional foods. The International Life Sciences Institute of North America has a similar definition: "foods that, by virtue of physiologically active food components, provide health benefits beyond basic nutrition" [12].

In the United States (US), the Food and Drug Administration (FDA) does not provide a legal definition for functional food. However, a number of working definitions have been developed by different organizations including the American Dietetic Association (ADA), the International Food Information Council (IFIC) and the Institute of Food Technologies (IFT). According to these organisations, a functional food is any food that moves beyond necessity (basic nutrition) to provide additional health benefits that may reduce disease risk and/or promote optimal health [13]. The Institute of Food Technologies defines functional food as, "foods and food components that provide a health benefit beyond basic nutrition (for the intended population). These substances provide essential nutrients often beyond quantities necessary for normal maintenance, growth, and development, and/or other biologically active components that impart health benefits or desirable physiological effects". The FDA does not have a formal definition for novel food. These foods are often referred to as new and emerging food/biotechnology products, and a premarket notification with the FDA is usually recommended by the FDA [14].

The Food and Nutrition Board (1994) limits functional foods to "those in which the concentrations of one or more food constituents have been manipulated to enhance their contributions to a healthful diet." The American Dietetic Association defines functional foods as inclusive of the entire range of whole foods and fortified, enriched, or enhanced foods (ADA, 1999). ADA further states that "the term 'functional' implies that the food has some identified value leading to health benefits, including reduced risk for disease, for the person consuming it" [15].

In Russia, food products with health benefits (functional food, and NHPs) are known as biologically active food supplements (BAFS). The Russian Federation defines BAFS as nutritive substances and minor food components used to ameliorate deficiencies, decrease risk of debilitating diseases, and improve the quality of life [16]. The food products and its components should be natural or identical to natural. They should be consumed along with food and incorporated into food. BAFS are not regarded as medicines and are not meant to treat or diagnose diseases. BAFS in Russia are equivalent to nutraceuticals in Canada, however, in Russia the definition used for NHPs also applies to functional foods and novel foods.

In India, functional foods are defined as food that encompasses potentially healthful products, including any modified food or food ingredient that may provide a health benefit beyond that of the traditional nutrients it contains [17]. Novel food on the other hand, refers to an article of food for which safety standards have not been specified but are not unsafe. These products should not contain any food or ingredient that is prohibited under the Food Safety Standards Act 2006 (Food Safety and Standards Act 2006, Chapter IV, article 22). India has no specified definition for natural health products.

The Food and Agriculture Organization (FAO) of the United Nations, in 2007, provided a working definition for functional food: "foods which are intended to be consumed as part of the normal diet and that, contain biologically active components which offer the potential of enhanced health or reduced risk of disease". The FAO does not have formal definitions for natural health products or novel foods [18].

While there is no universally accepted definition for functional food, some similarities in the terminology used across countries are apparent. The differences in the definitions are often cultural and regulatory in nature. All definitions require that the food provide benefits beyond that of conventional food, and that the food not be perceived as a drug, but should have some physiological functions. The absence of a uniform terminology for these food products/ingredients reflects differences in the underlying approaches to food regulation across different countries, as well perhaps a lack of scientific consensus on the presence, efficacy and measurement of the health benefits derived from these products. Differences in terminology raise costs and create uncertainty for firms marketing across international borders and as such can act as a brake on the expansion of international trade in these products.

On a whole, a functional food can be a natural whole food, a food to which a component has been added, or a food from which a component has been removed by technological or biotechnological means. It can also be a food, in which the nature of one or more components has been modified, or a food in which the bioavailability of one or more components has been modified, or any combination of these possibilities. A functional food may be targeted at the whole population or for particular groups, which may be defined, for example, by age or by genetic constitution. Table 1.2 presents various

Table 1.2: Terminology and acceptance in several jurisdictions

Country	Functional foods	Dietary supplements	Nutraceuticals	Medical foods	Dietary use	Health food
European Union	At EU level, currently no legal definition. Some member states have working definitions used by industry.	Products resembling those in U.S. are not subject to EU directive.	This term has met with limited success in Europe and appears to be out of favour as a descriptive term.	N.A.	N.A.	"Health Food" is a marketing, not a legal term, normally denoting product sold in specialty stores.
United States	No legal status or general acceptance in U.S. but accepted definition in dietetic profession.	Formal definition exists under the dietary supplements health education Act.	Accepted definitions in nutrition/science community but not embodied in law or regulation to date.	Regulated by FDA Office of Special Nutrition on a case by case basis as enteral foods or foods for sick infants.	N.A.	N.A.
Canada	No accepted regulated definition. All ingestible products sold either as a food or drug as defined in Act.	A food sold or represented as a supplement to a diet that may be inadequate in energy and essential nutrients.	Commonly used in nutrition and science community.	Food and drug regulations may be revised to define and regulate "foods for special medical purposes."	Eleven defined categories of Foods for Special Dietary Uses.	No separate definition exists for "health food" but it is a widely used term.
Japan	Functional foods recognized as Foods for Specified Health Use in 1991 (FOSHU), replacing "functional foods".	N.A.	N.A.	N.A.	N.A.	"Health food" is a widely accepted term used by consumers and industry in Japan.

N.A.: Not available

terminologies related to functional foods, dietary supplements, medical foods, health foods, etc. in different countries. A successful functional food or beverage must contain at least one active component with a proven benefit in a delivery system that appeals to consumers in its form, taste and price.

Nutraceuticals

The term nutraceutical is a hybrid or contraction of nutrition and pharmaceutical. More than a decade has passed since the term nutraceuticals was coined by Dr. Stephen DeFelice, founder of the Foundation for Innovation in Medicine. A nutraceutical is a product isolated from foods but sold in powders, pills and other medicinal forms that are not generally associated with food and is demonstrated to have physiological benefits or provide protection against chronic diseases [19]. Several nutraceuticals are known and can be categorized as flavonoids, antioxidants such as gamma-linolenic acid, beta carotene and anthocyanins, etc. With the dietary supplement health and education act (DSHEA) several other compounds were added to the list of supplements originally mentioned in FDA notifications. Newer items such as ginseng, garlic oil, etc. are being categorized in nutraceuticals now.

The Nutrition Business Journal states that it uses the term nutraceutical for anything that is consumed primarily or particularly for health reasons. Based on that definition, a functional food would be a kind of nutraceutical. On the other hand, Health Canada states that nutraceuticals are products that are "prepared from foods, but sold in the form of pills or powders or in other medicinal forms not usually associated with foods. A nutraceutical is demonstrated to have a physiological benefit or provide protection against chronic disease." Based on this definition and how functional foods are characterized, as noted previously, nutraceuticals would be distinct from functional foods [20].

Need of Functional Foods

Consumer awareness about relation of diet and health

Consumer interest in the relationship between diet and health has increased substantially now-a-days. There is much greater recognition today that people can help themselves and their families to reduce the risk of illness and disease and to maintain their state of health and well being through a healthy lifestyle, including the diet. Ongoing support for the important role of foods such as fruits and vegetables and wholegrain cereals in disease prevention and the latest research on dietary antioxidants and combinations of protective substances in plants has helped to provide the impetus for further developments in the functional food market. There is already a wide range of foods available to today's consumer but now the impetus is to identify those functional foods that have the potential to improve health and well-being, reduce the risk from, or delay the onset of, major diseases such as cardiovascular disease (CVD), cancer and osteoporosis. Combined with a healthy lifestyle, functional foods can make a positive contribution to health and well being.

Now-a-days, obesity has become a major problem not only in the developed countries but also in developing countries. So, consumers are trying everything that they can do to overcome this problem and functional foods are very promising in this direction. It is seen experimentally that increase in the amount of intake of calcium in diet results in weight reduction which is the main driving force behind the growing market for the calcium rich or calcium fortified functional food products, for example, calcium-fortified juices [21]. There are various other experiments done to show the benefits of functional foods. For example, the relation of calcium and osteoporosis, potassium and reduced risk of blood pressure and stroke. Hence, these foods are very promising in improving the health.

Several studies have indicated that consumption of oat bran lowers blood cholesterol and this effect has been attributed specifically to oat bran's soluble fiber (β-glucan). Concentrated β-glucan from oat is now commercially available as a dietary supplement. Oat β-glucan is a natural polymer comprised of individual glucose molecules that are linked together by α series of α-(1–3) and α-(1–4) linkages. [22-24]. Also, many studies shows that consumption of *Trigonella foenum graecum* (fenugreek) seed powder have potential antidiabetic effects. Fenugreek seeds are high in soluble fibre, which helps lower blood sugar by slowing down digestion and absorption of carbohydrates. This suggests they may be effective in treating people with diabetes [25-27].

These functional food products result from technological innovations at the processing level, such as cholesterol lowering spreads, xylitol-sweetened chewing gum, and dairy products fermented with specific lactic acid bacteria; technological innovation at the production level such as genetically modified crop seeds (e.g., golden rice) and planting materials derived through conventional breeding (e.g., orange fleshed sweet potato); or crops that naturally contain components that have a physiological function (e.g., soy bean and oats, which are good for cardio-vascular health).

Changes in lifestyle, increase of chronic diseases and willingness to pay

Increased affluence and urbanisation is also linked to a lifestyle where our daily routine requires less physical activity and there is greater access to foods with higher energy densities. We now have a real challenge to balance energy intake and expenditure as most of us are becoming overweight. For much of its history, nutrition science has focused on the role of essential nutrients in preventing deficiencies. However, there is now a need to ensure unaltered levels of key nutrients or functional components in the context of declining energy expenditures. Further, the prevalence of chronic non-communicable diseases such as cardio vascular disease (CVD), high blood pressure and type 2 diabetes has increased, leading to concerns about escalating costs of healthcare and reduced quality of life. Modern lifestyles and longer life are also linked to various mental health problems such as depression, poor concentration and loss of memory. It is seen that consumers are willing to pay premium prices for functional foods. Functional and novel foods with base products that have already been accepted by consumers for their nutritional content or healthiness are most likely to be accepted and stimulate a higher willingness to pay. The reward (health benefits) associated with consumption of functional food is a major determinant of willingness to pay. In a study, Markosyan et al. (2009) measured US consumers' responses to a specific functional food (apples with enriched antioxidant coatings). Antioxidants may protect cells from damage from free radicals which are believed to lead to cancer. A contingent valuation (CV) survey instrument was administered in face-to-face interviews with 730 US consumers [28].

The results indicated that consumers value and are willing to pay for the health attributes of functional foods. Consumers were responsive towards health claims, underlining the importance of credible information dissemination. Anxiety over the utilization of a new technology that was considered unnatural and risky was the primary reason for rejection by some respondents of apples with enriched antioxidant coatings [28]. In another study, UK consumers were examined for willingness to pay for enhanced foods. Attitudes toward a number of products were examined, including strawberries and lettuce with higher antioxidant levels, and lamb chops with higher levels of unsaturated fatty acids. The results indicate a viable market for functional foods. Highly skilled professionals in urban areas, women, and smaller households on lower incomes were more likely to be willing to pay for enhanced products [29]. Many other studies have reported that consumers are willing to pay more for foods with additional health benefits [30-32].

Classification of Functional Foods

The number of purported nutraceutical substances is in the hundreds, and some of the more recognizable substances include isoflavones, tocotrienols, allyl sulphur compounds, fibre, and carotenoids. In light of a long and growing list of nutraceutical substances, organization systems are needed to allow for easier understanding and application. This is particularly true for academic instruction, as well as product formulation by food companies.

Depending upon one's interest and/or background, the appropriate organizational scheme for nutraceuticals can vary. For example, cardiologists may be most interested in those nutraceutical substances that are associated with reducing the risk factors of heart disease. Specifically, their interest may lie in substances purported to positively influence hypertension and hypercholesterolemia and to reduce free radical or platelet-dependent thrombotic activity. Nutraceutical factors such as omega-3 fatty acids, phytosterols, quercetin, and grape flavonoids would be of particular interest. Meanwhile, oncologists may be more interested in those substances that target anti-carcinogenic activities. These substances may be associated with augmentations of microsomal detoxification systems and antioxidant defences, or they may slow the progression of existing cancer. Thus, their interest may lie in both chemo preventive or potential adjunctive therapy. On the other hand, the nutraceutical interest of food scientists working on the development of a functional food product will not only include physiological properties, but also stability and sensory properties, as well as issues of cost efficiency. To demonstrate this point, the anti-carcinogenic triterpene-limonin is lipid-soluble and intensely bitter, somewhat limiting its commercial use as a functional food ingredient [33]. However, the glucoside derivative of limonin, which shares some of the ant carcinogenic activity of limonin, is water soluble and virtually tasteless, thereby enhancing its potential use as an ingredient [34].

Classifying functional food on the basis of type of food

There are several ways of classifying functional foods. One way is according to the type of food (Table 1.3).

Table 1.3: Different categories of functional foods [6]

Basic foods	Processsed foods with added ingredients	Foods enhanced to have more functional component	Isolated, purified preparations of active food ingredients
Carrots (containing the natural level of the antioxidant β-carotene)	Calcium enriched fruit juice	Tomatoes with higher levels of lycopene (biotechnology intervention) Eggs with ω-3	Isoflavones from soy ω-3 from fish oils (DHA and ALA)

Another means of classifying nutraceuticals is by their mechanism of action as shown in Table 1.4. This system groups nutraceutical factors together, regardless of food source, based upon their proven or purported physiological properties.

Table 1.4: Classification on the basis of mechanism of action [20]

Anticancer	Positive influence on blood lipid	Antioxidant activity	Anti inflammatory	Osteogenetic or bone protective
Capsaicin	β-glucan	CLA	Linolenic acid	CLA
Genestein	Tocotrienols	Ascorbic acid	EPA	Soy protein
Daidzein	Resveratrol	Carotene	DHA	Genestein
Tocotrienol	Tannins	Tocopherols	GLA	Daidzein
CLA	Quercetin	Tocotrienols	gamma-linolenic acid	FOS
Lactobacillus acidophilus	Pectin	Ellagic acid	Capsaicin	Calcium
Sphingolipids	Guar	Glutathione Gingerol	Quercetin	Casein phosphopeptides
Liminene	Saponins Polyphenols	Hydroxytyrosol	Curcumin	Inulin
Diallyl sulphide	Sitosterol	Luteolin		Fructo oligosaccharides
Ajoene, Equol	MUFA	Lycopene		
Tocopherol	ω-3 Fatty acids	Tannins		
Glycyrrhizin		Chlorogenic acid		

Among the classes are included antioxidants, antibacterial, antihypertensive, antihypercholesterolemic, antiaggregate, antiinflammatory, anticarcinogenic, osteoprotective, and so on. This model would be helpful to an individual who is genetically predisposed to a particular medical condition or to scientists trying to develop powerful functional foods for just such a person. This developer might consider the ingredients listed in several categories to develop a product that would reduce blood pressure, LDL cholesterol level, and inflammation.

Classifying nutraceutical factors based on chemical nature

Another method of grouping nutraceuticals is based upon their chemical nature. This approach allows nutraceuticals to be categorized under molecular/ elemental groups. This preliminary model includes several large groups, which then provide a basis for sub-classification or sub-groups, and so on. One way to group nutraceuticals grossly is as shown in Table 1.5.

Table 1.5: Classification of nutraceutical factors based on chemical nature [20]

Isoprenoid	Phenolic compounds	Fatty acids	Carbohydrate	Protein	Minerals	Microbes
Carotenoids	Coumarinns	ω-3PUFA	Ascorbic acid	Amino acids	Ca	Probiotic
Saponins	Tannins	CLA	Oligo - saccharides	Allyl-S comps.	Se	Prebiotic
Tocotrienols	Lignins	MUFA	Non-starch Poly-saccharides	Capsaicinoid	K	
Tocopherols	Flavonols	Sphingolipids		Isothiocyanates	Cu	
Simple terpenes	Flavonones	Lecithin		Indoles	Zn	
	Isoflavones			Folate, Choline		
	Anthro- cyanins			Choline		

Isoprenoid derivatives (terpenoids)

Isoprenoids and terpenoids are terms used to refer to the same class of molecules. These substances are one of the largest groups of plant secondary metabolites and are basis of many plant-derived nutraceuticals. Under this large umbrella are many popular nutraceutical families such as carotenoids, tocopherols, tocotrienols, and saponins. This group is also referred to as isoprenoid derivatives because the principal building block molecule is isoprene. The basic chemical structure of isoprene molecule is shown in Fig. 1.1.

Isoprene itself is synthesized from acetyl coenzyme A (CoA), in the well-researched mevalonic acid pathway, and the glycolysis associated molecules pyruvate and 3-phosphoglycerate in a lesser understood metabolic pathway

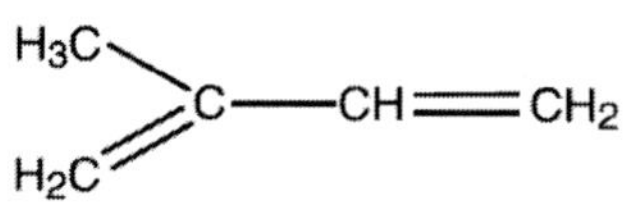

Isoprene

Fig. 1.1: Isoprene structure

[35]. This series of reactions result in the formation of monoterpenes, diterpenes, sesquiterpenes, triterpenes, etc. Most plants contain so called essential oils, which contain a mixture of volatile monoterpenes and sesquiterpenes. The chemical structure of monoterpenes is shown in Fig. 1.2. Limonene is found in the essential oils of citrus peels, whereas menthol is the chief monoterpene in peppermint essential oil. Two potentially nutraceutical diterpenes in coffee beans are kahweol and cafestol [36, 37]. Both of these diterpenes contain a furan ring. The furan-ring component might be very important in yielding some of the potential antineoplastic activity of these compounds.

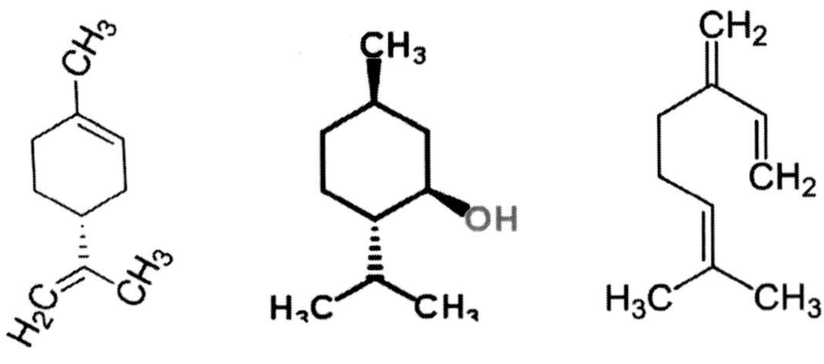

Limonene Menthol Myrcene

Fig. 1.2: Structure of some monoterpenes

Several triterpenes have been reported to have nutraceutical properties. These compounds include plant sterols; however, some of these structures may have been modified to contain fewer than 30 carbons. One of the most recognizable triterpene families is the limonoids. These triterpenes are found in citrus fruit and impart most of their bitter flavour. Limonin and nomilin are two triterpenoids that may have nutraceutical application, limonin more so than nomilin. Both of these molecules contain a furan component. In citrus fruit limonoids can also be found with an attached glucose, forming a limonoids glycoside [39]. Saponins are also triterpene derivatives, and their nutraceutical potential is attracting interest [40, 41].

Phenolic compounds

Like the terpenoids, phenolic compounds are also considered secondary metabolites. The base for this very diverse family of molecules is a phenol structure. From this structure, larger and interesting molecules are formed such as anthocyanins, coumarins, phenylpropamides flavonoids, tannins, and lignin. Phenolic compounds perform a variety of functions for plants including defending against herbivores and pathogens, absorbing light, attracting pollinators, reducing the growth of competitive plants, and promoting symbiotic relationships with nitrogen-fixing bacteria [42]. The flavonoids are one of the largest classes of phenolic compounds in plants. The basic carbon structure of flavonoids contains 15 carbons and is endowed with 2 aromatic rings linked by a 3- carbon bridge as shown in Fig. 1.3 [43, 44].

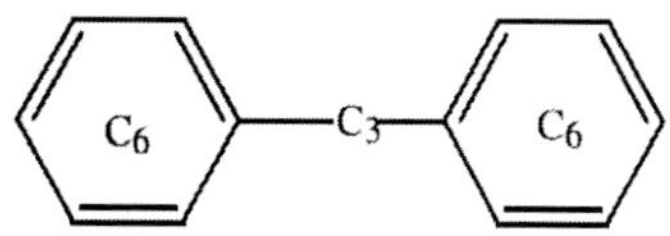

Fig. 1.3: Basic flavonoid structure

Fruits and vegetables are the main dietary sources of flavonoids for humans, along with tea and wine. Many flavonoids are shown to have antioxidative activity, free-radical scavenging capacity, coronary heart disease prevention, and anticancer activity, while some flavonoids exhibit potential for anti–human immunodeficiency virus functions [45]. Most of the literature on sources of plant phenolics focuses on fruits and vegetables; phenolic compounds are found in cereals and grains as well. Infact some phenolics like avenanthramides are present uniquely in oats [45-48].

There are a couple of biosynthetic pathways that form phenolic compounds. The predominant pathways are the shikimic acid pathway and the malonic acid pathway [35]. The malonic acid pathway begins with acetyl CoA. Meanwhile, in the shikimic pathway, simple carbohydrate intermediates of glycolysis and the pentose phosphate pathway (PPP) are used to form the aromatic amino acids phenylalanine and tyrosine. A third aromatic amino acid, tryptophan, is also a derivative of this pathway. As animals do not possess the shikimic acid pathway, these aromatic amino acids are diet essentials. These amino acids are considered primary metabolites or products. Thus, it is the reactions beyond the formation of these amino acids that are of greater importance to the production of secondary metabolites. Once formed, phenylalanine can be used to generate flavonoids.

Plant phenols are mostly products of the phenyl propanoid pathway and comprise a large variety of compounds viz cinnamic acids, benzoic acids, flavonoids, proanthocyanidins, stilbenes, coumarins, lignans and lignins. They are strong antioxidants and might prevent oxidative damage to biomolecules such as DNA, lipids and proteins which play a role in chronic diseases such as cancer and cardiovascular disease. Plant phenols may interfere with all stages of the cancer process, potentially resulting in a reduction of cancer risk [49]. The structure of some phenolic compounds is shown in Fig. 1.4.

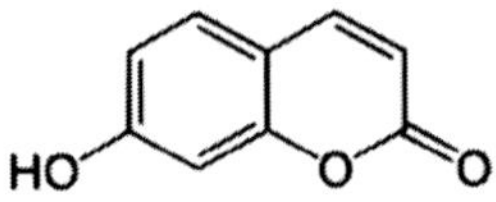

Umbelliferone

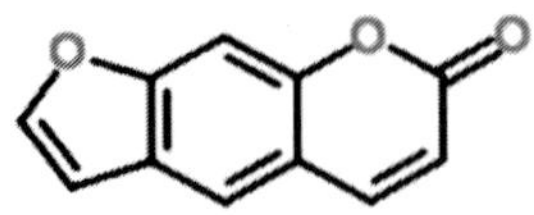

Psoralens

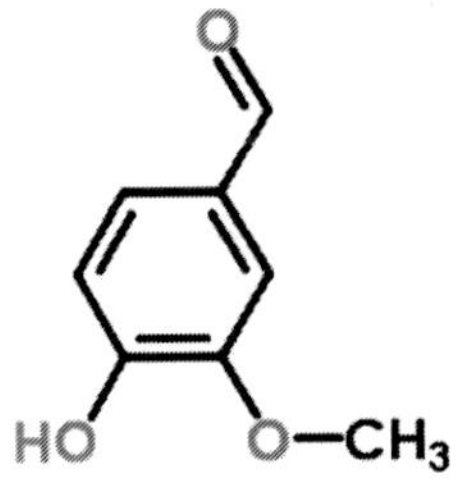

Vanillin

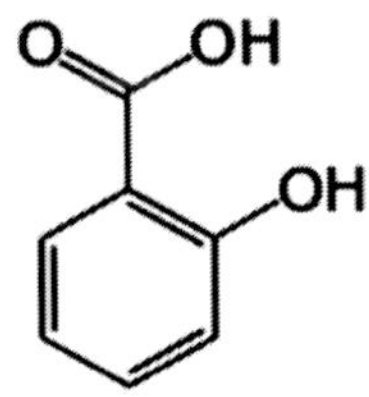

salicylic acid

Fig.1.4: Some phenolic compounds

Whereas the simpler phenolic compounds and lignin-building blocks result from the shikimic pathway and are phenyl alanine derivatives, formation of the flavonoids requires some assistance from both the shikimic pathway and the malonic acid pathway. The majority of naturally occurring flavonoids are actually glycosides, meaning a sugar moiety is attached. The attachment of hydroxyl groups and sugars will increase the hydrophilic properties of the flavonoid molecule, while attachment of methyl esters or modified isopentyl units will increase the lipophilic character.

Carbohydrates

The glucose derivative ascorbic acid (vitamin C) is perhaps one of the most recognizable nutraceutical substances and is a very popular supplement. Ascorbic acid functions as a nutraceutical compound, primarily as an antioxidant. Meanwhile, plants produce some oligosaccharides that appear to function as prebiotics substances. Prebiotics are nondigestible food ingredients that benefit the host by selectively stimulating the growth or activity of one or a limited number of bacteria in the colon [50]. Several plant polysaccharide families are not readily available energy sources for humans as they are resistant to secreted digestive enzymes. These polysaccharides are grouped together along with the phenolic polymer compound lignin to form one of the most recognizable nutraceutical families – fibres [51]. By and large the role of fibres is structural for plants. For example, cellulose and hemicellulose are major structural polysaccharides found within plant cell walls. Beyond providing structural characteristics to plant tissue, another interesting role of certain fibers is in tissue repair after trauma, somewhat analogous to scar tissue in animals. The non-starch polysaccharides can be divided into homogeneous and heterogeneous polysaccharides, as well as into soluble and insoluble substances. Cellulose is a homogeneous non-starch polysaccharide as it consists of repeating units of glucose monomers. These polysaccharides are found in plant cell walls as microfibril bundles. Hemicellulose is found in association with cellulose within plant-cell walls and is composed of a mixture of both straight-chain and highly branched polysaccharides containing pentoses, hexoses, and uronic acids. Pentoses such as xylans, mannans, galactans, and arabicans are found in relatively higher abundance. Hemicelluloses are somewhat different from cellulose in that they are not limited to glucose, and they are also vulnerable to hydrolysis by bacterial degradation. Another homopolysaccharide is pectin where the repeating subunits are largely methylgalacturonic acid units. It is a jelly-like material that acts as the cellular cement in plants. The carboxyl groups become methylated in a seemingly random manner as fruit ripen. Chemically related to pectin is chitin. Chitin is not a plant polysaccharide but is found within the animal kingdom, although not necessarily in humans. It is a homo-polymer of N-acetyl-glucosamine found in shells or exoskeletons of insects and crustacean [52, 53]. Chitin has recently surfaced as a dietary supplement for weight loss.

Another family of polysaccharides that is worthy of discussion is glycosaminoglycans (GAGs). While these compounds are found in animal connective tissue, they are important to this discussion as they are potential components of functional foods. At present, GAG and chondroitin sulphate are popular nutrition supplements being used by individuals recovering from joint injuries and suffering joint inflammatory disorders. Glycosaminoglycans are often referred to as mucopolysaccharides [54, 55]. They are characterized by their content of amino sugars and uronic acids, which occur in combination with proteins in secretions and structures. These polysaccharides are responsible for the viscosity of body-mucus secretions and are components of extracellular amorphous ground substances surrounding collagen and elastin fibers, and cells of connective tissues and bone. Some examples of glycosaminoglycans are hyaluronic acid and chondroitin sulphate. Hyaluronic acid is a component of the ground substance found in most connective tissue including the synovial fluid of joints. It is a jelly-like substance composed of repeating disaccharides of β-glucuronic acid and N-acetyl-D-glucosamine. Hyaluronic acid can contain several thousand disaccharide residues and is unique from the other glycosaminoglycans in that it will not interact with proteins to form proteoglycans [52, 56]. Chondroitin sulphate is composed of β-glucuronic acid and N-acetylgalactosamine sulphate. This molecule has a relatively high viscosity and ability to bind water. It is the major organic component of the ground substance of cartilage and bone. Both of these polysaccharides have β (1-3) linkage between uronic acid and acetylated amino sugars but are linked by β (1-4) covalent bonds to other polysaccharide units. Unlike hyaluronic acid, chondroitin sulphate will bind to proteins to form proteoglycans [57, 58].

Fatty acids and structural lipids

At present, there are several fatty acids and their derivatives that have received the interests of researchers for their functional potential. These include the ω-3 Polyunsaturated fatty acids (PUFA) found in higher concentrations in plants, fish, and other marine animals, and conjugated linoleic acid (CLA) produced by bacteria in the rumen of grazing animals such as cattle. The formation of CLA probably serves to help control the vitality of the released bacterial population in the rumen, whereas plants and fishes use ω-3 fatty acids for their properties in membranes. Some plants also use ω-3 PUFA in a second messenger system to form jasmonic acid when plant tissue is under attack (i.e., by insect feeding). Fish oil ω-3 PUFAs comprise EPA, DPA, and DHA. Numerous studies have shown beneficial effects in coronary heart disease (CHD), in particular a reduction in the risk of fatal CHD, and in inflammatory and immunological diseases [59].

The CLA precursor, linoleic acid, and ω-3 PUFA are produced largely in plants. In processes very similar to those found in humans, plants construct fatty acids using two-carbon units derived from acetyl CoA. Some of the major fatty acids produced by plants include palmitic acid (16:0), oleic acid (18:1 ω-9), linoleic acid (18:2 ω-6), and linolenic acid (18:3 ω-3) [60]. Grazing animals ingest linoleic acid which is then metabolized to CLA by rumen bacteria. Herbivorous fish also ingest these fatty acids when they consume algae and other seaweeds and phytoplankton. Carnivorous fish and marine animals then acquire these PUFA and derivatives from the tissue of other fish and marine life. Fish will further metabolize the PUFA to produce longer and more unsaturated fatty acids such as DHA and EPA [61].

Two other types of lipids in food products are structured lipids and diglycerides. Structured lipids are triglycerides that have undergone hydrolysis and re-esterification under conditions that resulted in triglycerides with new combinations of fatty acids [62]. For example, a mixture of medium chain triglycerides and fish oil taken through this process results in triglycerides that can contain medium-chain fatty acids, EPA and DHA. The basic process results in the free fatty acids being randomly re-esterified to the glycerol backbones. However, the process can be manipulated to place specific fatty acids in preferred positions on the glycerol molecule. This option is quite expensive and thus has not been adopted by the food industry to any degree. However, the random re-esterification process has been used to produce structured triglycerides designed to facilitate the absorption of both medium-chain and long-chain ω-3 fatty acids [63]. Diglycerides have been used as emulsifying agents in manufactured food products for many years.

Amino acids

This group include intact protein (i.e., soy protein), polypeptides, amino acids, and nitrogenous and sulphur amino acid derivatives. Now-a-days, a few amino acids are also being investigated for their nutraceutical potential like arginine, ornithine, taurine, and aspartic acid. Arginine has been speculated to be cardio protective in that it is a precursor molecule for the vasodilating substance nitric oxide (NO) [64]. Also, arginine may reduce atherogenesis. The non-protein amino acid taurine may also have blood pressure-lowering properties as well as antioxidant roles. However, the research in these areas is still inconclusive, and the effect of supplementation of these amino acids on other aspects of human physiology is unclear. Several plant molecules are formed via amino acids. A few of the most striking examples are isothiocyanates, indole-3-carbinol, allyl sulphur compounds, and capsaicinoids. Another nutraceutical amino acid-derived molecule is folic acid, which is believed to be cardio protective in its role of minimizing homocysteine levels [20]. Other members of this group would include the tripeptide glutathione and choline.

Microbes (Probiotics)

Where the other groupings of nutraceuticals involve molecules or elements, probiotics includes intact microorganisms. The word 'probiotic' is derived from the Greek meaning 'for life' and has had several different meanings over the years. It was first used by Lilley and Stillwell in 1965 to describe substances secreted by one microorganism which stimulated the growth of another. Probiotics are defined as selected, viable microbial dietary supplements that, when introduced in sufficient quantities, beneficially affect human organism through their effects in the intestinal tract [65-68]. The FAO/WHO has adopted the definition of probiotics as "Live microorganisms which when administered in adequate amounts confer a health benefit on the host" [69].

There are a large number of probiotics currently used and available in dairy fermented foods, especially in yogurts. This group largely includes bacteria, and its criteria are that a microbe must be resistant to acid conditions of the stomach, bile, and digestive enzymes normally found in the human gastrointestinal tract; able to colonize the human intestine, be safe for human consumption, and lastly, have scientifically proven efficacy. Among the bacterial species recognized as having functional food potential are *Lactobacillus acidophilus, L. plantarum, L. casei, Bifidobacterium bifidum, B. infantis*, and *Streptococcus salvarius* subspecies *thermophilus*. Some yeasts were noted as well, including *Saccharomyces boulardii*.

Lactic acid bacteria constitute a diverse group of organisms providing considerable benefits to humankind, some as natural inhabitants of the intestinal tract and others as fermentative lactic acid bacteria used in food industry, imparting flavor, texture and possessing preservative properties. Beyond these, some species are administered to humans as live microbial supplements, which positively influence our health mainly by improving the composition of intestinal microbiota. For this reason, they are called probiotics. Some selected strains of *Lactobacillus, Bifidobacterium, Streptococcus, Lactococcus* and Saccharomyces have been promoted in food products because of their reputed health benefits [65, 70, 71].

The physiological effects related to probiotic bacteria include the reduction of gut pH, production of digestive enzymes and vitamins, antibacterial substances such as organic acids, bacteriocins, hydrogen peroxide, diacetyl, acetaldehyde, lactoperoxidase system, lactones and other unidentified substances; reconstruction of normal intestinal microflora after disorders caused by diarrhoeas, antibiotic therapy and radiotherapy; reduction of cholesterol level in the blood, stimulation of immune functions, suppression of bacterial infections, removal of carcinogens, improvement of calcium absorption as well as the reduction of faecal enzyme activity [67, 72].

Minerals

Several minerals have been recognized for their nutraceutical potential and thus become components for functional food recipes.Among the most obvious is calcium with relation to bone health, colon cancer, hypertension and cardiovascular disease. *In-vitro* and *in-vivo* laboratory studies have shown that calcium may affect the risk of developing cardio vascular diseases (CVD) through multiple mechanisms including blood cholesterol, insulin secretion and sensitivity, vasodilation, inflammatory profile, thrombosis, obesity, and vascular calcification [73, 74].

In-vivo and *in-vitro* studies suggest that dairy products, calcium, and dietary vitamin D inhibits the development of colorectal cancer (CRC). High calcium intake has been shown to have a greater protective effect against tumors of the distal colon and rectal cancer vs. proximal colon. Higher consumption of milk/ dairy products reduces the risk of colon cancer; high calcium intake reduces the risk of CRC [75].

Potassium has also been reported to reduce hypertension and thus improve cardiovascular health. A couple of trace minerals have also been found to have nutraceutical properties. These include copper, selenium, manganese, and zinc. Their nutraceutical potential is usually discussed in relation to antioxidation. Copper, zinc, and manganese are components of superoxide dismutase (SOD) enzymes, whereas selenium is a component of glutathione peroxidase. Certainly more investigation is required in the area of trace elements in light of their metabolic relationships to other nutrients and the potential for toxicity [76].

Important Definitions

Bioactive compounds are the naturally occurring chemical compounds contained in, or derived from, a plant, animal or marine source, that exert the desired health/wellness benefit e.g. ω-3 fatty acids in flax or fish oils and β-glucan from oats and barley.

Functional ingredients are the standardized and characterized preparations, fractions or extracts containing bioactive compounds of varying purity, that are used as ingredients by manufacturers in the food (human and pet) and preparations, fractions or extracts containing bioactive compounds of varying purity, which are used as ingredients by manufacturers in the cosmetics and pharmaceutical sectors.

Herbs are the leaves, roots and flowers of plants grown and processed for culinary, cosmetic, industrial, medicinal, landscaping, decorative and fragrant purposes. Much of the early interest in functional foods and nutraceuticals was based on the medicinal uses of herbs.

Industrial ingredients are the standardized and characterized preparations, fractions or extracts of agricultural commodities of varying purity, that are used as ingredients by manufacturers of non-food products.

Natural health products **(NHP)** include homeopathic preparations, substances used in traditional medicines, minerals or trace elements, vitamins, amino acid, essential fatty acids, or other botanical, animal, microorganism derived substances. These products are generally sold in medicinal or "dosage" form to diagnose, treat, or prevent disease; restore or correct function, or to maintain or promote health. As a product group, NHPs include nutraceuticals.

Novel foods are defined by 'Health Canada' as products that have never been used as food; foods that result from a process that has not previously been used for food, or foods that have been modified by genetic manipulation. These categories of foods are genetically modified foods.

Spices are seeds, root, bark and flowers of plants that are grown, harvested and processed for use as food or beverage flavouring. Examples include caraway, coriander, dill and mustard. Recently there has been interest in bioactive compounds identified in spices.

Traditional food ingredients are the standardized and characterized preparations, fractions or extracts of agricultural commodities of varying purity, that originate from plant, animal or marine sources and are used as ingredients by manufacturers in the food (human and pets) and NHP sectors. They are used for a variety of reasons in food products including consistency, adding flavour or colouring, modifying texture or stabilizing mixtures. They are not marketed on the basis of any bioactive compounds that they may contain.

Traditional processed foods are the conventional foods that have been manufactured by the traditional food processing industry and sold to the public through established distribution systems for generations. These foods are consumed as part of a usual diet and are consumed primarily for basic nutritional purposes. Examples are processed meats, bottled fruit juice, yogurt, and breakfast cereal. They are not marketed on the basis of any bioactive compounds that they may contain.

Traditional whole foods are the conventional foods that have been grown by agricultural producers for generations. They are subjected to minimal processing (e.g. sorting, cleaning and bulk packaging) before being sold to the public through established distribution systems. These foods are consumed as part of a usual diet and are consumed primarily for basic nutritional purposes. Examples include raw vegetables (broccoli and spinach), fresh fruit (blueberries and apples) and fresh meat (tray packed beef, pork or chicken).

Designer foods were coined in 1989 by the National Cancer Institute to describe foods that naturally contain or are enriched with cancer-preventing substances such as phytochemicals.

Functional foods are defined by the Institute of Medicine of the US National Academy of Sciences as foods that encompass potentially healthful products, including any modified food or food ingredient that may provide a health benefit beyond the nutrients it contains.

Medical foods are special dietary food intended for use solely under medical supervision to meet nutritional requirements in specific medical conditions. Although a legal classification of foods under the jurisdiction of the FDA and technically confined to enteral formulas used to feed hospitalized patients and people with rare diseases, more recently spin-offs of these formulas have begun to be marketed to the public, especially to the elderly, as a source of supplemental nutrition.

Nutraceuticals is a term popularized by the Foundation for Innovation in Medicine that refers to "any substance that may be considered a food or part of a food and provides medical or health benefits including the prevention and treatment of disease.

Phytochemical/Phytonutrient are plant components that have health-promoting properties; originally this term was limited to substances found in edible fruits and vegetables that appeared to be protective against cancer. Now the term is much broader and applies to any plant component that has health-enhancing benefits.

Summary

There is no doubt that functional food development has great interest among consumers, industries, governments and universities. From the research and development point of view functional foods represent an opportunity to obtain innovative products that satisfy considerably this demand that already exists. The technologies mainly used as it is possible has changed considerably over the years. Thus, while traditional techniques are the most commonly used for development of functional foods, from years 2000 until 2010 the techniques aimed towards personalized nutrition have grown greatly. There is a group of technologies, which have grown significantly, that form a structure aimed to prevent the deterioration of physiologically active compounds. In this way, it is important to note the importance of the structure-property relationship. The functional effect of a food or food component depends on the active component gaining access to the functional target site. However, foods are mostly complex mixtures of macro- and micro-components that can trap active compound, modulate its release or inhibit its activity [6]. Thus, the food matrix both in its raw state, after storage or culinary preparation can have a significant influence on the activity or release on the key components. Selection and development of an appropriate food vehicle that maintain the active molecular form until the time of consumption, and deliver this form to the physiological target within the organism, is an important step to the success of a functional food. As an example, it has been demonstrated that plant sterol efficacy differs across various matrices, the milk matrix being almost three times more effective than in bread or cereal.

References

1. Hasler, C.M., *Functional foods: their role in disease prevention and health promotion.* Food Technology 1998. **52**: 63-147.
2. Saito, M., *Role of FOSHU (food for specified health uses) for healthier life.* Yakugaku zasshi: Journal of the Pharmaceutical Society of Japan, 2007. **127**(3): 407-416.

3. Mamtani, R. and R. Mamtani, *Ayurveda and yoga in cardiovascular diseases.* Cardiology in Review, 2005. **13**(3): 155-162.
4. Siro, I., *et al., Functional food. Product development, marketing and consumer acceptancea - A review.* Appetite, 2008. **51**(3): 456-467.
5. Sloan, A.E., *Top ten trends to watch and work on for the millennium.* Food Technology, 1999.**53:** 38-39
6. Arvanitoyannis, I.S. and M. Van Houwelingen-Koukaliaroglou, *Functional foods: a survey of health claims, pros and cons, and current legislation.* Critical Reviews in Food Science and Nutrition, 2005. **45**(5): 385-404.
7. Kwak, N.S. and D.J. Jukes, *Functional foods. Part 2: the impact on current regulatory terminology.* Food Control, 2001. **12**(2): 109-117.
8. Shimizu, T., *Health claims on functional foods: the Japanese regulations and an international comparison.* Nutrition Research Reviews, 2003. **16**(02): 241-252.
9. Bailey, R., *Functional Foods in Japan: FOSHU (Foods for Specified Health Uses) and Foods with Nutrient Function Claims.* Regulation of functional foods and nutraceuticals: A Global Perspective, 2008: **247**.
10. Diplock, A.T., *et al., Scientific concepts in functional foods in Europe: Consensus document.* British Journal of Nutrition (United Kingdom), 1999.
11. Ohama, H., H. Ikeda, and H. Moriyama, *Health foods and foods with health claims in Japan.* Toxicology, 2006. **221**(1): 95-111.
12. Clydesdale, F.M., *ILSI North America food component reports.* Crit Rev Food Sci Nutr, 1999. **39**(3): 203-316.
13. Hasler, C.M., *et al., Position of the American Dietetic Association: functional foods.* Journal of the American Dietetic Association, 2004. **104**(5): 814-826.
14. Clydesdale, F., *Functional foods: opportunities and challenges.* Food Technology, 2004. **58**(12): 35-40.
15. Lucchina, L.A., *Improving the success of functional foods.* Food Technology, 2003.
16. Zawistowski, J., *Regualation of Functional Food in Selected Asian Countries in the Pacific Rim*, In Debasis Bagchi. Nutraceutical and Functional Food Regulations in the United States and Around the World. USA: Elsevier 2008.
17. Anonyms , *Government of India.* 2010 [cited 2014 January]; Available from: http://wcd.nic.in.
18. FAO, *Report on Functional Foods*, in *Food and Agriculture Organization of the United Nations,* 2007.
19. Bagchi, D., *Nutraceutical and functional food regulations in the United States and around the world,* Academic Press 2008.
20. Wildman, R.E.C., *Handbook of nutraceuticals and functional foods*: CRC press, London 2006.
21. Zemel, M.B., *et al., Calcium and dairy acceleration of weight and fat loss during energy restriction in obese adults.* Obesity Research, 2004. **12**(4): 582-590.
22. Braaten, J.T., *et al., Oat beta-glucan reduces blood cholesterol concentration in hypercholesterolemic subjects.* European Journal of Clinical Nutrition, 1994. **48**(7): 465-474.
23. Queenan, K.M., *et al., Concentrated oat beta-glucan, a fermentable fiber, lowers serum cholesterol in hypercholesterolemic adults in a randomized controlled trial.* Nutrition Journal, 2007. **6**(1): 6.
24. Kerckhoffs, D.l.A.J.M., G. Hornstra, and R.P. Mensink, *Cholesterol-lowering effect of beta-glucan from oat bran in mildly hypercholesterolemic subjects may decrease when beta-glucan is incorporated into bread and cookies.* The American Journal of Clinical Nutrition, 2003. **78**(2): 221-227.

25. Gupta, A., R. Gupta, and B. Lal, *Effect of Trigonella foenum-graecum (fenugreek) seeds on glycaemic control and insulin resistance in type 2 diabetes mellitus: a double blind placebo controlled study.* The Journal of the Association of Physicians of India, 2001. **49**: 1057-1061.
26. Sharma, R.D., T.C. Raghuram, and N.S. Rao, *Effect of fenugreek seeds on blood glucose and serum lipids in type I diabetes.* European Clinical nutrition Journal, 1990. **44**(4):
27. 301-6.Raju, J., *et al.*, *Trigonella foenum graecum (fenugreek) seed powder improves glucose homeostasis in alloxan diabetic rat tissues by reversing the altered glycolytic, gluconeogenic and lipogenic enzymes.* Molecular and cellular biochemistry, 2001. **224** (1-2): 45-51.
28. Markosyan, A., J.J. McCluskey, and T.I. Wahl, *Consumer response to information about a functional food product: apples enriched with antioxidants.* Canadian Journal of Agricultural Economics/Revue canadienne d'agroeconomie, 2009. **57**(3): 325-341.
29. M.H. Arnoult, *et al.*, *Consumers' Willingness to Pay for Functional Agricultural Foods*, 2007. The University of Reading.Workpackage No. 2, Report No. 05;1-21.
30. Ares, A. Gimanez, and R. Deliza, *Influence of three non-sensory factors on consumer choice of functional yogurts over regular ones.* Food quality and preference, 2010. **21**(4): 361-367.
31. Krystallis, A., G. Maglaras, and S. Mamalis, *Motivations and cognitive structures of consumers in their purchasing of functional foods.* Food Quality and Preference, 2008. **19**(6): 525-538.
32. Marette, S., *et al.*, *Functional food, uncertainty and consumers choices: A lab experiment with enriched yoghurts for lowering cholesterol.* Food Policy, 2010. **35**(5): 419-428.
33. Manners, G.D., *Citrus limonoids: analysis, bioactivity, and biomedical prospects.* Journal of agricultural and food chemistry, 2007. **55**(21): 8285-8294.
34. Fong, C.H., *et al.*, *Limonoid glucosides in commercial citrus juices.* Journal of Food Science, 1989. **54**(6): 1505-1506.
35. Taiz, L. and E. Zeiger, *Plant Physiology.* Ed. Sinauer, Sunderland, Massachussets, 1998. **2**: 111-143.
36. Wattenberg, L.W. and L.K. Lam, *Protective effects of coffee constituents on carcinogenesis in experimental animals.* Banbury Rep, 1984. **17**: p. 137-145.
37. Miller, E.G., *et al.*, *Kahweol and cafestol: inhibitors of hamster buccal pouch carcinogenesis.* 1991.
38. Miller, E.G., *et al.*, *Inhibition of oral carcinogenesis by citrus flavonoids.* Nutrition and cancer, 2007. **60**(1): 69-74.
39. Hasegawa, S., *et al.*, *Limonoid glucosides in citrus.* Phytochemistry, 1989. **28**(6): 1717-1720.
40. Lee, S.-J., *et al.*, *Antitumor activity of a novel ginseng saponin metabolite in human pulmonary adenocarcinoma cells resistant to cisplatin.* Cancer Letters, 1999. **144**(1): 39-43.
41. Craig, W.J., *Health-promoting properties of common herbs.* The American Journal of Clinical Nutrition, 1999. **70**(3): 491s-499s.
42. Robards, K., *et al.*, *Phenolic compounds and their role in oxidative processes in fruits.* Food Chemistry, 1999. **66**(4): 401-436.
43. Wildman, R.E.C., *Nutraceuticals and functional foods*, in *Handbook of nutraceuticals and functional foods*, R.W. Robert E.C. Wildman, Taylor C. Wallace, Editor 2006, CRC Press : 560-570.
44. Hollman, P.C.H. and M.B. Katan, *Dietary flavonoids: intake, health effects and bioavailability.* Food and Chemical Toxicology, 1999. **37**(9): 937-942.

45. Yao, L.H., *et al., Flavonoids in food and their health benefits.* Plant Foods for Human Nutrition, 2004. **59**(3): 113-122.
46. Balasundram, N., K. Sundram, and S. Samman, *Phenolic compounds in plants and agri-industrial by-products: Antioxidant activity, occurrence, and potential uses.* Food Chemistry, 2006. **99**(1): 191-203.
47. Collins, F.W., *Oat phenolics: avenanthramides, novel substituted N-cinnamoylanthranilate alkaloids from oat groats and hulls.* Journal of Agricultural and Food Chemistry, 1989. **37**(1): 60-66.
48. Dimberg, L.H., O. Theander, and H. Lingnert, *Avenanthramides-a group of phenolic antioxidants in oats.* Cereal Chemistry, 1993. **70**: 637-637.
49. Hollman, P.C.H., *Evidence for health benefits of plant phenols: local or systemic effects.* Journal of the Science of Food and Agriculture, 2001. **81**(9): 842-852.
50. Roberfroid, M.B., *Prebiotics and probiotics: are they functional foods?* The American Journal of Clinical Nutrition, 2000. **71**(6): 1682s-1687s.
51. Elleuch, M., *et al., Dietary fibre and fibre-rich by-products of food processing: Characterisation, technological functionality and commercial applications: A Review.* Food Chemistry, 2011. **124**(2): 411-421.
52. Dumitriu, S., *Polysaccharides: structural diversity and functional versatility* : CRC Press, 2012.
53. Cunha, A.G. and A. Gandini, *Turning polysaccharides into hydrophobic materials: a critical review. Part 2. Hemicelluloses, chitin/chitosan, starch, pectin and alginates.* Cellulose, 2010. **17**(6): 1045-1065.
54. Gandhi, N.S. and R.L. Mancera, *The structure of glycosaminoglycans and their interactions with proteins.* Chemical Biology & Drug Design, 2008. **72**(6): 455-482.
55. Jackson, R.L., S.J. Busch, and A.D. Cardin, *Glycosaminoglycans: molecular properties, protein interactions, and role in physiological processes.* Physiological Review, 1991. **71**(2): 481-539.
56. Laurent, T.C., U.B.G. Laurent, and J.R.E. Fraser, *The structure and function of hyaluronan: An overview.* Immunology and cell biology, 1996. **74**(2): A1-A7.
57. Igarashi, N., *et al., Effect of Molecular Sizes of Chondroitin Sulfate on Interaction with L-Selectin.* International Journal of Carbohydrate Chemistry, 2013.
58. Gargiulo, V., *et al., Structural analysis of chondroitin sulfate from Scyliorhinus canicula: A useful source of this polysaccharide.* Glycobiology, 2009. **19**(12): 1485-1491.
59. Oliver, M.F., *It is more important to increase the intake of unsaturated fats than to decrease the intake of saturated fats: evidence from clinical trials relating to ischemic heart disease.* The American Journal of Clinical Nutrition, 1997. **66**(4): 980S-986S.
60. Qi, B., *et al., Production of very long chain polyunsaturated omega-3 and omega-6 fatty acids in plants.* Nature Biotechnology, 2004. **22**(6): 739-745.
61. O'Quinn, P.R., *et al., Conjugated linoleic acid.* Animal Health Research Reviews, 2000. **1**(01): 35-46.
62. Babayan, V.K., *Medium chain triglycerides and structured lipids.* Lipids, 1987. **22**(6): 417-420.
63. Tso, P., T. Lee, and S.J. Demichele, *Lymphatic absorption of structured triglycerides vs. physical mix in a rat model of fat malabsorption.* American Journal of Physiology-Gastrointestinal and Liver Physiology, 1999. **277**(2): G333-G340.
64. Niittynen, L., *et al., Role of arginine, taurine 4 and homocysteine in cardiovascular diseases.* Annals of medicine, 1999. **31**(5): 318-326.
65. Ziemer, C.J. and G.R. Gibson, *An overview of probiotics, prebiotics and synbiotics in the functional food concept: perspectives and future strategies.* International Dairy Journal, 1998. **8**(5-6): 473-479.

66. Holzapfel, W.H. and U. Schillinger, *Introduction to pre-and probiotics.* Food Research International, 2002. **35**(2): 109-116.
67. Zubillaga, M., *et al.*, *Effect of probiotics and functional foods and their use in different diseases.* Nutrition Research, 2001. **21**(3): 569-579.
68. Vaughan, E.E., B. Mollet, and W.M. deVos, *Functionality of probiotics and intestinal lactobacilli: light in the intestinal tract tunnel.* Current opinion in biotechnology, 1999. **10**(5): 505-510.
69. F.A.O., *Working Group Report on Drafting Guidelines for the evaluation of probiotics in food,* 2002.
70. Puupponen Pimis, R., *et al.*, *Development of functional ingredients for gut health.* Trends in Food Science & Technology, 2002. **13**(1): 3-11.
71. Sanders, M.E., *Overview of functional foods: emphasis on probiotic bacteria.* International Dairy Journal, 1998. **8**(5-6): 341-347.
72. Ouwehand, A.C., *et al.*, *Probiotics: mechanisms and established effects.* International Dairy Journal, 1999. **9**(1): 43-52.
73. Iso, H., *et al.*, *Prospective study of calcium, potassium, and magnesium intake and risk of stroke in women.* Stroke, 1999. **30**(9): 1772-1779.
74. Wang, L., J.E. Manson, and H.D. Sesso, *Calcium Intake and Risk of Cardiovascular Disease.* American Journal of Cardiovascular Drugs, 2012. **12**(2): 105-116.
75. Huncharek, M., J. Muscat, and B. Kupelnick, *Colorectal cancer risk and dietary intake of calcium, vitamin D, and dairy products: a meta-analysis of 26,335 cases from 60 observational studies.* Nutrition and cancer, 2008. **61**(1): 47-69.
76. Ascherio, A., *et al.*, *Intake of potassium, magnesium, calcium, and fiber and risk of stroke among US men.* Circulation, 1998. **98**(12): 1198-1204.

Section II: Functional Food Products: Ingredients and Functionality

2

Plant Based Ingredients

Introduction

Vast proportions of the population of the world are dependent on plants as their principal source of food either directly or indirectly. Recent decades have also seen increased numbers of people turning to alternative medical therapy based upon traditional folk medicine or food as a form of medicine. These alternative therapies emphasise the importance of a 'good diet' in maintaining and restoring health. Under these circumstances, vast majority of population may see diet therapy, dietary supplements and non-invasive alternative therapies as a safer, less technological and more natural way of maintaining health and treating and preventing individuals from illness. These alternatives may be seen as empowering the consumer and shifting the locus of control away from the medical establishment. This chapter covers many crucial topics to understand the chemistry and bioactivity of the active compounds of plant origin. It also aims to cover the latest research on the bioactive compounds.

Functional Benefits of Plant Based Ingredients

Plants have been a source of medicines in past centuries and even now scientists and the general public recognize their value as a source of new or alternative medicinal products. A significant percentage of the active compounds of currently prescribed medicines in western countries continue to be derived from higher plants. Large numbers of plant preparations and extracts as well as a few from animal sources are used as traditional or 'herbal' medicines and many of these are also marketed as dietary supplements.

Over the past decades interest in medicinal plants has grown enormously, from the use of herbal products as natural cosmetics and for self-medication by the general public to the scientific investigation of plants for their biological effects in human being [1, 2]. Even traditional herbal medicinal products offer unique possibilities to develop a self-medication approach and at the same time to boost research projects. As a matter of fact the traditional use of herbal medicine may serve as an initial screening of the biological activity of the product and of its compounds. Natural plant products are not only used for primary healthcare in rural areas but in developing countries also, where modern medicines are predominantly used. Beyond this pharmaceutical approach to plants, there is a wide tendency to utilize herbal products to supplement the diet, mainly with the intention of improving the quality of life and preventing the diseases of elderly people [3]. Numerous plant foods or physiologically active ingredients derived from plants have been investigated for their role in disease prevention and health. However, only a small number of these have had substantive clinical documentation of their health benefits. An even smaller number have surpassed the rigorous standard of "significant scientific agreement" required by the FDA for authorization of a health claim, which will be discussed in further detail below. Those plant foods currently eligible to bear an FDA-approved health claim include oat soluble fiber [4], soluble fiber from psyllium seed husk [5], soy protein [6] and sterol- and stanol-ester–fortified margarine [7].

Some plant-based foods or food constituents currently do not have approved health claims, but have growing clinical research supporting their potential health benefits, and thus would be described as having moderately strong evidence. In the last decade there has been a research interest and understanding of the functional bioactive ingredients and their health benefits. Prior to this period bioactive, secondary metabolites of food plants were considered to be undesirable. For example, in soy isoflavones are shown to prolong the oestrus cycle in women. Low dose exposures may result in benefits since there is a lower incidence of hormonally related cancers in women living in societies with a high consumption of soya [8]. Food plants are also the source of delivery of functional compounds. These functional compounds, which also act as secondary metabolites have specific functions within the plants. For example, anthocyanins acts as a pollinator attractant, carotenoids act as a UV-protectants and phytoalexins acts as infestation control. Schematic view of the diversity of the secondary metabolites formation is shown in Fig. 2.1.

Plant Natural Product Formation

↓

(1) Photosynthesis (2) Glycolysis (3) Citrate Cycle (4) Shikimate Pathway (5) DOXP/MEP

↓

Formation of Metabolites

↓

Ligation, Cyclisation

↓

Alkaloids, Phenylpropanoids, Terpenoids, Polyketides

↓

(1) Glycosylation, (2) Methylation, (3) Acylation, (4) Prenylation, (5) Oxidation

↓

>200, 000 Derived Plant Products

Fig. 2.1: Schematic view of diversity of the secondary metabolites formation

The biosynthetic pathway of the production of these bioactive compounds is under regulatory control and any attempt to modify the pathway can result in the adverse effect. For example, the symptoms of oxidative stress in the plant can be result of the overexpression of the glutathione synthesis [9]. In vast majority, plants use two principal biosynthetic pathways – the phenylpropanoid and isoprenoid pathways along with some common links through which majority of phytochemical of interest are synthesised. The phenylpropanoid pathway leads to the production of lignins and their phenolic ester precursors, the flavones and related compounds, and isoflavones. Phenylpropanoids contribute to all aspects of plant responses towards biotic and abiotic stimuli. They are not only indicators of plant stress responses upon variation of light or mineral treatment, but are also key mediators of the plants resistance towards pests [10]. The isoprenoid pathway leads to the formation of terpenes, sterols, carotenoids, and the tocopherols.

Examples of plant based ingredients

Modern knowledge of health is being used to improve food. In recent years, scientific evidence has revealed that bioactive dietary components benefit health in ways that extend beyond meeting basic nutritional needs. This chapter is confined to the discussion of specific functional ingredients originated from the plant. There is also an attempt to understand the chemistry and biological activity of different plant based functional ingredients.

Phytosterols and Phytostanols

Chemical structure and properties

Plant membranes contain several types of phytosterols that are similar in structure to cholesterol but are metabolized differently and include a methyl or ethyl group at C-24. The chemical structure of some of the phytosterols are presented in Fig. 2.2.

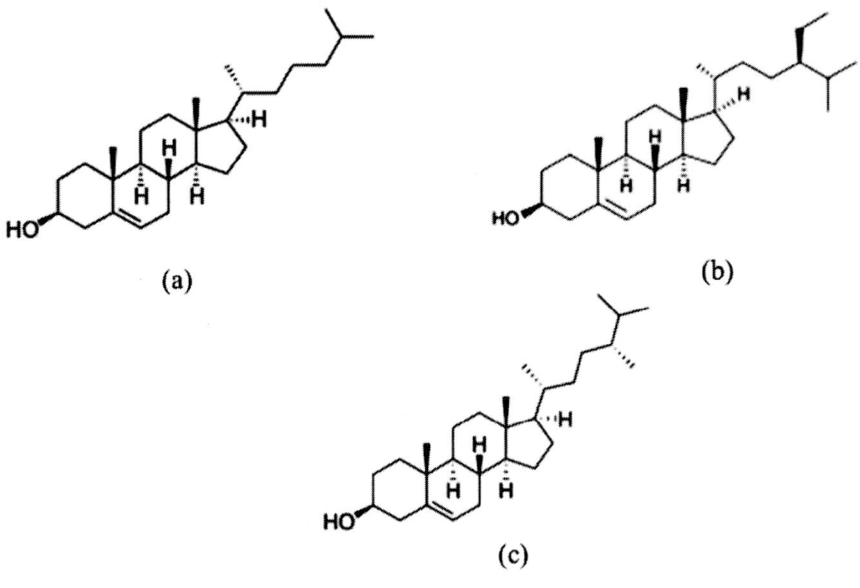

Fig. 2.2: Chemical structure of (a) cholesterol (b) sitosterol and (c) campesterol

These compounds are commonly known as phytosterols, β-sitosterol and campesterol. The scientific names of phytosterols are given to them according to the number of C-atoms in the C-17 side chain, the number and the position of the double bond in the ring system and the side chain. Their scientific names are usually very complex, so the most common phytosterols are referred to by their trivial names. The trivial and the scientific names of the most important phytosterols are given in Table 2.1 [11].

Table 2.1: Trivial and scientific names of selected sterols from the sterol biosynthetic pathway

Trival Name	Scientific Name	Sterol Class
Cycloartenol	9beta,19-cyclo-24-lanosten-3beta-ol	4,4-dimethyl
24-Methylene	4-alpha-Methyl-5-alpha-ergosta-7,24-dien-3-beta-ol	4-methyl
Avenasterol	24-ethylcholesta-5,24(28)Z-dien-3,ol	4-desmethyl
Cholesterol	cholest-5-en-3,-ol	4-desmethyl
Campesterol	24α-methyl-5-cholestern-3,-ol	4-desmethyl
Brassicasterol	5,22-cholestadien-24β-methyl-3,-ol	4-desmethyl
Sitosterol	24α-ethylcholest-5-en-3,-ol	4-desmethyl
Stigmasterol	5,22-cholestadien-24α-ethyl-3,ol	4-desmethyl

Phytosterols are triterpenes and are one of the most important structural components of the plant membrane. These are the most abundant phytosterols in the human diet. They make up over 80% of the total phytosterol intake, typically 100–300 mg/day, similar to the average daily intake of cholesterol. Other phytosterols include stigmasterol and the sterol ergosterol which is made by yeast and other fungi and which can also act as a precursor for vitamin D if irradiated with ultraviolet light.

Phytosterols (plant sterols) are members of the triterpene family of natural products, which includes more than 100 different phytosterols and more than 4000 other types of triterpenes [12, 13]. All triterpenes are synthesized via a pathway that starts with reduction of 3-hydroxy-3-methylglutaryl-coenzyme A (HMG-CoA) (six carbons) to mevalonate (five carbons). Six mevalonate units are then assembled into two farnesyl diphosphate molecules, which are combined to make squalene (30 carbons or three terpenes). Enzymatic ring closure steps then form cycloartenol (also 30 carbons), and additional enzymatic reactions form common plant triterpenes such as phytosterols, triterpene alcohols, and brassino steroids. The biosynthesis of phytosterols and other triterpenes are shown in Fig. 2.3.

Phytosterols are largely hydrophobic having one polar hydroxyl group at 3rd C-atom, making them amphiphilic. According to one of the sterol classifications phytosterols with free 3β- hydroxyl group, are named free phytosterols and they are the major end product of biosynthetic pathway of phytosterols. However, phytosterols also occur as steryl esters where 3β-hydroxyl group of free phytosterols is esterified with a long chain of saturated or unsaturated fatty acids , mainly linoleic and oleic, or with phenolic acids. Steryl glycosides are formed when the 3β-hydroxyl group is linked with monosaccharides, usually glucose, at the first C position. When this monosaccharide is at the 6-C position esterified with fatty acid, then so-called acylated steryl glycosides are formed [14].

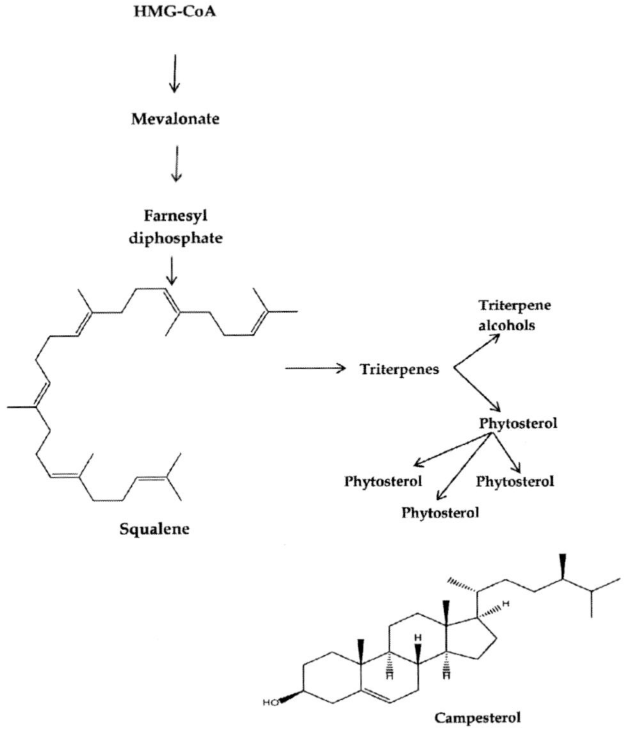

Fig. 2.3: Biosynthesis of phytosterols and other triterpenes

Phytosterols are in general poorly absorbed from the gut and phytostanols are especially poorly absorbed. For instance less than 1% of ingested β-sitostanol is absorbed in the body. It has been known since decades that, the phytosterols are poorly absorbed from the gut, but are able to inhibit the absorption of dietary cholesterol and the re-absorption of biliary cholesterol, probably by displacing cholesterol from micelles in the gut. Micelles are the minute suspension particles from which lipids and fat soluble vitamins are absorbed in the gut. Indeed β sitosterol was first described as a therapeutic agent for hypercholesterolemia in 1951[15]. They thus have the potential to lower blood cholesterol by increasing the amount of dietary and biliary cholesterol that is lost in the faeces [16]. There is recent renewed interest in phytosterols as a possible novel coronary artery disease (CAD) risk factor. This interest has been heightened by the availability of an agent, ezetimibe, which interferes with the absorption of both cholesterol and phytosterols [17, 18].

Plants have a variety of more than 40 well identified and studied sterols [19], which are termed phytosterols and are predominantly present in oilseed plants. The amount of phytosterol content in different food is shown in Table 2.2 [20].

Cereals are recognised as significant sources of phytosterols as well, whereas phytosterol content in vegetables and nuts is considerably low [21, 22]. The most abundant phytosterols are: sitosterol, campesterol and stigmasterol. Other phytosterols like avenasterol and cycloartenol are synthesised earlier in the biosynthetic pathway and as sterol precursors they usually occur in relatively smaller amounts. Phytosterols are, with respect to their physiological function and their chemical structure, similar to the major and only animal produced sterol cholesterol.

Table 2.2: Phytosterol content of different types of foods

Food	Phytosterol Content (mg/100g of edible portion)
Corn oil	952
Sunflower oil	725
Safflower oil	444
Soybean oil	221
Olive oil	176
Almonds	143
Beans	76
Corn	70
Wheat	69
Palm oil	49
Lettuce	38
Banana	16
Apple	12
Tomato	7

Biological function in plants

Phytosterols in plant are the integral part of membrane component as well as they have metabolic functions. They determine the characteristics of plasma membrane, endoplasmic reticulum and mitochondria membranes. Phytosterols also regulate the metabolic function of membrane permeability, fluidity, signal transduction events for cell division and even activity of membrane-bound enzyme [23]. It has been postulated that sitosterol and campesterol are most efficient in membrane permeability and fluidity regulation and there are evidence that stigmasterol play an important role in cell proliferation, but has reduced ordering effect [23]. During the germination of seed, it has been found that, the phytosterol content increases. The increase in the phytosterol is associated with the intensive membrane biosynthesis in the plant. Phytosterols, that are accumulated in seeds and meristematic tissue, play an important role in cellular proliferation and differentiation. Phytosterol synthesis will appear mostly during the seed formation and germination and phytosterols themselves will provide a supply for the growth of new cells and young shoots. By the time when seeds are already mature or tissues ages, the intensity of biosynthesis will decline[11]. Phytosterol amount in different plant tissues could be explained by the anatomical structure. For instance, flower heads of broccoli and cauliflower (*Brassica oleracea* L.) have higher proportion of membrane-rich meristematic tissue, which results in higher content of total phytosterols in this species in comparison with other vegetables, fruits or berries [24]. Moreover, free phytosterols are precursors of bioactive steroids, growth factors and substrates for synthesis of numerous secondary plant metabolites [24].

Biological function in human body

According to many conducted experiments phytosterols decreases serum total and LDL cholesterol levels [24, 25]. It was suggested that phytosterols compete at the same time, in the micellar phase of the small intestine, for the limited space with cholesterol. Micelles are the essentially small aggregates, which carry a mixture of lipids and bile salts in intestinal lumen.

Sitostanol esters, 5α-saturated sitosterol esters, were recognised as the most efficient for reducing the serum cholesterol concentration and the intestinal cholesterol absorption [25]. As it seems they could be easily produced by sterol hydrogenation and trans-esterification with polyunsaturated fatty acids. Sitostanol esters apparently result similar in prevention of cholesterol absorption but, in contrast to phytosterol esters, they do not increase their own absorption. In addition to this fact, it has been confirmed that relatively high campesterol content, in some phytosterol complexes like soy phytosterols for example, can increase the campesterol proportion in serum, what was not acknowledged for plant stanols including campestanol (5α-saturated campesterol).

Legislation and labelling aspect

In the USA plant sterols, stanols and their esters were given generally recognized as safe (GRAS) status. On the basis of this recognition, the US Food and Drug Administration (FDA) approved fat spreads containing up to 20% of either steryl or stanyl esters [26]. In Europe in 2000, the EU Scientific Committee on Food (SCF) approved the use of phytosterol esters in yellow fat spreads (maximum level of 8% of free phytosterols) [27, 28]. SCF and its successor, the European Food Safety Authority (EFSA) through its Scientific Panel on Dietetic Products, Nutrition and Allergies, have recommended that plant sterol containing food stuffs should not be consumed in amounts resulting in total phytosterol intakes exceeding 3g/day [29].

In 2003, the FDA published the interim final rule (IFR) authorizing the use of a health claim for plant sterol/stanol esters and reduced risk for coronary heart disease on food labels. The IFR authorizes the use of a health claim relating between plant sterol/stanol esters and reduced risk of coronary heart disease on labelling of spreads and dressings for salad containing at least 0.65 g of plant sterol esters per serving and spreads, dressings for salad, snack bars, dietary supplements in soft gel form containing at least 1.7 g plant stanol esters per serving [30]. In 2004, the EU Commission published Regulation 608/2004/EC concerning the labeling of foods and food ingredients with added phytosterols and/or phytosterol esters, requiring such products to be labelled with additional information including the words "with added plant sterols/plant stanols". In England, the food with added phytosterols (labelling) Regulations 2004 were published to provide for the enforcement of Regulation 608/2004/EC [31].

Flavonoids

Occurrence and chemical structure

Flavonoids are a large class of secondary metabolites encompassing more than 10,000 structures. The word flavonoid comes from Latin *flavus*, which means yellow, and included at the beginning only yellow colored group of compounds with a flavone nucleus [32]. However, the term also includes colourless compound (flavans-3-ol) to less coloured compounds (Flavanone) as well as red and blue compound like anthocyanins, commonly occurring in the plants. Their wide occurrence, complex diversity and manifold functions have made flavonoids a very attractive system for chemical, genetic and enzymological studies, and for research on a molecular biological level. The coloration of flowers and other plant organs due to flavonoids, in particular the red/bluish colours from anthocyanins, other flavonoid compound bring white to yellow colour to plants. Flavonoid can form precipitate in the presence of proteins, thus forming insoluble structures in the mouth. This property provides astringency associated with flavonoids. Additional important functions include protection to plants from deteriorating UV radiation from the sun, to protect against parasite attack, to regulate certain enzyme reactions, to function as signal substances and to provide flavour to various plant products used for human consumption [32].

Flavonoids are plant phytochemicals that cannot be synthesized by humans. The six classes of flavonoids (flavanones, flavones, flavonols, isoflavonoids, anthocyanins, and flavans) vary in their structural characteristics around the heterocyclic oxygen ring. Flavanones occur predominantly in citrus fruits, flavones in herbs, isoflavonoids in legumes, anthocyanins and catechins in fruits and flavonols in all fruits and vegetables. Most anthocyanins and flavonols occur as glycosides in plants. The glycosides thus consist of the flavonoid unit (the aglycon) and one or more monosaccharides. The most common substitutions occur in the 3- and the 5-positions of the flavans structure for the anthocyanidins and in the 3-position for the flavonols. The sugar moiety may have one or more aliphatic or aromatic acids attached in an acyl linkage. Food preparation and processing of fresh fruits and vegetables may decrease flavonoid content by 50% owing to leaching into water or removal of portions of the plant that are rich in them. Grains and oilseeds have flavonoids, but processing removes or reduces them. Other plant food groups contain differing amounts of flavonoids. Honey, chocolate, and sweets that contain some plant constituents have flavonoids. Flavonoids are not present in animal foods. Dietary intake estimates vary from 23 mg/day to 1000 mg/day but the number of compounds and classes assessed vary, and most estimates currently are incomplete. In vitro and animal studies have demonstrated that flavonoids have antioxidant and antimutagenic activities. Case control studies suggest that flavonoids may reduce the risk of cardiovascular disease and stroke. Flavonoid classes vary in their absorption and their metabolism is still obscure. They are conjugated in the liver or kidney and excreted into bile or urine. Colonic bacteria split the heterocyclic ring and degrade the flavonoids to phenyl acids which may be absorbed, conjugated, and excreted or metabolized further by colonic bacteria. In the plant kingdom, different plant families have characteristic patterns of flavonoids and their conjugates. All these compounds play important biochemical and physiological roles in the various cell types or organs where they accumulate. Different classes of flavonoids and their conjugates have numerous functions during the interactions of plant with the environment, both in biotic and abiotic stress conditions [33]. Additionally, flavonoid conjugates, because of their common presence in plants, are important components of human and animal diet. Due to the different biological activities of plant secondary metabolites, their regular consumption may have serious consequences for health, both positive and negative [34].

Flavonoids as functional food

Flavonoids have many functions in the biochemistry, physiology and ecology of plants, and they are important in both human and animal nutrition. Flavonoids have functions in protecting against UV light (UV-B screening pigments), in warding off pathogenic microorganisms (phytoalexins) or pests (antifeedants), in the fertility and germination of pollen, in activating bacterial nodulation genes (nitrogen fixation) and in regulating plant growth and enzyme activity. Plant coloration is not only attractive for pollinators and seed distribution, but also provides aesthetically valuable characteristics for humans. The major nutraceuticals ingredients in plants are flavonoids. Being a phenolic compound they can act as a potent antioxidant as well as metal chelators.

It has been demonstrated, reactive oxygen species (ROS) contribute to cellular aging, utagenesis, carcinogenesis, and coronary heart disease possibly through the destabilization of membranes, DNA damage, and oxidation of low-density lipoprotein (LDL) [35]. Flavonoids have demonstrated peroxyl radical scavenging abilities, which contribute to inhibiting lipid peroxidation and oxidation of LDL [36]. Since LDL oxidation is implicated in the pathogenesis of coronary heart diseases through its ability to decrease the susceptibility of LDL to oxidation, a number of researches have undertaken investigations examining the activity of dietary agents rich in flavonoids in inhibiting LDL oxidation [37, 38].

Flavonoids are regarded as anti-allergenic, anti-inflammatory, anti-viral, anti-proliferant and anti-carcinogenic and act in certain metabolic pathways in mammals [39]. A mixture of two flavonoid compounds, called citrin, was previously considered to have a vitamin-like activity. The compounds were called vitamin P and were known to decrease capillary permeability and fragility and to help under vitamin C deficit conditions. Later, the vitamin classification of citrin was abandoned. The antioxidant and lipid peroxidation inhibition potential of flavonoids predominantly resides in their radical-scavenging capacity rather than in their metal-chelating potential. Most flavonoids are effective radical scavengers. This property by itself does not imply a beneficial effect, because after scavenging a flavonoid radical is formed. A very reactive flavonoid radical would propagate rather than interrupt the deleterious events initiated by the radical attack. However, a flavonoid radical with high stability will not readily react. As a consequence, this flavonoid will act as an antioxidant. Green tea is a rich source of flavonoids, primarily catechins and flavonols. In black tea, as a consequence of the fermentation process, catechins are oxidized to complex condensation products, the theaflavins and thearubigins. Tea polyphenols show strong antioxidative effects and provide powerful scavengers against superoxide, hydrogen peroxide, hydroxyl radicals, nitric oxide, and peroxynitrite produced by various chemical and biological systems. Many studies demonstrated that topical application or oral feeding of a polyphenolic fraction from tea extract or of individual catechin derivatives had anti-carcinogenesis effects in animal skins and other organs [40]. The bioavailability of tea constituents is apparently a key factor determining the effectiveness of tea in inhibiting tumor formation. In this respect, the oral cavity and digestive tract, which have direct contact with orally administrated tea, may represent good targets for chemoprevention.

The recent studies focussing on the health-benefiting compounds create the necessity to search for their exact function in the human body [41]. From a health point of view, the main interest at present is concentrated on the ability of the flavonoids to act as free radical scavengers as well as bioavailability aspects in the body. Methodologies for the assessment of these and other functions are currently being developed.

Lycopene

Chemical structure

Lycopene is an oil soluble lipophilic carotenoid pigment widely distributed in fruits and vegetables. Carotenoids are also found in algae, bacteria, fungi, and in animals, birds, fish and invertebrates, which rely exclusively on plant sources of carotenoids. Carotenoid pigments consist of two major classes; i.e., the hydrocarbon carotenes (including lycopene) and the oxygenated xanthophylls. The molecular formula of lycopene is $C_{40}H_{56}$ (Fig. 2.4). All chemical structures of the carotenoids may be formally derived from the acyclic $C_{40}H_{56}$ of structure of lycopene. The 11 conjugates bonds in the lycopene are arranged in the linear array, making it longer than any other carotenoids. From the plant products commonly consumed by humans, more than 600 different carotenoids (the number includes cis-trans isomeric forms) have been isolated to date. Lycopene is dispersible in edible oils and soluble in polar organic solvents. In aqueous systems, lycopene tends to aggregate and to precipitate as crystals; this behavior is suspected to inhibit the bioavailability of lycopene in humans. In fresh tomatoes, the crystalline form of lycopene is responsible for the typical bright red colour of the ripe fruits [42, 43].

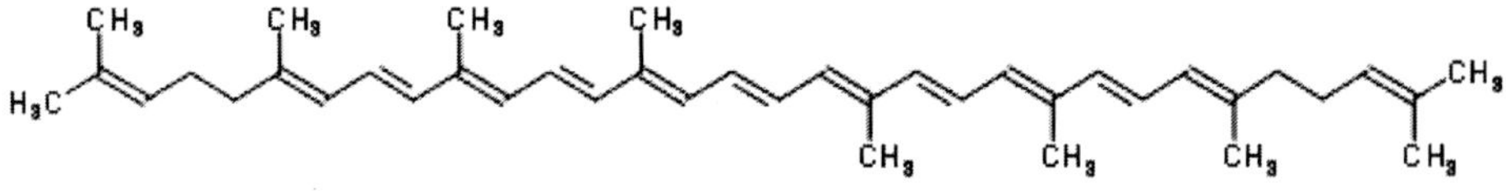

Fig. 2.4: Chemical structure of lycopene

Lycopene as functional food

Many of the putative biologic effects and health benefits of lycopene and other carotenoids are hypothesized to occur via protection against oxidative damage [44, 45]. Studies suggest that lycopene is a more potent scavenger of oxygen radicals than other major dietary carotenoid. Gastric cancer remains a major problem in many parts of the world, and several studies have reported protective effects for diets rich in tomato products. Lycopene intake is also associated with the reduction in skin cancer [44, 46]. There has been a great deal of interest in exploring the relationship between carotenoids, including lycopene, and risk of skin cancer. Exposure of the skin to ultraviolet (UV) radiation leads to bums, photosensitivity, premature aging, and increased risk of cancer. UV exposure is associated with the local production of singlet oxygen and free radicals. The role of carotenoids in plants in quenching UV-induced oxidative products has been proposed to be analogous to human skin, because UV light acts as a photo sensitizer, inducing the generation of free radicals, singlet oxygen, and other reactive specie. However, there is inadequate information on potential risks, dose response relationships, appropriate timing of intake relative to stages of disease processes, pharmacokinetics or mechanisms of action for purified lycopene preparations, and which disease processes may be inhibited by pharmacologic lycopene supplements.

Increasing clinical evidence supports the role of lycopene as a micronutrient with important health benefits, because it appears to provide protection against a broad range of epithelial cancers. Dietary intakes of tomatoes and tomato products containing lycopene have been shown to be associated with decreased risk of chronic diseases such as cancer and cardiovascular diseases in several recent studies. Serum and tissue lycopene levels have also been inversely related with the chronic disease risk. Although the antioxidant properties of lycopene are thought to be primarily responsible for its beneficial properties, evidence is accumulating to suggest other mechanisms such as modulation of intercellular gap junction communication, hormonal and immune system and metabolic pathways may also be involved [47]. One study has reported on tomato intake in relation to total cancer risk. In this prospective study, based on 42 cancer deaths among 1271 elderly persons, individuals who were in the top half of tomato consumption had a lower risk of all cancers combined compared with those in the bottom half ($R^2 = 95\%$, CI = 0.3-0.8). Other items, including green and yellow vegetables and strawberries, were also associated with a decreased risk of total cancer. Carrots and squash were unrelated to risk. There were too few cancers to allow examination of specific cancer sites [48].

Twelve case-control studies from a variety of populations, including the United States, Japan, Spain, Poland, Belgium and Sweden have reported data on tomato or lycopene intake and stomach cancer risk. Inverse associations between tomato consumption and risk of gastric cancer were observed in all these diverse populations. An ecologic study in Japan showed that examined plasma levels of various nutrients in samples of populations in various regions found that regions high in plasma lycopene had the lowest gastric cancer rates and regions low in lycopene had the highest rates. While other fruits and vegetables have frequently been inversely associated with gastric cancer, inverse associations with tomatoes have been among the most consistent and strongest [49, 50]. Although no prospective studies of tomato intake and gastric cancer were identified, the consistent inverse association observed in diverse populations strongly suggests a protective effect of tomato or lycopene consumption on gastric cancer [46].

Limonene

Citrus fruits (particularly oranges, mandarins, lemons and limes) contain significant quantities of limonene in the peel and smaller quantities in the pulp. Citrus is one of the most heavily produced crops in the world (103,639,732 metric tonnes in 2000). Recently, the industrialized countries have seen a marked increase in the level of processed citrus fruits, which has generated a large number of by-products derived from peel. These include pectin, concentrated hesperidin and naringin, dried peel, molasses, alcohol, carotenoids, seed oil and citric acid and essential oils, which contain a great proportion of limonene [51].

Chemical structure and occurrence

Limonene is a monocyclic monoterpene [1-methyl-4-(1-methylethenyl) cyclohex-1-ene] that appears to have been formed by the union of two isoprene molecules. Because C-4 in limonene is asymmetric, this terpene naturally occurs in an optically active form as the d- or l-limonene. Limonene's molecular structure may be altered chemically, by oxidation yielding alcohols, aldehydes, ketones or acids, or reduction and isomerization, resulting in the formation of a great variety of simple terpenes or metabolites of the essential oils as is shown in Fig. 2.5.

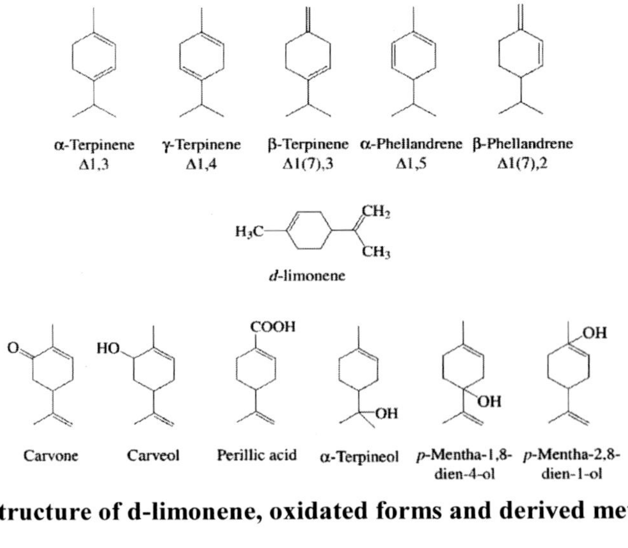

Fig. 2.5: Structure of d-limonene, oxidated forms and derived methadienes

Oil yield, properties and *d*-limonene content are affected by many aspects such as seasonal factors, ripening, soil types, climatic conditions, cultural practices, cultivar and varietal factors, scion and rootstock, etc. Variation of oil content was also partially correlated with fruit size or peel thickness. Nitrogen fertilization increases the oil yield, while potassium fertilization had no significant effect on peel oil content [52].

Limonene as functional food

Human consumption of *d*-limonene occurs mainly through citrus fruits, juices and flavored beverages. The daily per capita consumption from both its natural occurrence in food and as a flavouring agent is estimated to be 0.27 mg/kg body weight/day [53]. Nevertheless, the main physiological action reported for limonene concerns cancer prevention and therapy. In fact, there is some evidence to suggest that *d*-limonene may afford protection from chemical tumor formation. Limonene has been shown to block and suppress carcinogenic events, because certain biochemical pathways in tumor tissues may be sensitive to the inhibitory action of *d*-limonene and certain terpene constituents of essential oils [43]. Endogenous antioxidant systems such as glutathione *S*-transferase (GST) are involved in the detoxification pathways of carcinogens and, therefore, in the prevention of carcinogenesis. The effects of *d*-limonene on rat mammary tumors were investigated. There was a highly significant increase in the regression of the first tumors in the rats fed *d*-limonene. In addition, *d*-limonene given at this time inhibited the formation of subsequent multiple tumors in these rats [54].

Limonene has demonstrated efficacy in pre-clinical models of breast and colon cancers. Seventy samples of various citrus juice products were analysed by gas chromatography. Preliminary results suggest that sources and preparation or storage practices have major influences on *d*-limonene levels. Many commonly consumed citrus beverages contain no *d*-limonene. The largest sources of *d*-limonene (mg/L), identified to date, are fresh Mediterranean-style lemonade (1027) and limeade (402), followed by commercial orange juices in cans (51.9–85), in glasses (35.4–60.3); from concentrates (27.5–90) and grapefruit juices in cans (13.5–50) [55]. Several mechanisms may account for the antitumorigenic effects of monoterpenes. For example, many monoterpenes inhibit the post-translational isoprenylation of cell growth-regulatory proteins. In addition, monoterpene-treated, regressing rat mammary tumors exhibit increased expression of transforming growth factor concomitant with tumor remodeling/redifferentiation to a more benign phenotype. Monoterpenes are effective, nontoxic dietary antitumor agents which act through a variety of mechanisms of action and hold promise as a novel class of antitumor drugs for human cancer [56].

In-vitro, *d*-limonene dissolved human gallstones within two hours. In animals, infusion of d-limonene into the gallbladder dissolved and disintegrated gallstones, which were excreted through the common bile duct [57]. In 14 in-patients post gallstone surgery, infusion of 20 mL *d*-limonene every other day dissolved gallstones overlooked during surgery. In some patients gallstone dissolution occurred after only three infusions. A study with 200 patients reported a direct infusion of 20-30 mL *d*-limonene (97% solution) completely or partially dissolved gallstones in 141 patients. Stones completely dissolved in 96 cases (48%); partial dissolution was observed in 29 cases (14.5%); and in 16 cases (8%) complete dissolution was achieved with the inclusion of hexamethaphosphate (HMP), a chelating agent that can dissolve bilirubin calcium stones. All the stones were between 0.5 and 1.5 cm with an average diameter of 1.0 cm. The duration of the treatment ranged from three weeks to four months [58]. The *d*-limonene has also been shown to be effective in relieving occasional heartburn and gastroesophageal reflux disorder (GERD) [59].

Anthocyanins and Betalains

Anthocyanins and betalains play important roles both in plant physiology, visual attraction for pollinators and seed dispersers, but also in food mainly defining its aesthetic value. Since anthocyanin and betalain structures allow to predict only part of their appearance, additional chemical and anatomical functions are required to modulate the appearance of plants and coloured food. Physiological effects that the same pigments exert in plants are supposedly similar to those which they show in humans following ingestion of coloured food. Therefore, anthocyanins and betalains both in fresh and also processed fruit and vegetables serve two major functions. They improve the overall appearance, and also contribute to consumers' health and well-being.

Structure and occurrence

The anthocyanins are a sub-group within the flavonoids characterised by a C_6–C_3–C_6-skeleton. Eighteen different aglycones were reported, among which the six most common are pelargonidin, cyanidin, delphinidin, peonidin, petunidin and malvidin as shown in Fig. 2.6 [60]. These structures differ in their hydroxylation and methoxylation patterns producing shades from orange-red (pelargonidin) to blue-violet (delphinidin) at a non-physiological pH around 1.0.

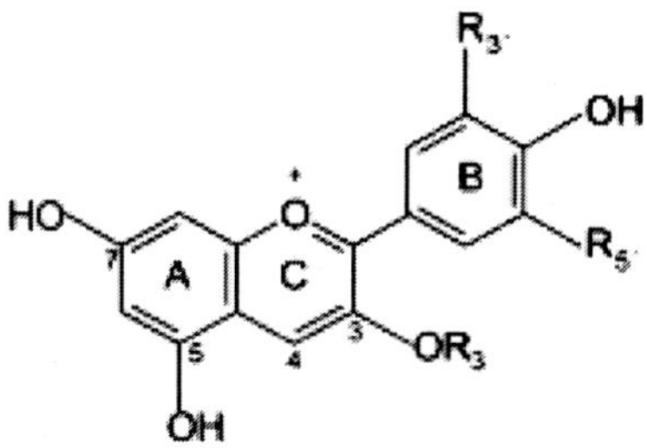

Anthocyanin	R_3	R_{3-}	R_5	λ_{max} [nm]
Pelargonidin	H	H	H	520
Cyanidin	H	OH	H	535
Delphinidin	H	OH	OH	546
Peonidin	H	OCH_3	H	532
Petunidin	H	OCH_3	OH	543
Malvidin	H	OCH_3	OHC_3	542
Pelargonidin-3-glucoside	Gle	H	H	516
Cyanidin-3-gluocside	Gle	H	H	530
Delphinidin-3-glucoside	Gle	OH	OH	543
Peonidin-3-glucoside	Gle	OCH_3	H	536
Petunidin-3-glucoside	Gle	OCH_3	OH	546
Malvidin-3-glucoside	Gle	OCH_3	OCH_3	546

Fig. 2.6: Structure of anthocyanin

With the exception of one study where betacyanins were deposited as crystals in vacuolar spherical bodies [61], betalains are usually dissolved in the vacuolar sap as bis-anions [62]. Structurally, the betaxanthins are condensation products of betalamic acid and various amino compounds, whereas the betacyanins are characterized by a cyclo-dopa structure with additional substitutions through varying glycosylation and acylation patterns (Fig. 2.7)

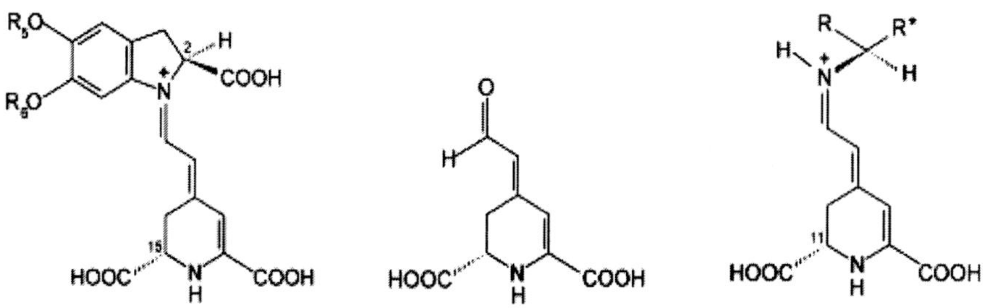

Fig. 2.7: Structure of betacyanin

Anthocyanins and betalins as functional food

Apart from their physiological roles in plants, anthocyanins are regarded as important components in human nutrition [63] which is supported by numerous studies that report a high positive correlation of fruit or vegetable pigment content and antioxidant capacities [64-67]. Moreover, structural variations in anthocyanins seem to influence antioxidant activities [68, 69]. Since anthocyanin mixtures exerted higher radical scavenging potencies than purified pigments on the same quantitative level [69], synergistic effects have been assumed suggesting that whole extracts are biologically more active than isolates [70, 71].

Little has been published with respect to the physiological role of betalains in mammals. In earlier reports, it was found that red beet neither exerted hepatotoxic nor mutagenic action [72-74]. Kapadia et al. (1996) showed a significant inhibitory effect of beetroot towards skin and lung cancer in mice. In other animal feeding trials, purple cactus pear [75] and garambullo [76] proved to be devoid of toxicity and their pigments did not provoke any allergic reactions [77]. The excretion of red urine after ingestion of beetred in 10–14% of the population, the so-called beeturia, was classified as an idiosyncratic reaction being dependent on individual's physiological constitutions rather than representing a pathological event or a genetic defect [78]. A very recent study investigated structure–activity relationships of various betaxanthins and betacyanins from members of the Amaranthaceae with respect to free radical scavenging capacities [79]. The antioxidant potential was found to be related to structural features of the respective betalains. In betaxanthins, an increasing number of hydroxy and imino residues improved free radical scavenging. In betacyanins, glycosylation reduced activity while acylation generally raised the antioxidant potential. Furthermore, 5-O-glycosylated structures produced lower antioxidant values than 6-O-glycosylated betacyanins. Future studies will have to ascertain the alleged promising physiological and pharmacological effects of betalains on humans published only very recently [80].

Putting anthocyanins and betalains into this nutritional context, caution is warranted to judge their actual bioactivities. Besides, factors that influence their release from the respective food matrix, interactions with the same will also modulate their absorption [81, 82]. In future studies, the exact pharmacokinetics and parameters governing gastrointestinal absorption as suggested by Ekmekcioglu (2002) as well as the tissue distribution of ingested compounds need to be assessed. Charge alterations, structural transformations with changing pH and the formation of degradation products among others, will also have to be considered to assess the benefit of these plant-derived secondary compounds [83, 84].

Secondary Plant Metabolites

Plants produce many thousands of so-called secondary metabolites, which are substances that are additional to those involved in the primary processes of respiration, photosynthesis, growth and development. Primary metabolites are distributed throughout the plant kingdom and often throughout much of the animal kingdom as well. In contrast to the ubiquitous presence of primary metabolites, each plant produces its own complement of secondary metabolites that serve a variety of diverse functions. Some of these compounds, such as several of the carotenoids, are widespread in the plant kingdom but many others may be confined to just a small number of plants. This is one reason why consumers are also advised to consume a variety of fruits and vegetables. Some of these secondary metabolites of plants have well-established medicinal properties such as:

- The salicylates (aspirin-like substances) from willow bark.
- The cardiac glycosides (including digoxin) found in the purple foxglove, *Digitalis purpurea*, and used in the treatment of heart failure.
- The anti-malarial drug quinine extracted from the bark of *Cinchona succiruba.*

Crozier (2003) classified these secondary metabolites of plants into three major groupings based upon their synthetic origins in plants rather than their dietary effects in animals and people [85]. Details of the biosynthetic origins and chemical structures of these secondary plant metabolites may be found in this source. The three major groupings used by Crozier are:

- The terpenoids
- The phenols and polyphenols
- Sulphur-containing compounds and the nitrogen-containing alkaloids

Plant metabolites as functional food

Plants synthesize certain compounds called secondary metabolites which are not naturally produced by humans but can play vital roles in protection or destruction of human health. One such group of metabolites is phytoestrogens found in nuts, oilseeds, soy. Phytoestrogens are a group of compounds found in plants that are structurally or functionally similar to the hormone estradiol, and act as selective estrogen receptor modulators. Epidemiological studies show an inverse correlation between consumption of a high phytoestrogen-rich diet, as seen in traditional Asiatic societies, and the incidence of certain hormone-dependent cancers such, as breast and prostate cancer. Phytoestrogens may protect against hormone-dependent cancers through their ability to lower the level of endogenous estrogens and their genotoxic metabolites, activation of estrogen receptor beta and peroxisome proliferator activated receptor gamma, inhibiting cell signaling pathways of growth factors, upregulating expression of antioxidant genes, tumor suppressor genes and immunomodulatory effects [86, 87].

The potential mechanisms by which bioactive substances, including plant secondary metabolites act to reduce the risk of chronic degenerative diseases such as cancer and heart disease are explained [88]. For example, it has been suggested that certain isothiocyanates derived from dietary glucosinolates (e.g. from brassica vegetables) may inhibit this metabolic activation of potential carcinogen. The sulphur-containing compounds found in the Allium (onion) genus, as well as some isothiocyanates, may have detoxification effect which involves conversion of foreign chemical to harmless metabolites. This has led to suggestions that garlic and broccoli, for example, may have anti-cancer properties. Many plant substances with antioxidant activity such as the antioxidant vitamins, carotenoids and many flavonoids may reduce high blood cholesterol concentration, and more specifically an elevated level of low density lipoprotein (LDL) cholesterol in plasma is causally associated with increased risk of atherosclerosis and coronary heart disease. Some plant chemicals may reduce plasma cholesterol concentration and thus exert a protective effect against those cardiovascular diseases that are precipitated by atherosclerosis. Some plant sterols that are structurally similar to cholesterol may reduce blood cholesterol by inhibiting its absorption from the gut. Sulphur compounds in garlic and other members of the Allium genus are also claimed to lower blood cholesterol levels in high doses. High levels of several carotenoids are found in parts of the human retina where they may protect it from light-induced free radical damage by virtue of their antioxidant effects. It has been suggested that carotenoids may help to reduce the degenerative changes in the retina seen in age-related macular degeneration [88].

Many of the claims about the cancer preventing or heart disease preventing properties of particular foods or supplements are also based upon such studies. Evidence from such studies should be regarded with extreme scepticism unless the safety and efficacy of any supplement has been substantiated in large well-designed clinical trials.

Summary

Plant foods contain many bioactive compounds in addition to those which are traditionally considered as nutrients, such as vitamins and minerals. These physiologically active compounds, referred to simply as phytochemicals, are produced via secondary metabolism in relatively small amounts. Some groups of phytochemicals which have or appear to have significant health potentials are: carotenoids, phenolic compounds (flavonoids, phytoestrogens, phenolic acids), phytosterols and phytostanols, tocotrienols, organosulfur compounds and dietary fiber. Overwhelming evidences from epidemiological, *in-vivo*, *in-vitro*, and clinical trial data indicate that a plant based diet can reduce the risk of chronic diseases such as cancer and heart diseases. Scientists have identified thousands of different phytochemicals found in vegetables and fruits. Eating lots of plant foods rich in phytochemicals may help to prevent at least one in every five cases of cancer as well as other serious ailments such as heart diseas.

References

1. Aburjai, T. and F.M. Natsheh, *Plants used in cosmetics.* Phytotherapy Research, 2003. **17**(9): 987-1000.
2. Ross, I.A., *Medicinal plants of the world, Volume 3: Chemical constituents, traditional and modern medicinal uses*. 2007 Springer.U.K.
3. Nahler, G. and G. Nahler, *Traditional herbal medicinal product.* Dictionary of Pharmaceutical Medicine, 2009: 184-184.
4. US-FDA, *Department of Health and Human Services*, in *Food labeling: health claims; Oats and Coronary Heart disease* Federal Register.1997, 3584-3601.
5. US-FDA, *Department of Health and Human Services*, in *Food labeling: health claims; soluble fiber from certain foods and coronary heart disease*, 1998, Federal Register. p. 8103-8121.
6. US-FDA, *Department of Health and Human Services*, in *Food labeling: health claims; soy protein, and coronary heart disease* , Federal Register : 1999, 57700-57733.
7. US-FDA, *Department of Health and Human Services*, in *Food labeling: health claims; plant sterol esters and coronary heart disease* 2000, Federal Register. p. 54686-54739.
8. Setchell, K.D., et al., *Nonsteroidal estrogens of dietary origin: possible roles in hormone-dependent disease.* Americal J Clinnical Nutrition, 1984, **40**(3): 569-78.
9. Mullineaux, P.M. and G.P. Creissen, *Opportunities for the genetic manipulation of antioxidants in plant foods.* Biochemical Society Transactions, 1996. **24**(3): 829-35.
10. La Camera, S., et al., *Metabolic reprogramming in plant innate immunity: the contributions of phenylpropanoid and oxylipin pathways.* Immunol Review, 2004. **198**: 267-84.
11. Grunwald, C., *Steroids* in *Encyclopedia of Plant Physiology*, M.H.Z. A. Pirson Editor 1980. 221-239.
12. Goad, J.L., ed. *Methods in plant biochemistry*. ed. D.V.B. B.V. Charlewood. Vol. 7. 1991, Academic Press: London. 369-434.
13. J.D. Connolly and R.A. Hill, *Methods in plant biochemistry*, ed. D.V.B. B.V. Charlewood 1991, London Academic Press. pp 331-368
14. Gül, M.K. and S. Amar, *Sterols and the phytosterol content in oilseed rape (Brassica napus L.).* Journal of Cell and Molecular Biology, 2006. **5**: 71-79.
15. Peterson, D.W., *Effect of soybean sterols in the diet on plasma and liver cholesterol in chicks.* Proceeding of Society for Exp Biology 1951. **78**(1): 143-7.
16. Jones, P.J., et al., *Dietary phytosterols as cholesterol-lowering agents in humans.* Can J Physiology Pharmacology, 1997. **75**(3): 217-27.
17. Sudhop, T., et al., *Inhibition of Intestinal Cholesterol Absorption by Ezetimibe in Humans.* Circulation, 2002. **106**(15): 1943-1948.

18. Salen, G., et al., *Ezetimibe Effectively Reduces Plasma Plant Sterols in Patients With Sitosterolemia.* Circulation, 2004. **109**(8): 966-971.
19. Law, M., *Plant sterol and stanol margarines and health.* British Medical Journal, 2000. **320**: 861-864.
20. Weihrauch, J.L. and J.M. Gardner, *Sterol content of foods of plant origin.* Journal of the American Dietetic Association, 1978. **73**(1): 39-47.
21. Piironen, V., et al., *Plant sterols: biosynthesis, biological function and their importance to human nutrition.* Journal of the Science of Food and Agriculture, 2000. **80**(7): 939-966.
22. Normen, L., et al., *Plant sterols in vegetables and fruits commonly consumed in Sweden.* European Journal of Nutrition, 1999. **38**(2): 84-9.
23. Hartmann, M.-A., *Plant sterols and the membrane environment.* Trends in plant science, 1998. **3**(5): 170-175.
24. Piironen, V., et al., *Plant sterols in vegetables, fruits and berries.* Journal of the Science of Food and Agriculture, 2003. **83**(4): 330-337.
25. Nissinen, M., et al., *Micellar distribution of cholesterol and phytosterols after duodenal plant stanol ester infusion.* Am J Physiol Gastrointest Liver Physiol, 2002. **282**(6): p. 1009-15.
26. Qullez, J., P. Garcla-Lorda, and J. Salas-Salvado, *Potential uses and benefits of phytosterols in diet: present situation and future directions.* Clinical Nutrition, 2003. **22**(4): 343-51.
27. Tasan, M., et al., *Phytosterols as functional food ingredients.* Journal of Tekirdag Agricultural Faculty, 2006.
28. Anonymous, *EU Scientific Committee on Food. General view on the long-term effects of the intake of elevated levels of phytosterols from multiple dietary sources, with particular attention*, 2002.
29. Jones, P.J.H. and S.S. AbuMweis, *Phytosterols as functional food ingredients: linkages to cardiovascular disease and cancer.* Current Opinion in Clinical Nutrition & Metabolic Care, 2009. **12**(2): 147-151.
30. Anonymous, *Functional foods fact sheet: Plant stanols and sterols*, 2003. 21-45
31. Anonymous, *Commission Regulation (EC) No.608/2004 of March 2004 concerning the labeling of foods and food ingredients with added phytosterols, phytosterols esters, and phytostanols and/or phytostanol esters*, 2004.
32. S.V Jovanovic, et al., *Antioxidant properties of flavonoids: Reduction potentials and electron transfer reactions of flavonoid radicals*, in *Flavonoids in Health and Disease*, C.R.-E. Lester Packer, Catherine Rice-Evans, Editor 1998, Marcel Dekker, Inc: New York. p. 137-161.
33. Dixon, R.A. and C.L. Steele, *Flavonoids and isoflavonoids - a gold mine for metabolic engineering.* Trends in plant science, 1999. **4**(10): 394-400.
34. Nestel, P., *Isoflavones: their effects on cardiovascular risk and functions.* Current Opinion in Lipidology, 2003. **14**(1): 3-8.
35. Takabe, W., et al., *Oxidative stress promotes the development of transformation: involvement of a potent mutagenic lipid peroxidation product, acrolein.* Carcinogenesis, 2001. **22**(6): 935-41.
36. Castelluccio, C., et al., *Antioxidant potential of intermediates in phenylpropanoid metabolism in higher plants.* FEBS Letters, 1995. **368**(1): 188-192.
37. Nigdikar, S.V., et al., *Consumption of red wine polyphenols reduces the susceptibility of low-density lipoproteins to oxidation in vivo.* The American Journal of Clinical Nutrition, 1998. **68**(2): 258-65.
38. Ishikawa, T., et al., *Effect of tea flavonoid supplementation on the susceptibility of low-density lipoprotein to oxidative modification.* The American Journal of Clinical Nutrition, 1997. **66**(2): 261-6.

39. Hollman, P.C.H. and M.B. Katan, *Absorption, metabolism and health effects of dietary flavonoids in man.* Biomedicine & Pharmacotherapy, 1997. **51**(8): 305-310.
40. Yang, G.-y., et al., *Characterization of Early Pulmonary Hyperproliferation and Tumor Progression and Their Inhibition by Black Tea in a 4-(Methylnitrosamino)-1-(3-pyridyl)-1-butanone-induced Lung Tumorigenesis Model with A/J Mice.* Cancer Research, 1997. **57**(10): 1889-1894.
41. Rodriguez-Mateos, A., et al., *Bioavailability, bioactivity and impact on health of dietary flavonoids and related compounds: an update.* Archives of toxicology, 2014. **88**(10): 1803-1853.
42. Nguyen, M.L. and S.J. Schwartz, *Lycopene: chemical and biological properties.* Food technology (USA), 1999.
43. Mazza, G., *Functional foods: biochemical and processing aspects.* Vol. 1. 1998: CRC Press,London
44. Gerster, H., *The potential role of lycopene for human health.* Journal of the American College of Nutrition, 1997. **16**(2): 109-126.
45. Stahl, W. and H. Sies, *Lycopene: a biologically important carotenoid for humans?* Archives of Biochemistry and Biophysics, 1996. **336**(1): 1-9.
46. Giovannucci, E., *Tomatoes, tomato-based products, lycopene, and cancer: review of the epidemiologic literature.* Journal of the National Cancer Institute, 1999. **91**(4): 317-331.
47. Rao, A.V. and S. Agarwal, *Role of antioxidant lycopene in cancer and heart disease.* Journal of the American College of Nutrition, 2000. **19**(5): 563-569.
48. Colditz, G.A., et al., *Increased green and yellow vegetable intake and lowered cancer deaths in an elderly population.* The American Journal of Clinical Nutrition, 1985. **41**(1): 32-36.
49. Modan, B., H. Cuckle, and F. Lubin, *A note on the role of dietary retinol and carotene in human gastro-intestinal cancer.* International Journal of Cancer, 1981. **28**(4): 421-424.
50. Buiatti, E., et al., *A case-control study of gastric cancer and diet in Italy.* International Journal of Cancer, 1989. **44**(4): 611-616.
51. Chiralt, A., et al., *Limonene from citrus*, in *Functional foods: biochemical and processing aspects*, G.M. John Shi, Marc Le Maguer, Editor 1998, CRC Press. p. 169 - 170.
52. Sinclair, W.B., *Biochemistry and physiology of the lemon and other citrus fruits.* 1984.
53. Schrankel, K.R., *Safety evaluation of food flavorings.* Toxicology, 2004. **198**(1): 203-211.
54. Elegbede, J.A., et al., *Regression of rat primary mammary tumors following dietary d-limonene.* Journal of the National Cancer Institute, 1986. **76**(2): 323-325.
55. Hakim, I., T. McClure, and D. Liebler, *Assessing dietary d-limonene intake for epidemiological studies.* Journal of Food Composition and Analysis, 2000. **13**(4): 329-336.
56. Crowell, P.L., A.S. Ayoubi, and Y.D. Burke, *Antitumorigenic effects of limonene and perillyl alcohol against pancreatic and breast cancer*, in *Dietary Phytochemicals in Cancer Prevention and Treatment*1996, Springer. p. 131-136.
57. Igimi, H., T. Hisatsugu, and M. Nishimura, *The use of d-limonene preparation as a dissolving agent of gallstones.* The American Journal of Digestive Diseases, 1976. **21**(11): 926-939.
58. Igimi, H., et al., *Medical dissolution of gallstones.* Digestive diseases and sciences, 1991. **36**(2): 200-208.
59. Ueno, R., *Method for treating gastrointestinal disorder*, 2005, Google Patents.
60. Mazza, G. and E. Miniati, *Anthocyanins in fruits, vegetables, and grains*1993: CRC press.
61. Iwashina, T., S. Ootani, and K. Hayashi, *On the pigmented spherical bodies and crystals in tepals of cactaceous species in reference to the nature of betalains or flavonols.* The botanical magazine, 1988. **101**(2): 175-184.

62. Wyler, H., *Die betalaine.* Chemie in unserer Zeit, 1969. **3**(5): 146-151.
63. Ross, J.A. and C.M. Kasum, *Dietary flavonoids: bioavailability, metabolic effects, and safety.* Annual review of Nutrition, 2002. **22**(1): 19-34.
64. Abuja, P.M., M. Murkovic, and W. Pfannhauser, *Antioxidant and prooxidant activities of elderberry (Sambucus nigra) extract in low-density lipoprotein oxidation.* Journal of Agricultural and Food Chemistry, 1998. **46**(10): 4091-4096.
65. Cao, G., E. Sofic, and R.L. Prior, *Antioxidant capacity of tea and common vegetables.* Journal of Agricultural and Food Chemistry, 1996. **44**(11): 3426-3431.
66. Halvorsen, B.L., et al., *A systematic screening of total antioxidants in dietary plants.* The Journal of nutrition, 2002. **132**(3): 461-471.
67. Moyer, R.A., et al., *Anthocyanins, phenolics, and antioxidant capacity in diverse small fruits: Vaccinium, Rubus, and Ribes.* Journal of Agricultural and Food Chemistry, 2002. **50**(3): 519-525.
68. Rice-Evans, C.A., N.J. Miller, and G. Paganga, *Structure-antioxidant activity relationships of flavonoids and phenolic acids.* Free radical biology and medicine, 1996. **20**(7): 933-956.
69. Stintzing, F.C., et al., *Color and antioxidant properties of cyanidin-based anthocyanin pigments.* Journal of Agricultural and Food Chemistry, 2002. **50**(21): 6172-6181.
70. Southon, S., R. Faulks, and W. Jongen, *Health benefits of increased fruit and vegetable consumption.* Fruit and vegetable processing: Improving quality, 2002: 5-22.
71. Vinson, J.A., et al., *Phenol antioxidant quantity and quality in foods: fruits.* Journal of Agricultural and Food Chemistry, 2001. **49**(11): 5315-5321.
72. Schwartz, S.J. and H. Joachim, *Identification of betanin degradation products.* Zeitschrift für Lebensmittel-Untersuchung und Forschung, 1983. **176**(6): 448-453.
73. Elbe, J.H. and S.J. Schwartz, *Absence of mutagenic activity and a short-term toxicity study of beet pigments as food colorants.* Archives of toxicology, 1981. **49**(1): 93-98.
74. Kapadia, G.J., et al., *Chemoprevention of lung and skin cancer by Beta vulgaris (beet) root extract.* Cancer letters, 1996. **100**(1): 211-214.
75. Krifa, F., et al. *Pourpre de barbarie»: nouveau colorant naturel extrait d'Opuntia stricta.* in *Annales des falsifications, de l'expertise chimique et toxicologique*. 1994. Société des experts-chimistes de France.
76. Reynoso, R., T. Giner, and E.G. de Mejia, *Safety of a filtrate of fermented garambullo fruit: biotransformation and toxicity studies.* Food and chemical toxicology, 1999. **37**(8): 825-830.
77. Kuramoto, Y., et al., *Effect of natural food colorings on immunoglobulin production in vitro by rat spleen lymphocytes.* Bioscience, biotechnology, and biochemistry, 1996. **60**(10): 1712-1713.
78. Mitchell, S., *Food idiosyncrasies: beetroot and asparagus.* Drug Metabolism and Disposition, 2001. **29**(4): 539-543.
79. Cai, Y., M. Sun, and H. Corke, *Antioxidant activity of betalains from plants of the Amaranthaceae.* Journal of Agricultural and Food Chemistry, 2003. **51**(8): 2288-2294.
80. Wettasinghe, M., et al., *Phase II enzyme-inducing and antioxidant activities of beetroot (Beta vulgaris L.) extracts from phenotypes of different pigmentation.* Journal of Agricultural and Food Chemistry, 2002. **50**(23): 6704-6709.
81. Arts, M.J., et al., *Interactions between flavonoids and proteins: effect on the total antioxidant capacity.* Journal of Agricultural and Food Chemistry, 2002. **50**(5): 1184-1187.
82. Murkovic, M., et al., *Analysis of anthocyanins in plasma for determination of their bioavailability.* Journal of Food Composition and Analysis, 2000. **13**(4): 291-296.
83. Ekmekcioglu, C., *A physiological approach for preparing and conducting intestinal bioavailability studies using experimental systems.* Food Chemistry, 2002. **76**(2): 225-230.

84. Balimane, P.V., S. Chong, and R.A. Morrison, *Current methodologies used for evaluation of intestinal permeability and absorption.* Journal of Pharmacological and Toxicological Methods, 2000. **44**(1): 301-312.
85. Crozier, A., *Classification and biosynthesis of secondary plant products: an overview.* Plants: Diet and Health: The Report of a British Nutrition Foundation Task Force, 2003: p. 27-48.
86. Mathew, B., et al., *Phytoestrogens and cancer.* Journal of Institute of Medicine, 2009. **31**(1): 11-19.
87. Adlercreutz, H., *Phytoestrogens: epidemiology and a possible role in cancer protection.* Environmental Health Perspectives, 1995. **103**(Suppl 7): 103.
88. Jackson, M., *Potential mechanisms of action of bioactive substances found in foods.* Plants: Diet and Health: The Report of a British Nutrition Foundation Task Force, 2003: p. 65-75.

3

Dairy Ingredients

Introduction

Dairy ingredients are of significant importance in the formulation of many food products. Additionally, the incorporation of dairy ingredients, comprehensively recognized by the consumer as "natural" intensifies the odds of success of packaged foods and adds functional properties of the dairy ingredients in the targeted product. Dairy ingredients can be extracted from fluid milk in the form of cream, butter, condensed milk, dry milk, cheese, and whey products [1]. They provide beneficial functionality to foods, such as delivery of key nutrients, water management, fat-holding capacity, emulsification capability, viscosity creation, gel formation, and foam generation apart from providing desirable attributes of texture and flavour. Various ingredients and claims associated with dairy foods are shown in Table 3.1.

The dairy industry is in an excellent position to develop and exploit the functional food market. Growth of the dairy industry is set to continue with total world milk production of 780 million tonnes during the year 2013.

Table 3.1: Ingredients and claims associated with functional dairy foods

Ingredient	Sources	Examples of Claim Areas
Minerals	Calcium, Casein peptides	Optimum growth and development, dental health, osteoporosis
Fatty acids	CLA	Heart disease, cancer prevention, weight control
Prebiotics/ carbohydrates	Galacto-oligosaccharides Lactulose Lactose	Digestion, pathogen prevention, gutflora balance, immunity, lactose intolerance
Probiotics	Lactic acid bacteria Bifidobacteria	Digestion, immunity, vitamin production, heart disease, antitumor activity, remission of inflammatory bowel disease, prevention of allergy, alleviation of diarrhea
Proteins/peptides	Caseins,whey proteins, immunoglobulins, lactoferrin, glycoproteins, specific peptides	Immunomodulation, growth, antibacterial activity, dental health, hypertension regulation (angiotensin inhibitors)

Example of Functional Dairy Ingredients

Whey based ingredients

According to the Food and Drug Administration Code (Title 21 CFR Section 184.1979), whey is the liquid substance obtained by separating the coagulum from milk, cream, or skim milk in cheese making. Whey also may be obtained from curd formation by the direct acidification of milk. Whey is a dilute liquid containing approximately 6% solids, of which approximately 5.4% is lactose and 0.69 % is protein. The whey protein is a complex mixture of β-lactoglobulin (β-lg, approximately 55%), α-lactalbumin (α-Ig approximately 24%), bovine serum albumin (BSA, approximately 5%), immunoglobulins (Ig approximately 15%), and several other minor proteins [3]. Whey also can include more than 20% glycomacropeptide (GMP) if it is from the cheese manufacture [4].

The proteins found in whey streams include globular proteins (β-lg α-lg, BSA), immunoglobulins (Ig -G, Ig -A, and Ig -M), some minor proteins (lactoperoxidase, lactoferrin, and lysosyme), and polypeptides (proteose -peptone 3, and proteose -peptone 5) and glycomacropeptides (present in rennet casein or cheese whey only) [1]. The functional properties of whey ingredients are defined by the physico-chemical properties of the whey proteins that contribute to the desired characteristics in food products [5]. The conformation and functional properties of whey proteins are interrelated and governed by changes in the globular folded structure of the molecule [6]. β-lactoglobulin comprises more than 50% of the whey proteins in bovine milk. Because of its prevalence in bovine milk, the properties of β-lactoglobulin determine the properties of whey protein concentrates [7]. β-Lactoglobulin contains two disulfide bonds and one sulfhydryl group. It is through thiol/ disulfide interchanges that protein aggregation is initiated. It is also through heat induced sulfhydryl and disulfide interactions of α-lactalbumin with β-lactoglobulin that the heat sensitivity of α-lactalbumin is increased [7]. β-Lactoglobulin is not present in human milk. Treatments to reduce the allergenicity of bovine milk generally involve the hydrolysis of β-lactoglobulin.

Whey protein as functional food

Whey protein, with its high protein quality score and high percentage of branched chain amino acids (BCAA), has long been popular in the exercise industry as a muscle building supplement. However, research suggests it may have far wider applications as a functional food in the management of conditions such as cancer, hepatitis B, HIV, cardiovascular disease, osteoporosis and even chronic stress. [8,9].

Whey protein, a by-product recognized as a valuable food ingredient with important nutritional and functional properties is gaining acceptance as functional food ingredient. Commercial whey proteins are considered a GRAS substance for food product applications. It constitutes about 85–90% of the volume of the milk used for transformation into ripened cheese, and it retains about 55% of the milk nutrients. Much attention is focused on whey proteins. Not only is their biological value superior to most other proteins, but whey proteins also have a high content of sulfur containing amino acids, which support antioxidant functions. Hydrolyzed whey protein based formulations are beneficial for infants intolerant to cow's milk protein. During hydrolysis proteins are broken down into peptides of different sizes and free amino acids as a result of cleavage of peptide bonds.

Whey is made up of a number of proteins including β-lactoglobulin, α-lactalbumin, bovine serum albumin (BSA) and glycomacropeptide (GMP). Collectively, these contain a full spectrum of amino acids including the BCAA leucine, isoleucine and valine. BCAA are required for tissue growth and repair and leucine in particular plays a key role in the translation-initiation of protein synthesis. The sulphur-containing amino acids cysteine and methionine are also found in high concentrations in whey protein, contributing to enhanced immune function through intracellular conversion to glutathione. Interestingly, GMP, although a source of BCAAs, lacks the aromatic amino acids phenylalanine, tryptophan and tyrosine. This makes it a viable protein option for individuals with PKU (Phenylketonuria) [8].

Whey protein products possess a number of functional and nutritional properties to varying degrees. However, no single whey protein can be used in a wide spectrum of food products because very few proteins exhibit the needed multiple functionalities. Whey proteins contain high levels of the branched chain amino acids leucine, isoleucine, and valine. These amino acids can be metabolized as an energy source and thus are considered to be useful in sport drinks, where they assist both as an energy source and as a protein source, and for muscle building or sparing [10]. The putative biological activity of whey proteins and their derived peptides, the rapidly expanding functional food market, and the growing consumer demand for high-quality protein combine to provide an attractive foundation for the development of whey protein and whey protein fractions as functional food ingredients. To strengthen this foundation and to provide evidence for health claims associated with functional foods containing whey proteins, scientific substantiation of the putative physiological activity of these proteins will be essential in both animal and human trials. The health benefits associated with different components of whey are shown in Figure 3.1.

For hundreds or perhaps thousands of years, various cultures and societies have used whey in the prevention and treatment of disease as part of folk medicine. Indeed, there are references to various health benefits of whey in the Italian literature dating back to the early 17th century [11]. During this period of history, whey was often prescribed for the treatment of numerous ailments, including acute septic conditions and gastrointestinal infections. Rather than rely on folk medicine, however, careful scientific investigation of the physiological functionality of the whey protein components and their possible role in disease prevention have been pursued. For these studies much of the attention has been focussed on colorectal cancer or bowel cancer which is one of the most common neoplastic diseases affecting men and women in western countries. There is increasing evidence that components of the diet in these countries, particularly red meat and animal fat, are linked in a causative fashion to the high incidence of colon cancer [12]. Dairy foods are usually included in animal based diets, rich in red meat and animal fat, and are capable of contributing a significant amount of dietary protein and/or fat, although the fat issue is being addressed through the increasing availability of low-fat dairy products.

Whey's antioxidant and detoxifying activity is most likely linked to its contribution to the synthesis of glutathione (GSH). Cysteine (which contains an antioxidant thiol group) combines with glycine and glutamate to form GSH1 [13]. GSH is the major endogenous antioxidant produced by cells, providing protection for RNA, DNA and proteins via its redox cycling from GSH (the

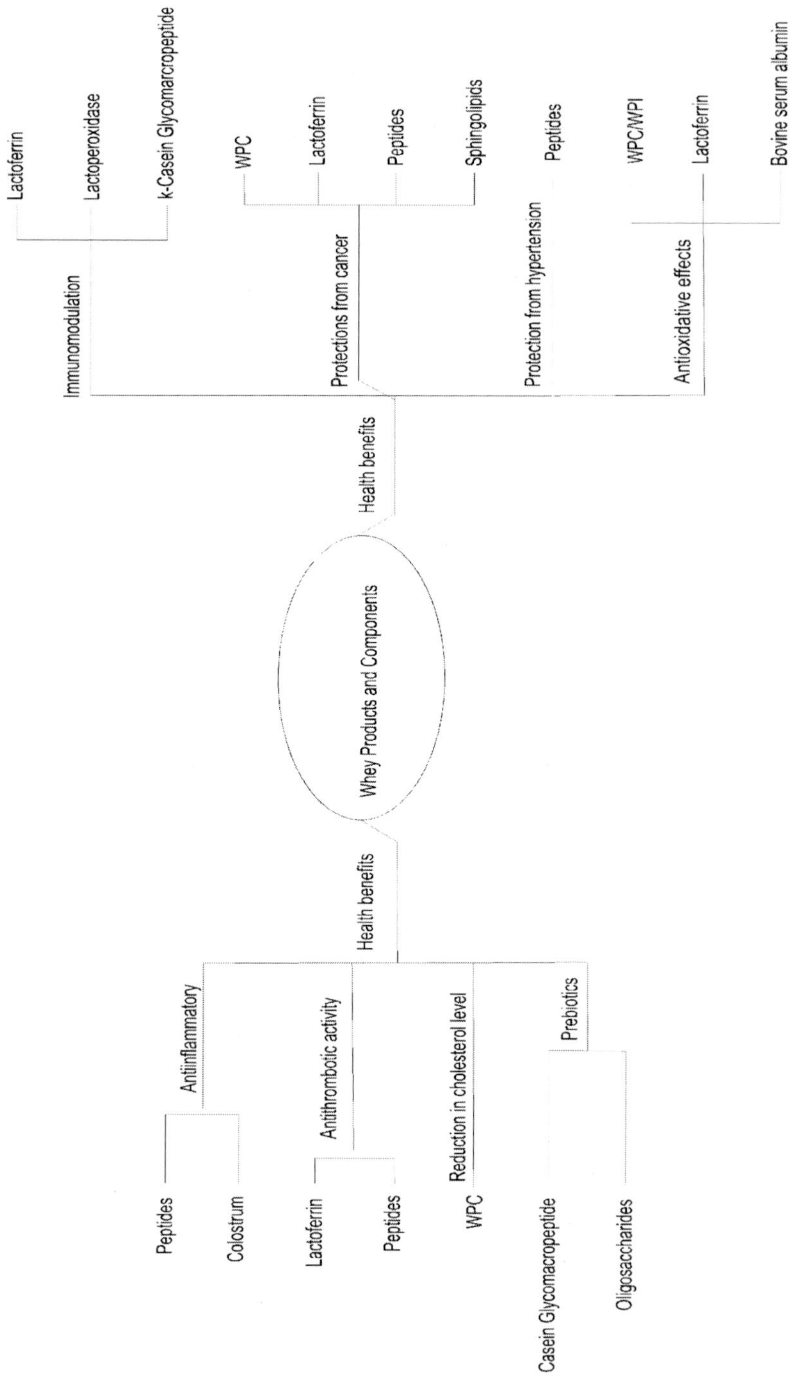

Fig. 3.1: Whey products and components health benefits

reduced form) to GSSH (the oxidised form). Through direct conjugation, GSH detoxifies a host of both endogenous and exogenous toxins including toxic metals, petroleum distillates, lipid peroxides, bilirubin and prostaglandins [14]. Lactoferrin's antioxidant and antimicrobial effects have already been touched on briefly. Due to its ability to chelate iron, organisms requiring this metal to replicate would seem to be particularly vulnerable to lactoferrin's effects. Lactoferrin also demonstrates an ability to stimulate immune responses involving natural killer (NK) cells, neutrophils and macrophage cytotoxicity. Furthermore, a mouse study concluded that lactoferrin acts as an anti-inflammatory by regulating levels of tumour necrosis factor (TNF) and interleukin 6 (IL-6) [15]. The protein β-lactoglobulin contains anti-hypertensive peptides which act as significant angiotensin converting enzyme (ACE) inhibitors. Cholesterol-lowering effects have also been noted as a result of changes in micellar cholesterol solubility in the intestine [8].

Bioactive Peptides

Milk proteins have received increasing attention as potential ingredients of health-promoting functional foods targeted at diet-related chronic diseases, such as cardiovascular disease, diabetes type 2 and obesity. To this end, growing interest has been focused on physiologically active peptides derived from milk proteins. These peptides are inactive within the sequence of the parent protein molecule and can be liberated by gastrointestinal digestion of milk, fermentation of milk with proteolytic starter cultures or hydrolysis by proteolytic enzymes. Milk protein-derived peptides have been shown under *in-vitro* and *in-vivo* conditions to exert a number of activities affecting the digestive, endocrine, cardiovascular, immune and nervous systems. A great variety of naturally formed bioactive peptides have been found in fermented dairy products, such as yoghurt, sour milk and cheese. Recently, industrial-scale technologies suitable for the production of bioactive milk peptides have been developed. In addition, a few commercial food products supplemented with milk protein derived bioactive peptides have been launched on limited markets. Some of these products carry clinically documented benefits, in particular for reduction of mild hypertension. The multi-functional properties of milk peptides appear to offer considerable potential for the development of many similar products in the near future [16]. The list of bioactive peptides identified in cheese varieties and fermented milk products are shown in Table 3.2 [16].

Table 3.2: Bioactive peptides identified in cheese varieties and fermented milk products

Product	Examples of identified bioactive peptide	Bioactivity
Cheese type		
Cheddar	αs1 - and β-casein fragments	Several phosphopeptides
Italian Mozzarellazola	αs1-cn f(1–9) β-cn f(60–68)	ACE inhibitory
Gouda	αs1-cn f(1–9) β-cn f(60–68)	ACE inhibitory
Festivo	αs1-cn f(1–9), f(1–7), f(1–6)	ACE inhibitory
Emmental	αs1- and β-casein fragments	Immunostimulatory, several phosphopeptides, antimicrobial
Manchego	Ovine αs1-, αs2- and β-casein fragments	ACE inhibitory
Emmental	Active peptides not identified	ACE inhibitory
Hard, semi-hard and soft	Val-Pro-Pro, Ile-Pro-Pro	ACE inhibitory
Cheddar	αs1-cn f(1–6), f(1–7), f(1–9), f(24–32), f(102–110), β-cnf(47–52), f(193–209)	ACE inhibitory
Fermented milks		
Sour milk	β-cn f(74–76, f(84–86) j-cnf(108–111)	Antihypertensive
Yoghurt	Active peptides not identified	ACE inhibitory
Dahi	Ser-Lys-Val-Tyr-Pro	ACE inhibitory

The production and properties of milk protein-derived bioactive peptides have been reviewed in many articles [17-20]. Upon oral administration, bioactive peptides may affect the major body systems – namely, the cardiovascular, digestive, endocrine, immune and nervous systems. For this reason, the potential of distinct dietary peptide sequences to promote human health by reducing the risk of chronic diseases or boosting natural immune protection has aroused increasing scientific and commercial interest over the past decade [19, 21].

The beneficial health effects may be attributed to numerous known peptide sequences exhibiting antimicrobial, antioxidative, antithrombotic, antihypertensive, immunomodulatory and opioid activities, among others [22]. The activity of these peptides is based on their inherent amino acid composition and sequence. The size of active sequences may vary from 2 to 20 amino acid residues; many peptides are known to reveal multi functional properties. For above reasons, milk-derived bioactive peptides are considered as prominent candidates for various health-promoting functional foods targeted at heart, bone and digestive system health as well as improving immune defence, mood and stress control. Recent studies suggest that bioactive milk peptides may also be beneficial in reducing the risk of obesity and development of type 2 diabetes.

Bioactive peptides as a functional food

Biologically active peptides can be produced from precursor milk proteins in the various ways such as: (a) enzymatic hydrolysis by digestive enzymes, (b) fermentation of milk with proteolytic starter cultures, (c) proteolysis by enzymes derived from microorganisms or plants. In many studies, a combination of above methods has proven effective in generation of short functional peptide. Bioactive peptides have attracted increasing interest as prominent candidates for various health-promoting functional foods.

At present, milk proteins are the best known source of such ingredients, but until recently the commercial production of milk derived bioactive peptides has been limited by a lack of suitable large scale technologies. Membrane separation techniques seem to provide the best technology available for the enrichment of peptides with a specific molecular weight range [23, 24]. Nanofiltration and ultrafiltration techniques are now employed industrially to produce ingredients which contain specific bioactive peptides based on casein or whey protein hydrolysates. Such preparations are commercially available and are being introduced into different consumer products, such as dairy and fruit based drinks, confectionery, chewing gum, pastilles and capsules. Currently marketed products contain peptides with anticariogenic, antihypertensive, mineral-binding, stress-relieving and satiety inducing properties. Examples of these products are listed in Table 3.3.

Only a few among the large number of milk peptides identified as antihypertensive under *in-vitro* conditions have so far proven clinically effective in animal and human studies as reviewed in recent articles [25-28]. Fig. 3.2 shows the potential health benefits of peptides derived from milk protein. In most of the human studies moderate or significant reduction of blood pressure has been observed after consumption of specific milk protein hydrolysates or fermented dairy products. The potential health benefits of bioactive milk peptides have been a subject of growing commercial interest in the context of health promoting functional foods. So far, antihypertensive, mineral-binding and anticariogenic peptides have been most studied for their physiological effects. A few commercial products have been launched on the market and this trend is likely to continue alongside with increasing knowledge about the functionalities of the peptides. It can be anticipated that more peptide-based products will be developed to manage various risk factors of the metabolic syndrome, e.g., hypertension, serum lipid levels, glucose balance and body mass index. In view of the current global trend of increasing prevalence of

Table 3.3: Examples of products containing milk peptides

Brand name	Type of product	Claimed functional bioactive peptides	Health/function claims	Manufacturer
Calpis	Sour milk	Val-Pro-Pro, Ile-Pro-Pro, derived from β-casein and κ-casein	Reduction of blood pressure	Calpis Co., Japan
Evolus	Calcium enriched fermented milk drink	Val-Pro-Pro, Ile-Pro-Pro, derived from β-casein and κ-casein	Reduction of blood pressure	Valio Oy, Finland
BioZate	Hydrolysed whey protein isolate	β-Lactoglobulin fragments	Reduction of blood pressure	Davisco, USA
BioPURE-GMP	Whey protein isolate	β-casein f (106–169) (Glycomacropeptide)	Prevention of dental caries, influence the clotting of blood, protection against viruses and bacteria	Davisco, USA
PRODIET F200 /Lactium	Flavoured milk drink, confectionery, capsules	α -casein f (91–100) (Tyr-Leu-Gly-Tyr-Leu- Glu-Gln-Leu-Leu-Arg)	Reduction of stress effects	Ingredia, France
Festivo	Fermented low-fat	α_{s1}-casein f (1–6), (1–7), (1–9) hard cheese	No health claim	MTT Agrifood Research Finland
Cysteine Peptide	Ingredient/hydrolysate	Milk protein derived peptide	Aids to raise energy level and sleep	DMV International, The Netherlands
C12	Ingredient/hydrolysate	Casein derived peptide	Reduction of blood pressure	DMV International, The Netherlands
Capolac	Ingredient	Caseinophosphopeptide	Helps mineral absorption	Arla Foods Ingredients, Sweden
PeptoPro	Ingredient/hydrolysate	Casein derived peptide	Improves athletic erformance and muscle recovery	DSM Food Specialties, The Netherlands
Vivinal Alpha	Ingredient/hydrolysate	Whey derived peptide	Aids relaxation and sleep	Borculo Domo Ingredients (BDI), The Netherlands
Recaldent	Chewing gum	Calcium casein peptone-calcium phosphate	Anticariogenic	Cadbury Adams, USA

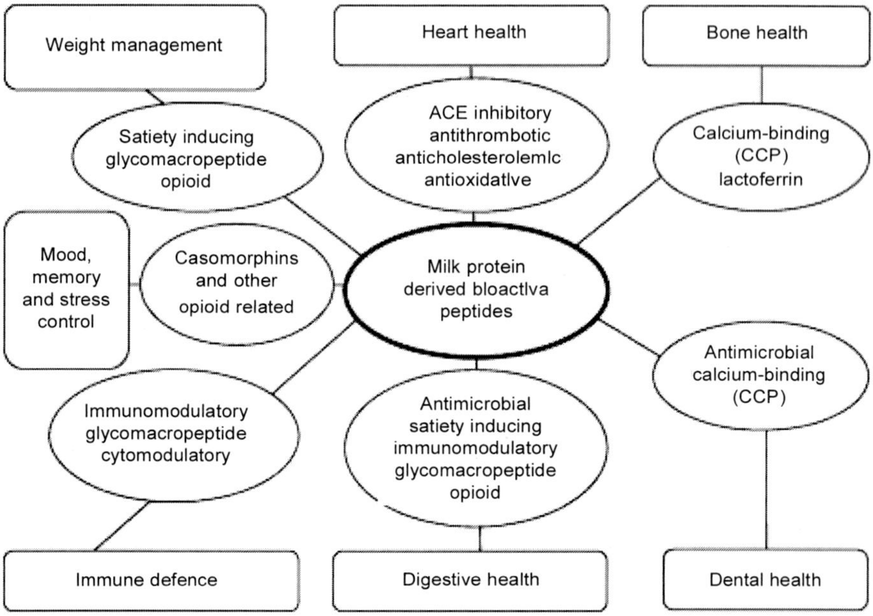

Fig. 3.2: Functionality of milk protein derived bioactive peptides and their potential health targets

obesity and related diseases, type 2 diabetes, in particular, more experimental research should be focused on natural dietary bioactive peptides which can induce satiety and reduce insulin resistance. Other new areas where more research about bioactive peptides is warranted are impairment of cognitive functions, memory-related diseases and mood control. In this context, antioxidative and opioid properties of many milk peptides may be worth further investigations. Also, potential to reduce oxidative stress in the body via oral administration of antioxidative peptides may be of considerable interest in view of the inflammatory events caused by radical oxygen species (ROS) in the living cells. It is now well established that persistent low grade inflammation in the body may lead to the development of various chronic diseases, such as atherosclerosis and type 2 diabetes.

Novel technologies, such as chromatographic and membrane separation techniques seem to be appropriate means to enrich active peptide fractions from the hydrolysates of various food proteins. Production of bioactive peptides from protein rich raw materials may be scaled up to industrial level using controlled fermentation in bioreactors with known lactic acid bacteria (LAB). To this end, recombined enzyme technology and specific production strains or peptidases isolated from suitable microorganisms are likely to be employed industrially in the future. In addition to enzymatic hydrolysis, microbial fermentation provides a natural technology applicable for the production of bioactive peptides either from animal or plant proteins. The potential of this approach is already well demonstrated by the presence of bioactive peptides in fermented dairy products [29]. The formation and stability of such peptides is, however, not controlled in traditional dairy products and, therefore, certain health benefits cannot be guaranteed. New manufacturing technologies such as supplementation of fermented milk with specific bioactive peptides concentrated from hydrolysates of milk proteins seem to provide a practical solution for the time being. Important research topics will continue to be the bioavailability and safety of bioactive peptides. To this end, the stability of chemical structures and biological activity of peptides in different food matrices and subsequent fate during the digestion process are to be taken into account in formulation of products containing bioactive peptides. Again, new emerging technologies, such as micro or nano encapsulation may offer feasible solutions for improving stability of peptides in various food products and during digestion. Furthermore, molecular studies are needed to assess the mechanisms by which bioactive peptides exert their activities. For this approach, it is necessary to employ new nutrigenomic techniques, for example proteomics and metabolomics. By developing such novel facilities it will be possible to study the impact of proteins and peptides on the expression of genes and hence optimize the nutritional and health effects of these compounds. This research area is currently considered highly challenging and will revolutionize the protein research in the near future. Also, biosafety aspects of novel peptides generated by new technologies have to be considered before entering into proper product formulation and marketing efforts. Further challenges with functional food products containing novel bioactive peptides are associated with regulatory issues if any health-related claims will be attached to the products. Currently, health claim regulations seem to vary greatly in different countries but regional and continental harmonization processes are in progress.

Caseinophosphopeptides (CPPs)

Caseinophosphopeptides (CPPs), as the name suggests, are casein-derived peptides that have phosphorus bound via monoester linkages to seryl residues. The term appears to have been first introduced by Mellander in 1950 to describe a group of phosphorylated casein peptides that enhanced bone calcification independently of vitamin D in rachitic children. Given their highly negatively charged structures arising from phosphorylation, CPPs have the ability to bind a range of macroelements such as calcium, magnesium and iron, and trace minerals such as zinc, barium, chromium, nickel, cobalt and selenium. It has long been thought that the enhanced dietary bioavailability of calcium from milk and dairy products may be attributed to the presence of CPPs. This is due to the observation that *in-vitro* CPPs possess the ability to bind and solubilise calcium at high pH, which corresponds to the alkaline conditions pertaining in the small intestine where maximal passive vitamin D independent calcium adsorption takes place [30].

The functional food ingredient role of CPPs is mainly linked to their potential ability to enhance the bioavailability of dietary calcium. A number of food ingredient companies currently market CPP containing/enriched milk protein hydrolysate products. A general challenge to the functional food ingredients sector is to demonstrate the efficacy and safety of food products or components therein that are claimed to have beneficial health effects. The current challenge in relation to CPPs is to unequivocally demonstrate their efficacy in enhancing dietary mineral bioavailability in humans. This chapter describes the structure, production, potential contribution to calcium bioavailability, safety and general ingredient roles of CPPs.

CPP as a functional food

The potential ingredient applications of CPPs have been previously reviewed [30] and are summarised in Table 3.4.

Table 3.4: Potential application of CPPs

Application	Rationale	Product
Prevention of osteoporosis	Increase absorption of Ca by Ca–CPP complexes including Ca–CPP as functional food ingredient for target populations, allowing maximum deposition of bone in early life	Ca–CPP complexes
Recalcification of bones after fracture	Increase serosal [Ca], aiding bone mineralisation	Ca–CPP complexes
Calcic addition during treatment of rickets	Increase serosal [Ca]	Ca–CPP complexes
Hypertension	Increased serosal [Ca] may aid reduction of hypertension	Ca–CPP complexes
Prevention of dental caries	Inclusion of Ca–CPP as ingredients in mouth washes, toothpastes	Ca–CPP complexes
Mg deficiency in pregnancy	Aid Mg supplementation	Mg-CPP complexes
Anaemia	Aid Fe supplementation,	Fe–CPP complexes
Humanisation of bovine milk	Increase phosphorus levels	CPPs
Oligoelement supplementation	Aid supplementation of Zn, Cu, Cr, Ni, Co, Mn and Se	Oligoelement–CPP complexes

The application for which CPPs are probably best linked is in the area of mineral supplementation. It is thought that CPPs may aid dietary mineral absorption by enhancing solubility. A whole range of minerals including calcium, magnesium and iron are required to maintain health [31-33]. Inadequate uptake of these minerals may lead to conditions such as osteoporosis, dental caries, hypertension and anaemia. Various foods have been cited in the patent literature as potential carriers for CPPs. These include confectionary products, breakfast cereals, juices, milk, dairy products and sports drinks [34, 35]. However, as already stated herein, there is a lack of extensive evidence linking consumption of CPPs and enhanced mineral bioavailability in humans.

The anticariogenic activity of CPPs is related to their potential ability to localise calcium phosphate at the tooth surface and as a consequence their ability to help prevent demineralisation and aid in remineralisation of tooth enamel. The anticariogenic effects of CPPs have been shown in animal and human experiments [36-38].

The functional food ingredient/nutraceuticals potential of CPPs revolves around their ability to bind and solubilize minerals such as calcium. Preparative scale production procedures for the generation of CPPs are available. Peptide preparations enriched in CPPs are available from multinational food ingredient suppliers. Furthermore, dental-care/hygiene products containing CPPs, targeted to have an anticariogenic effect, are currently available in the marketplace. A number of patents exist on the production and application of CPPs as functional food ingredients/nutraceuticals and as dental care/hygiene agents.

CPPs are casein-derived peptides that have phosphorous bound via monoester linkages to seryl residues. CPPs have the ability to bind to a range of macroelements such as calcium, magnesium, and iron and trace elements such as zinc, barium, chromium, nickel, cobalt, and selenium. Moreover, the enhanced dietary bioavailability of calcium from milk and dairy products may be attributed to the presence of CPPs. This is due to the fact that in vitro CPPs possess the ability to bind and solubilize calcium at high pH, corresponding to the alkaline conditions occurring in the small intestine where maximal massive vitamin D-independent calcium absorption takes place. Many companies market CPP containing/enriched milk protein hydrolysate products. CPPs are produced by enzymatic digestion of whole casein or fractions enriched in specific individual caseins. Hydrolysis is carried out at neutral to alkaline pH (7.0–8.0). CPPs present in the resulting clarified hydrolysate supernatants can be aggregated on inclusion of a range of mineral salts around neutral pH with calcium used as the major aggregating agent. The aggregated CPPs can be precipitated with hydrophilic solvents such as ethanol, methanol, propanol, butanol, and acetone and have been collected by centrifugation. Ultrafiltration through defined molecular mass cut-off membranes has been used at pilot scale to separate aggregated CPPs from nonphosphorylated peptides. Alternatively, CPPs have been isolated/enriched from calcium/ethanol-precipitated aggregates using many ion-exchange techniques. Fractions containing isolated/enriched CPPs are then subjected to spray drying or freeze drying.

CPPs bind and solubilize minerals such as calcium. The highly anionic phosphorylated regions in CPPs are responsible for mineral binding. However, the amino acid sequence around this anionic hydrophilic domain also seems to play a significant role in mineral binding [38]. The calcium-binding properties of CPPs have been studied using calcium selective electrodes, micro ultrafiltration methods in combination with atomic absorption, competitive assays with Chelex resin, and capillary zone electrophoresis.

CPPs show protective effects in cancer development. The most probable target site for the cytomodulatory action seems to be the gastrointestinal tract. The antiproliferative effect of milk protein derived peptides as well as other minor milk components in colon cancer cell lines suggests that they could have a role in the prevention of colon cancer by blocking hyperproliferation of the epithelium and by promoting apoptosis [20]. Cytochemical experiments carried out on CPP preparations using a variety of human cell culture model systems did not show a cytotoxic response of CPPs toward human cells, provided that the orally ingested doses correspond to reasonable concentrations for use in functional foods. It should also be noted that the allergenic potential of CPPs, following their consumption by the human population that shows allergenic reactions to caseins, should be considered in the use of CPPs as functional food ingredients/neutraceuticals. More specifically, IgE displays a higher binding affinity for phosphorylated caseins than the corresponding dephosphorylated derivatives. Potential applications of CPPs include prevention of osteoporosis, recalcification of bones after fracture, hypertension, prevention of dental caries, Mg deficiency in pregnancy and old age, anemia, oligoelement supplementation, and humanization of bovine milk by increasing phosphorus levels [20].

Conjugated Linoleic Acid (CLA)

CLA is usually found in dairy products and meat from ruminants [39]. Out of many positional and geometrical isomers of CLA, the 9-cis, 11-trans-octadecadienoic acid (9c,11t-18:2) predominates in food products. Milk and dairy products (butter, cheese, and yogurt) seem to contain the highest amounts of CLA [40]. For dairy products, CLA contents up to 30 mg/g fat were reported [41]. Plant oils or margarine contain only small amounts of CLA.

The CLAs are formed as a result of industrial processing (refining process, mainly deodorization) and catalytic process of hydrogenation to produce margarine. Otherwise they appear in oils due to high temperature treatments.

The naturally occurring CLA in milk is formed using two different pathways. In the first pathway, the bioconversion of polyunsaturated fatty acids (PUFA) takes place in the rumen. The ingested dietary unsaturated fatty acids, for example, linoleic acid, are metabolized by enzymes of different bacteria present in the rumen [42]. Various trans fatty acids appear in this biohydrogenation pathway as intermediates. The 9c,11t-18:2 is formed during this bacterial transformation. The endogenous synthesis in the mammary gland is very important as ~60% of CLA in milk fat is formed via this pathway in the lactating cow. Various factors influence the CLA content in milk such as the food of the ruminant, the season (decreased CLA content during winter, with the lowest level in March and increased level in the summer, positively correlated to the grazing period), the animal breeding type, the number of lactation times, and the current lactation state. The lactation stage can affect CLA content in milk since body fat stores of the cow are mobilized at the beginning of lactation. Pasture in spring time contains higher amounts of PUFAs, which caused a higher bacterial biohydrogenation in the rumen.

Commercial production of CLAs is carried out by alkali isomerization of linoleic acid producing complex CLA mixtures containing two predominant isomers. Dehydration of hydroxyl fatty acids has been used to prepare conjugated fatty acids. For example, dehydration of ricinoleic acid, an expensive starting material, can prepare 9c,11t-18:2. Zinc in aqueous n-propanol is very selective in the reduction of the acetylenic bond to the z-olefinic bond to prepare 9c,11t-18:2. Pure CLA isomers can also be produced by total stereo selective multiple step synthesis in small quantities; the isomeric purity can be higher than 98% [43].

Stereoselective synthesis guarantees the exact chemical structure of the final molecule and limits the formation of by-products due to several purification procedures carried out during the synthesis. These syntheses could allow the preparation of labeled molecules of CLA with deuterium, C13 or C14. Functional benefits of CLA include high anticarcinogenic effects, decreased LDL cholesterol in rabbits, decrease in atherosclerotic lesions, protection against immune induced cachexia (growth suppression or weight loss), and questionable antidiabetic effects. Human long term experiments need to be carried out. The only experiments carried out in humans have investigated the influence on body composition with controversial results. More experiments using isolated isomers to evaluate the effects of each CLA isomer are required. Also, toxicological risks need to be evaluated since little data on the safety aspects of CLAs are available [44].

Finally, the CLAs, due to the conjugated double bond system, are seen as more sensitive to oxidation, or isomerization during heat treatment than linoleic acid. It has been demonstrated that food processing methods such as frying, baking, boiling, or microwaving did not affect or change CLA content in grilled hamburger beef patties [45]. Only intensive heating for 15 min, using temperatures higher than 200°C, could lead to isomerization of CLA in milk [46]. The influence of storage on the CLA content was also tested in butter, yogurt, and sour cream with no modification of CLA content detected. The use of different fermentation cultures, processing temperatures, or ripening periods could affect the CLA level in the final food product

Hypoallergenic (HA) Hydrolysates

HA protein hydrolysates are used as functional ingredients in infant formulas and are of high nutritional and therapeutic value. HA cow's milk based formulas are administered during the first 6 months of life in infants at risk that are partially breast fed or not breast fed at all [47]. They are added to prevent CMA, mainly an IgE-mediated hypersensitivity reaction. They are classified as extensively hydrolyzed formulas (eHF) and partially hydrolyzed formulas (pHF) based on the degree of proteolysis. Depending on primary or secondary prevention of CMA, the HA formulas used are different [48]. When infants are already sensitized to cow's milk proteins (CMPs), low amounts of cow's milk may trigger an allergic reaction and only eHF or amino acids formulas should be used for infants with established CMA, whereas pHF should be used for primary prevention of CMA in nonsensitized at-risk infants. The pHF may be of benefit in long-term protection against CMA. Such formulas are sufficiently reduced in allergenicity [49].

These protein hydrolysates should be rich in low-molecular-weight peptides, especially di- and tripeptides, with the least possible quantity of free amino acids. Quality control of these products includes degree of hydrolysis (percentage of peptide bonds released by enzymatic hydrolysis), *in-vitro* characterization of peptide size, and determination of allergenicity. The safety and efficacy of these formulas can only be determined by clinical trials using appropriate standards. The presence of reactive epitopes in the hydrolysates, very important for the suitability of these HA products, is determined by *in-vitro* immunochemical assays such as enzyme-linked immunosorbent assay (ELISA), radioallergosorbent (RAST) and enzyme allergosorbent (EAST) tests, and Western blotting techniques. Enzymatic proteolysis and heat treatment are normally used to reduce the content of β-lactoglobulin and other intact proteins and therefore reduce the antigenicity of milk proteins in the production of pHF or eHF from cow's milk. These techniques are also used to prepare HA formulas based on protein hydrolysates from other sources. Ultrafiltration is used as a posthydrolysis process to remove residual high molecular weight peptides and proteins and the enzymes used in the digestion. The residual antigenicity of infant formulas will depend on the hydrolysis grade obtained, the enzyme specificity, and the filtration technique used. However, residual allergenicity has been reported in several commercial preparations, which could be due to inaccessibility of some epitopes to proteases even in the denatured protein. Most recently there has been great interest in the development of new HA hydrolysates with limited hydrolysis using high-pressure and microwave irradiation techniques. Excess hydrolysis might cause low palatability, off flavor, bitterness, and loss of nutritional value [50].

Lactose and Lactose Derivatives

Lactose has many uses in both food and pharmaceutical industries. Food applications include infant formulas, candy, baked goods, sauces, dressings, instantized food powders, and flavors [1, 51, 52]. This disaccharide is used in the pharmaceutical industry as a binder and solid diluent for many drugs. Lactose can be transformed into three products without breaking the 1,4-/3° linkage between glucose and galactose. Lactulose is produced by isomerization, lactitol by reduction, and lactobionic acid by oxidation. Lactitol is a nonnutritive sweetener equivalent to sorbitol but has not been used for food application until after toxicological and physiological studies [53]. Lactulose is known to promote growth of bifidobacteria and has important implications in infant nutrition. The inclusion of lactulose and bifidobacteria in cultured dairy products may be one mechanism to establish and maintain these organisms in the lower gastrointestinal tract. Lactose can be fermented to produce products such as alcohol, lactic acid, tartaric acid, single cell protein, and gums. Lactulose is not digested and could serve as a source of nutrients for bifidobacteria. Lactobionic acid is the glucose equivalent of gluconic acid and can be used to chelate iron, calcium, and magnesium in mineral supplementation of human diets and related products.

Oligosaccharides

According to IUB-IUPAC terminology, oligosaccharides have been defined as carbohydrates with a degree of polymerization up to 10. Oligosaccharides have also been defined as ranging from a degree of polymerization of 2 up to 20 and more [54]. Human milk contains considerable amounts of oligosaccharides showing that they play an important role during postnatal period. Human milk oligosaccharides contribute to the establishment of specific intestinal flora dominated by bifidobacteria, the postnatal stimulation of the immune system, the defense activity of human milk against viral and bacterial infections, and the enhancement of the bioavailability of minerals. Human milk contains 0.7–1.2 g oligosaccharides per litre, making the oligosaccharide fraction the major component of human milk [55]. The monomers of human milk oligosaccharides are *d*-glucose, *d*-galactose, N-acetylglucosamine, fucose, and sialic acid (N-acetyl neuraminic acid). All core molecules carry lactose at their reducing end with few exceptions. The activity of several fucosyl transferases and sialyl transferases results in an attachment at a different position in the core molecule in different linkages. Fucosylated oligosaccharides, representing the majority of oligosaccharides in human milk, could not be detected in cow's milk, goat's milk, or sheep's milk. On the other hand, 32- and 62-galactosyl lactose and 32- and 62-sialyl lactose could be detected in both animal milks and human milk. Moreover, oligosaccharides of more simple structures than human milk oligosaccharides have been used as components in several dietary products to mimic the beneficial effect of human milk oligosaccharides [50].

Summary

It has been substantiated that functional foods containing physiologically active components, either from plant or animal sources, may have a beneficial effect on health and longevity. However, functional foods are not a universal panacea for poor health habits. The perception of "deficient" versus "adequate" diets is gradually moving toward the concept of "optimal" nutrition. The evolving concept of functional foods is closely linked to that of optimal nutrition. Functional foods must be safe and claimed benefits must be real and their taste and convenience must not be compromised or they will not sell. The merit of such products can be appreciated only by educated consumers who must also be able to afford them. The regulations governing the scientific evidence required to support claims must protect the consumer while not obstructing innovation. It will be difficult to achieve a regulatory formula that pleases everyone, but the consumer's rights must have a priority over any other involved issue. Provided that these hurdles are eventually overcome, potential markets for functional foods are bound to become a success. Innovation in science results in more and more products with proven efficacy, helping improving and maintaining the health and vitality of the greater part of the world's population, not just the affluent few in developed countries. However, it is clear that a great deal of further research is required, including development of valid disease biomarkers and clinical trials, before most of these numerous benefits are properly measured and substantiated.

References

1. Playne, M.J., L.E. Bennett, and G.W. Smithers, *Functional dairy foods and ingredients.* Australian Journal of Dairy Technology, 2003. **58**(3): 242-264.
2. Dobson, W.D. and P. Christ, *Structural change in the US dairy industry: Growth in scale, regional shifts in milk production and processing, and internationalism*, 2000, University of Wisconsin, Agricultural and Applied Economics.
3. Fennema, O.R., S. Damodaran, and K.L. Parkin, *Food Chemistry* 2008: CRC, London.
4. Regester, G.O. and G.W. Smithers, *Seasonal changes in the β-lactoglobulin, α-lactalbumin, glycomacropeptide, and casein content of whey protein concentrate.* Journal of dairy science, 1991. **74**(3): 796-802.
5. McIntosh, G.H., et al., *Whey proteins as functional food ingredients.* International Dairy Journal, 1998. **8**(5-6): 425-434.
6. Morr, C.V. and E.Y.W. Ha, *Whey protein concentrates and isolates: processing and functional properties.* Critical Reviews in Food Science & Nutrition, 1993. **33**(6): 431-476.
7. De Wit, J.N., *Nutritional and functional characteristics of whey proteins in food products.* Journal of Dairy Science, 1998. **81**(3): 597-608.
8. Bayford, C., *Whey Protein: A Functional Food.* The Nutrition Practitioner , 2010.
9. De Wit, J., *Nutritional and functional characteristics of whey proteins in food products.* Journal of dairy science, 1998. **81**(3): 597-608.
10. Abd El-Salam, M.H., S. El-Shibiny, and A. Salem, *Factors affecting the functional properties of whey protein products: A review.* Food Reviews International, 2009. **25**(3): 251-270.
11. Baricellus, J., *The first little work concerning the qualities and use of milk.* Naples-azarum Scoriggium, 1603.
12. Sesink, A.L.A., *et al.*, *Red Meat and Colon Cancer The Cytotoxic and Hyperproliferative Effects of Dietary Heme.* Cancer research, 1999. **59**(22): 5704-5709.
13. Keri Marshall, N., *Therapeutic applications of whey protein.* Alternative Medicine Review, 2004. **9**(2): 136-156.
14. Liska, D. and J.S. Bland, *Clinical nutrition: a functional approach* 2004: Recording for the Blind & Dyslexic. 2004, Princeton, N.J
15. Machnicki, M., M. Zimecki, and T. Zagulski, *Lactoferrin regulates the release of tumour necrosis factor alpha and interleukin 6 in vivo.* International journal of experimental pathology, 1993. **74**(5): 433.
16. Korhonen, H., *Milk-derived bioactive peptides: From science to applications.* Journal of Functional Foods, 2009. **1**(2): 177-187.
17. Meisel, H., *Overview on milk protein-derived peptides.* International Dairy Journal, 1998. **8**(5-6): 363-373.
18. López-Fandiño, R., J. Otte, and J. Van Camp, *Physiological, chemical and technological aspects of milk-protein-derived peptides with antihypertensive and ACE-inhibitory activity.* International Dairy Journal, 2006. **16**(11): 1277-1293.
19. Erdmann, K., B.W. Cheung, and H. Schröder, *The possible roles of food-derived bioactive peptides in reducing the risk of cardiovascular disease.* The Journal of Nutritional Biochemistry, 2008. **19**(10): 643-654.
20. Hartmann, R. and H. Meisel, *Food-derived peptides with biological activity: from research to food applications.* Current opinion in biotechnology, 2007. **18**(2): 163-169.
21. Xu, R.J., *Bioactive peptides in milk and their biological and health implications.* Food Reviews International, 1998. **14**(1): 1-16.
22. Meisel, H. and R. FitzGerald, *Biofunctional peptides from milk proteins: mineral binding and cytomodulatory effects.* Current Pharmaceutical Design, 2003. **9**(16): 1289-1296.

23. Saxena, A., *et al.*, *Membrane-based techniques for the separation and purification of proteins: an overview.* Advances in colloid and interface science, 2009. **145**(1): 1-22.
24. Korhonen, H. and A. Pihlanto, *Technological options for the production of health-promoting proteins and peptides derived from milk and colostrum.* Current Pharmaceutical Design, 2007. **13**(8): 829-843.
25. Clare, D.A. and H.E. Swaisgood, *Bioactive milk peptides: a prospectus.* Journal of Dairy Science, 2000. **83**(6): 1187-1195.
26. Meisel, H. and W. Bockelmann, *Bioactive peptides encrypted in milk proteins: proteolytic activation and thropho-functional properties.* Antonie Van Leeuwenhoek, 1999. **76** (1-4): 207-215.
27. FitzGerald, R.J., B.A. Murray, and D.J. Walsh, *Hypotensive peptides from milk proteins.* The Journal of Nutrition, 2004. **134**(4): 980S-988S.
28. Raikos, V. and T. Dassios, *Health-promoting properties of bioactive peptides derived from milk proteins in infant food: a review.* Dairy Science & Technology, 2014. **94**(2): 91-101.
29. Hajirostamloo, B., *Bioactive component in milk and dairy product.* World Academy of Science and Engineering Technology, 2010. **72**: 162-6.
30. FitzGerald, R., *Potential uses of caseinophosphopeptides.* International Dairy Journal, 1998. **8**(5-6): 451-457.
31. Passmore, R., *Handbook on human nutritional requirements.*: Food & Agriculture Organization, 1974. Vol 4,WHO Geneva.
32. Osborne, C.G., *et al.*, *Evidence for the Relationship of Calcium to Blood Pressure.* Nutrition reviews, 1996. **54**(12): 365-381.
33. Buttriss, J., *The role of calcium in a balanced diet.* International Journal of Dairy Technology, 1990. **43**(1): 1-1.
34. FitzGerald, R.J., *Potential uses of caseinophosphopeptides.* International Dairy Journal, 1998. **8**(5-6): 451-457.
35. Korhonen, H. and A. Pihlanto, *Bioactive peptides: production and functionality.* International Dairy Journal, 2006. **16**(9): 945-960.
36. Aimutis, W.R., *Bioactive properties of milk proteins with particular focus on anticariogenesis.* The Journal of nutrition, 2004. **134**(4): 989S-995S.
37. Silva, S.V. and F.X. Malcata, *Caseins as source of bioactive peptides.* International Dairy Journal, 2005. **15**(1): 1-15.
38. McDonagh, D. and R.J. FitzGerald, *Production of caseinophosphopeptides (CPPs) from sodium caseinate using a range of commercial protease preparations.* International Dairy Journal, 1998. **8**(1): 39-45.
39. D'Orazio, N., *et al.*, *Conjugated linoleic acid: a functional food?* International journal of immunopathology and pharmacology, 2002. **16**(3): 215-220.
40. Benjamin, S. and F. Spener, *Conjugated linoleic acids as functional food: an insight into their health benefits.* Nutrition Metabolism, 2009. **6**(1): 36.
41. Belury, M.A., *Dietary Conjugated Linoleic Acid in Health: Physiological Effects and Mechanisms of Action 1.* Annual Review of Nutrition, 2002. **22**(1): 505-531.
42. Sampath, H. and J.M. Ntambi, *Polyunsaturated fatty acid regulation of genes of lipid metabolism.* Annu. Rev. Nutr., 2005. 25: p. 317-340.
43. Perfield II, J., *et al.*, *Conjugated Linoleic Acid Reduces Milk Fat Synthesis in Lactating Dairy Cows.* Journal of dairy science, 2007. **90**(5): 2211-2218.
44. Jahreis, G., J. Fritsche, and H. Steinhart, *Conjugated linoleic acid in milk fat: high variation depending on production system.* Nutrition Research, 1997. **17**(9): 1479-1484.
45. Shantha, N.C., A.D. Crum, and E.A. Decker, *Evaluation of conjugated linoleic acid concentrations in cooked beef.* Journal of Agricultural and Food Chemistry, 1994. **42**(8): 1757-1760.

46. Shantha, N.C., *et al.*, *Conjugated linoleic acid concentrations in dairy products as affected by processing and storage.* Journal of Food Science, 1995. **60**(4): 695-697.
47. Sampson, H.A., J.M. James, and J.A.N. Bernhisel-Broadbent, *Safety of an amino acid-derived infant formula in children allergic to cow milk.* Pediatrics, 1992. **90**(3): 463-465.
48. Halken, S. and A. HÃ¸st, *How hypoallergenic are hypoallergenic cow's milk based formulas?* Allergy, 1997. **52**(12): 1175-1183.
49. Kleinman, R.E., *et al.*, *Use of infant formulas in infants with cow milk allergy.* Pediatric Allergy and Immunology, 1991. **2**(4): 146-155.
50. Yildiz, F., *Development and manufacture of yogurt and other functional dairy products*, 2010, CRC Press.London
51. Zadow, J.G., *Whey and lactose processing.* 1992.Springer Australia
52. Gillies, M.T., *Whey processing and utilization. Economic and technical aspects*1974: Noyes Data Corp.
53. Ballongue, J., C. Schumann, and P. Quignon, *Effects of lactulose and lactitol on colonic microflora and enzymatic activity.* Scandinavian Journal of Gastroenterology. Supplement, 1996. **222**: 41-44.
54. Roberfroid, M.B., *Health benefits of non-digestible oligosaccharides*, in *Dietary Fiber in Health and Disease*1997, Springer. p. 211-219.
55. Kunz, C., *et al.*, *Nutritional and biochemical properties of human milk, Part I: General aspects, proteins, and carbohydrates.* Clinics in perinatology, 1999. **26**(2): 307-333.

4

Probiotics and Prebiotics

Introduction

The term probiotic is derived from Greek and means 'for life' as opposed to antibiotics which means 'against life'. The history of probiotics began with the consumption of fermented foods which was first observed in ancient Greeks and Romans [1]. The term, probiotic, as is used today was first used by Parker in 1974. Parker defined 'probiotic' as substances and organisms which contribute to intestinal microbial balance. In 1989, the term was modified further by Fuller. Thus, probiotic is a live microbial supplement which affects the host's health positively by improving its intestinal microbial balance. This definition was broadened to include mono or mixed culture of live microorganisms which benefits animals or man by improving the properties of the indigenous microflora [2]. Some authors have interpreted probiotics as microbial cell preparations or components of microbial cells that have a beneficial effect on the health and well-being of the host.

A prebiotic is a non-digestible food ingredient that beneficially affects the host by selectively stimulating growth, activity, or both, of one or a limited number of bacterial species already resident in the colon [2, 3]. To exhibit such effects, a prebiotic must neither be hydrolysed nor absorbed in the upper part of the gastrointestinal tract, and must be selectively fermented by one or a limited number of potentially beneficial bacteria residing in the colon [4]. The number of probiotics in the human gut tends to decrease with age. Two major strategies have been proposed to maintain a high level of probiotics to sustain beneficial health effects (continuous ingestion of probiotics containing foods), or supplementation of food with prebiotics [5]. These prebiotics are fermented by one or a limited number of potentially beneficial bacteria from the resident colonic microflora. A prebiotic is expected to improve the composition of the colonic microbiota and through this serve as beneficial to the host health [5].

A range of dietary compounds have been suggested as prebiotics, and these have been selected for their health benefits on the host. The popularity of inulin, fructooligosaccharides (FOS) and galctooligosaccharides (GOS) as health benefiting subtrates is well known [6]. In human studies the addition to bread of 7g of FOS has been shown to beneficially affect the dominant bifidobacteria as compared to common bread. The authors clearly proved that the use of FOS exerted a profound effect upon bifidobacteria. The only prebiotics for which sufficient data has been generated to allow an evaluation of their possible classification as functional food ingredients are the inulin type fructans, which include native inulin, enzymatically hydrolyzed inulin or oligofructose, and synthetic fructooligosaccharides [7, 8]. The two basic types of fermentations taking place in the gut are saccharolytic fermentation and proteolytic fermentation [9]. The main end products of carbohydrate metabolism are the short chain fatty acids such as acetate, propionate and butyrate. These may be further metabolised systematically or locally to generate energy for the host. The end products of the proteolytic fermentation include more or less toxic compounds as amines, ammonia and phenolic compounds. Fermentation in the gut can be modulated towards saccharolytic by prebiotic consumption. Much of the interest is aimed at non-digestible oligosaccharides and indeed more than 36000 plants worldwide contain FOS; some common sources of inulin are onion (2–6%), garlic (9–16%), leek (3–10%), banana (0.3–0.7%), asparagus (10–15%), Jerusalem artichokes (15–20%), chicory (13–20%), and wheat (1–4%). Consumption of more than 4 g of FOS daily is needed to induce changes in LAB levels in the gut, though estimated daily consumption differs in the US and Europe [6].

Mechanism of Action

With the understanding of the probiotic mechanism it could be demonstrated that many gastrointestinal (GI) disorders are based on malfunctions of intestinal microflora [10]. The human GIT is the habitat of a complex microbiota consisting of either facultative or obligate anaerobic microorganisms including streptococci, lactobacilli, and yeasts. These are settled in different quantities throughout the GIT and possess a wide spectrum of differences (Fig. 4.1). The human GIT could be considered as a traveling bioreactor, as the number of bacteria that colonize the human body is so large that researchers have estimated that the human body contains 10^{14}cells, of which only 10% are not bacteria [11]. Some of the probiotic mechanisms of action for the prevention of GI disorders are the suppression of harmful bacteria and viruses through the production of bacterocins, stimulation of local and systemic immunity, and the modification of gut microbial metabolic activity [12].

A positive feature of probiotic bacteria is their ability to produce organic acids at levels that have bactericidal and bacteriostatic effects that inhibit the growth of pathogenic organisms. Probiotic bacteria also compete against harmful pathogens for a position in the intestinal mucosa during the colonization of the GIT which prompts the state of well being in the host [13, 14]. Further, metabolic activities of the GIT microbiota that beneficially affect the host include continued degradation of food components, vitamin production, and production of short chain fatty acids (SCFAs) [15]. The degradation of food components is mainly by fermenting dietary carbohydrates that have escaped digestion in the upper GIT such as resistant starch, cellulose, hemicellulose, pectins, and gums [Fig. 4.1].

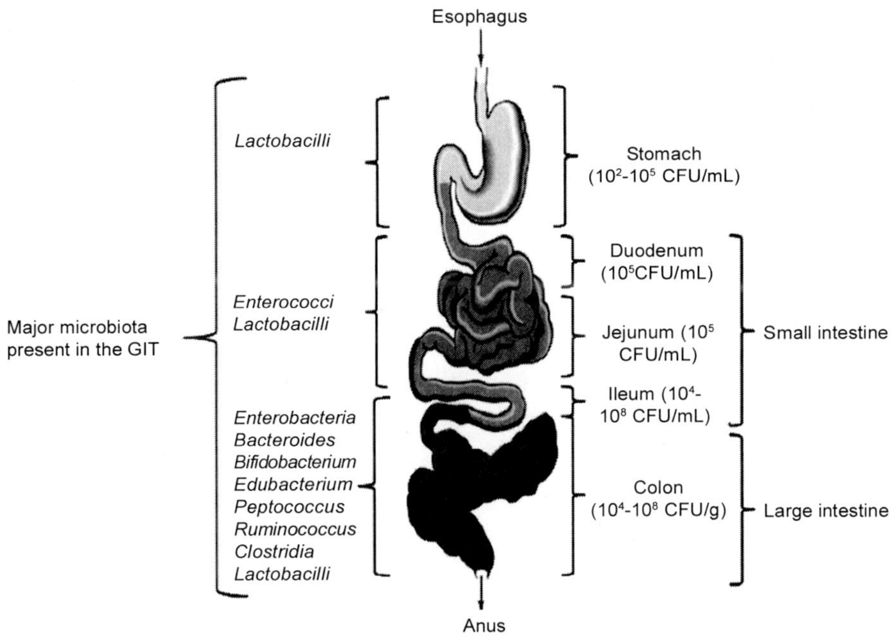

Fig. 4.1: Microbial colonization of the human gastrointestinal tract (GIT)

Food Application of Probiotics and Prebiotics

There are lots of studies on the health benef ts of fermented foods and probiotics. However, in most of these studies, researchers did not use sufficient test subjects or the used microorganisms were not identified definitely. So, while a number of reported effects have been only partially established, some can be regarded as well established and clinically well documented for specific strains. These health related effects can be considered as indicated below [3,16, 17].

- Managing lactose intolerance
- Improving immune system
- Prevention of colon cancer
- Reduction of cholesterol and triacylglycerol plasma concentrations
- Lowering blood pressure
- Reducing inflammation
- Reduction of allergic symptoms
- Beneficial effects on mineral metabolism, particularly bone density and stability

In the development of probiotic containing food products, the product concept must take into account the desire of consumers for health enhancing foods, but not at the expense of taste, convenience and the pleasurable nature of the product. The ability to communicate messages on health to the consumer is extremely important.For example, consumers can be told that the product promotes gastrointestinal health or supports the body's natural immune function, but the claims must be scientifically accurate and in accordance with actual legislation, which may differ from country to country. In addition to communications on the health benefits of probiotic products, another important area of communication is on viable count or active ingredients in probiotic products. However, statements as to content and counts of bacteria in commercial products are frequently not accurate, and may claim the presence of certain genera, species, or strain of probiotic, but rarely include any messages on probiotic potency. In countries where there is no legislation requiring this type of labeling, this leaves the consumer unable to make an informed choice based on a guaranteed probiotic content of a food product. In the long run, the failure of the industry to self-regulate in this regard may be its undoing. Consumers, unable to identify a potent product, may find a specific probiotic product ineffective and turn away from the entire product category.

Prebiotics are increasingly used in development of new food products, e.g., drinks, yoghurts, biscuits and table spreads [17]. Several prebiotics are available in Europe. The positive effects of prebiotic consumption are improvement of bowel habit, reduction of diarrhoea and constipation, modulation of lipid metabolism by normalizing cholesterol values, reduction of osteoporosis by improved mineral absorption, reduction of allergy risk through immune system modulation, and reduction of colon cancer risk. Many of these health claims require further research for validation.

Prebiotics Based Functional Ingredients

Fructooligosaccharides

Fructan prebiotics can be divided into inulin extracted from plant sources or lower molecular weight FOS. The FOS are either derived from inulin by partial enzymatic hydrolysis or are derived from sucrose by the transferase activity of β-fructofuranosidase enzymes. The commercial source of inulin for functional food application is chicory root; it is a linear chain of fructofuranose residues via linked β(2→1) linkages. A proportion of the chains terminates in a glucopyranosyl-residue linked β(1→2) (as in sucrose) and is thus nonreducing. The molecular weight distribution of plant extracted inulin is a disperse with a degree of polymerization (dp) of 2 to around 60. The general structure of oligosaccharides is presented in the Fig. 4.2.

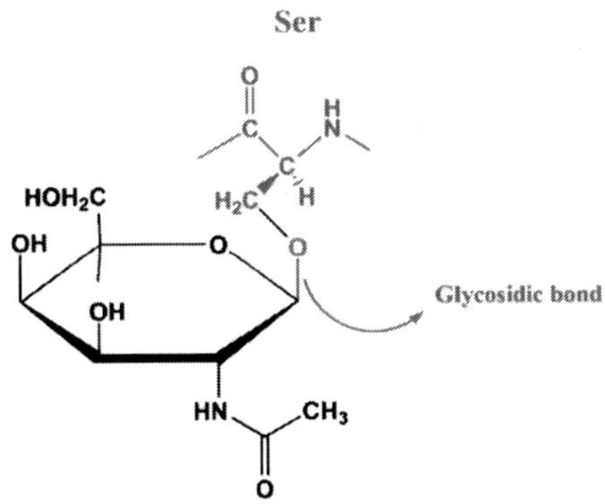

Fig. 4.2: Oligosaccharide general structure depicting O-glycosidic bond

The average dp is 12 with approximately 10% of the molecular species having a dp of 2 to 5. FOS derived from chicory inulin by hydrolysis largely terminates in fructofuranosyl residues but a low proportion terminates in glucosyl residues. The hydrolysis products have a dp range of 2 to 7 with an average of 4. Chicory inulin and inulin derived FOS are commercialized in Europe by Orafti NV (Tienen, Belgium) who manufacture a range of products with varying degrees of purity and varying technological properties. Of note is a product called Synergy which is a mixture of high molecular weight inulin chains prepared by fractionation with low molecular weight FOS. The FOS derived from sucrose terminate in glucosyl residues and are thus nonreducing. The synthetic products have a dp of 2 to 4 with an average of 3.6. Synthetic FOS is commercialized in Europe by Beghin Meiji (Thumeries, France).

The fructans are, by a large margin, the most thoroughly researched of the prebiotics. There is an abundance of literature on the biological properties of fructan prebiotics, including some well designed human trials using modern microbiological techniques. *In-vitro* studies have been carried out using pure bacterial cultures [18] and using mixed fecal inocula [19]. These have established that inulin and FOS can support the growth of Bifidobacterium spp. and to a lesser extent, Lactobacillus spp., in mixed fecal cultures at the expense of less desirable microbial genera such as Bacteroides and Clostridium. The conclusions of these investigations have been confirmed in several *in-vivo* trials in animals and humans. Studies in gnotobiotic ("germ-free") rats have shown a consistent selection for bifidobacteria and lactobacilli.

Galactooligosaccharides (GOS)

GOS are synthesized by galactosyl transfer reactions catalyzed by β-galactosidase utilizing lactose as a substrate. Lactose acts as both glycosyl donor and acceptor in such reactions, and a complex series of oligosaccharides are built up with dp ranging from 2 to 8 [20]. The commercial products are not pure and they typically contain around 50% GOS with 38% glucose and 12% lactose on a weight basis. The principal components are trisaccharides and tetrasaccharides and GOS contain mainly $\beta(1\rightarrow4)$ and $\beta(1\rightarrow6)$ linkages with lesser amounts of $\beta(1\rightarrow2)$ and $\beta(1\rightarrow3)$. The precise composition in terms of dp and linkage depends upon the enzyme used in manufacture. GOS are considered as nondigestible oligosaccharides (excluding the glucose and lactose components) on the basis of *in-vitro* tests, feeding studies in rats, and breath hydrogen tests in humans [21].

GOS have been extensively studied *in-vitro* and *in-vivo* in animal and human feeding trials. It is, however, probable that the early studies used GOS with different composition from later studies. GOS generally display a consistent prebiotic effect *in-vitro* in mixed fecal cultures. In comparative studies with other prebiotics using batch fecal cultures and molecular microbiological analyses, GOS displayed good selectivity for bifidobacteria and decreased clostridial populations [22]. GOS has also shown good selectivity in three-stage models of the human colon, particularly for lactobacilli [23]. GOS also reduced the β-glucosidase, β-glucuronidase, and arylsulphatase activities, considered to be undesirable in the colon. Inulin had little effect. The source of the enzyme used to manufacture the GOS can have an impact on the selectivity, at least in pure culture. Rabiu et al. used β-galactosidases from a panel of probiotic bacteria to synthesize GOS mixtures from lactose. These were then tested for their ability to support the growth of the probiotic panel [24]. The GOS mixtures had varying linkage profiles and supported different growth rates in the probiotic bacteria. Significantly, most of the GOS mixtures supported higher growth rates in the producing organism than in the others. Although not yet shown in mixed culture, this raises the exciting possibility of targeting prebiotics at particular species of probiotics. Tzortzis et al. studied the prebiotic properties of the novel GOS in a three-stage model of the human colon [25].

The GOS had a significant bifidogenic effect in vessels one and two, modeling the ascending and transverse colon, respectively. The total bacterial population and those of lactobacilli, bacteroides, and the *Clostridium histolyticum* group were also monitored, but no significant changes were seen. A feeding study was then carried out in pigs where a significant prebiotic effect was seen, to a higher level than with inulin. Using an *in-vitro* cell culture system, the novel GOS also inhibited the adhesion of strains of *E. coli* and *Salmonella enterica* serotype *Typhimurium*. This GOS is now being commercialized by Clasado. Trials of GOS in human volunteers have had mixed results.

Lactulose

Most prebiotics are extracted from plant tissues or are manufactured using enzymes. Lactulose is an exception, however, in that it is produced by alkaline isomerization of lactose to produce 4-O-β-galactopyranosyl-D-fructose. The commercial lactulose product is very pure and is available in crystalline and syrup formulations. These are currently targeted at medical applications, however, no food grade form of lactulose currently exists on the market in the EU. Although lactulose has a long history of use as a laxative and for treatment of hepatic encephalopathy, it is a prebiotic at sublaxative doses. It is used in Japan in several food vehicles and has FOSHU status.

Lactulose was actually one of the earliest recognized prebiotics and has been tested *in-vitro* and *in-vivo*. Earlier studies carried out comparative studies in static batch cultures with microbiology by FISH and found that lactulose was selective for bifidobacteria [26]. Lactulose also has good support for a prebiotic claim in human volunteer trials [28]; showed a significant increase in bifidobacteria with a reduction in bacteriodes, streptococci, *Clostridium perfringens*, and Enterobacteriaceae. This was achieved with a very low dose of only 3 g per day. More recent work has used more satisfactory study designs (placebo controlled and double blinded) and have shown consistent bifidogenic effects with reductions in less desirable bacterial groups [29, 30]. A significant study in one of the work, scientist used FISH to characterize the microflora changes occurring in a human trial at 10g/day [29]. They found significantly enhanced numbers of bifidobacteria and reduced numbers of clostridia. It seems clear that lactulose can be considered a prebiotic, although it is not used as such outside Japan.

Isomalto oligosaccharides

Isomalto oligosaccharides (IMO) consist of oligosaccharides with (largely) $\alpha(1\rightarrow6)$ linkages, although they also contain some panose (Glclcα $(1\rightarrow4)$ Glc). The molecular weight range is 2 to 6 kDa. They are commercial products manufactured from starch by a 2-stage process. This involves the use of α-amylase to hydrolyse starch to maltose and then transglycosylation catalyzed by α-glucosidase. This effectively converts the α (1→4) linkages to α (1→6) linkages. IMO do not strictly qualify as prebiotics as they are partially metabolized by the human small intestine. Some of the ingested material, however, does reach the colon where it has a selective metabolism.

There is very little evidence to support a prebiotic claim for IMO. The human trials have been small and poorly controlled [31]. They have reported a prebiotic effect using culture-based techniques. *In-vitro* work has supported a prebiotic claim for IMO. The comparative study using fecal batch cultures and FISH to enumerate bacterial groups, showed that IMO are selective for bifidobacteria [32]. The pH controlled fecal batch cultures with FISH also found good selectivity with IMO [33]. It is clear that good human trials with molecular microbiology techniques are overdue to support a prebiotic claim for IMO [27].

Soybean oligosaccharides

Soybean oligosaccharides (SOS) are extracted from soybean whey and consist predominantly of raffinose and stachyose. They are manufactured commercially. SOS have been investigated in small human feeding trials using culture based feeding techniques. Feeding 10g per day to 66 volunteers for 33 weeks demonstrated an increase in bifidobacteria and concomitant decrease in clostridia [34]. Another group of scientists found similar results in a study of 77 volunteers taking a low dose of 0.6 g/day for three weeks [34]. They also reported decrease in genotoxic enzymes and in toxic bacterial metabolites. An increase in bifidobacteria on feeding 1 to 2 g per day was observed [28]. The SOS have also been investigated in a comparative study in earlier work, who found that they increased populations of bifidobacteria but also produced a transient increase in bacteroides [32].

In an in vitro experiment, the addition of soybean oligosaccharides to a growth medium increased the counts of bifidobacteria, measured as a proportion of total viable counts in the culture [35]. In an experiment with young women who received 1.5 or 3 g/day of soy oligosaccharides for 30 days, the counts of bifidobacteria in faeces increased. Significantly higher concentrations of fecal short chain fatty acids, propionate and butyrate were also noted in this group.

Targeted prebiotics

With the advent of nutrigenomics and its promise of individually targeted nutrition, the generation of prebiotics targeted at particular probiotics is an interesting prospect. Virtually all of the fermentation studies to date have characterized the microflora changes at the genus level only. This is due to the technical difficulty of characterizing the complex colonic ecosystem at a species level. New DNA based techniques to achieve this are now becoming available. Two approaches can be considered in the generation of targeted prebiotics. One is the screening of as wide a range of candidate prebiotics as possible and using techniques such as quantitative PCR and DGGE to examine the resultant bacterial populations. An alternative is to utilize the enzymes of the target probiotics themselves for the synthesis of oligosaccharides. These enzymes should produce oligosaccharide mixtures that would be more readily metabolized by the producing organism, resulting in higher selectivity. Early results with this method are encouraging and novel GOS mixtures have been produced this way. Many of these mixtures resulted in higher growth rates for the producing species in pure culture, although the selectivity has not yet been described in mixed culture systems [36].

Persistent prebiotics

The majority of chronic gut diseases arise in the distal colon and spread proximally, the most important of which are colonic cancer and ulcerative colitis. Many of the metabolites elaborated by the nonprobiotic flora in the colon are genotoxic or are tumor promoters [37]. Although the precise role of these compounds in colon cancer is still not understood, it would seem sensible to reduce their levels in the colon. There is a growing body of evidence that prebiotics have a protective effect against the development of colon cancer. The precise mechanism behind the development of ulcerative colitis is still not clear but there is evidence that colonic bacteria, particularly the sulfate reducers such as *Desulfovibrio desulfuricans*, are involved in the etiology of the disease [38]. If a prebiotic is to have a protective effect, it must reach the distal colon and influence the microflora there. An approach to increasing persistence is regulation of the molecular weight distribution of the prebiotics. Most prebiotics are of relatively low molecular weight with the exception of inulin. High-molecular weight polysaccharides are not generally prebiotic but are nonselectively fermented. It seems that selectivity decreases with increasing molecular weight, consequently development of persistent prebiotics will be a compromise between persistence and selectivity. The most persistent of the current prebiotics is Synergie II, manufactured by Orafti (Tienen, Belgium). This is an inulin and FOS mixture with controlled chain length distribution. The long chain inulin exerts a prebiotic effect in more distal colonic regions and the lower molecular weight FOS is rapidly fermented in the proximal colon. This might be a more generic means of developing persistent prebiotics derived from polysaccharides.

Natural resistant starches as prebiotics

Resistant starch (RS) is defined as "the sum of starch and products of starch degradation not absorbed in the small intestine of healthy individuals" [39]. Resistant starch has been classified principally on a structural basis and most (with the exception of RS4) could be considered a natural source of RS. The four classes are simply referred to as RS1, RS2, RS3, and RS4. RS1 is starch trapped in or by a food material (e.g., whole grains). RS2 is found in native or physically processed starch granules. High amylose corn starches (HACS) including Hi Maize ® 260 are considered primarily RS2. Resistant starch is formed when starch containing foods are cooked and cooled such as in bread, cooked and chilled potatoes, or retrograded high amylose corn. The resistant structure that is formed can be degraded by microbial fermentation, but is not hydrolyzed by human alimentary enzymes. Novelose® 330 starch is a retrograded RS3 generated from annealed, enzyme-treated HACS (approximately 56% RS3). RS4 refers to chemically modified starches using standard starch chemical modification techniques, such as cross-linking, substitution, or a combination of the various chemistries [40].

In the large intestine, RS is fermented by intestinal bacteria to produce short chain fatty acids (SCFAs), particularly butyrate [41]. Some data suggest that the colonic microflora may adapt to produce more butyrate if given time and the proper substrate [41]. The fermentation of these RS led to *in-vitro* SCFAs levels (acetate, propionate, butyrate) of 2,000 to 2,500 µmol/g feces dry weight with butyrate contents of 30 to 60 mol% [42]. In human studies, RS2 and RS3 from HACS results in selective colonic microflora activity as well as increased fecal butyrate concentrations [41]. Thus, RS can be considered a prebiotic because it promotes health of the host through fermentation. Jacobasch et al. (2006) demonstrated that RS3 (Novelose ®330 starch) was well fermented in the cecum and proximal colon in rats, whereas the degradation of hydrothermally treated RS3 (hydrothermally treated Novelose) took place beyond the cecum and increased continuously through the colon to favor SCFA production in the distal colon [43].

Corresponding to the high rate of hydrothermally treated RS3 fermentation in the distal colon, the SCFA concentrations in the feces and the growth of bacteria increased significantly, resulting in a nearly twofold increase in wet content. An exchange of 10 % starch with a butyrogenic RS3 in the diet was proved to be sufficient to provide enough substrate for bacterial fermentation in the distal colon and rectum. As SCFA concentration increased, the pH decreased in the large bowel. Consumption of RS3 lowered the pH in the cecum and proximal colon to 6.5 to 6.6 from 7.5 (control diet) and intake of hydrothermally treated RS3 lowered the pH in the distal part of the colon to 6.3.

It was demonstrated that different amylomaize starches could generate desirable variation in gut microflora in mice. In this study the effects of HACS and modified (carboxymethylated and acetylated) HACS on the composition of colonic bacteria and the production of volatile fatty acids was investigated in mice [44]. All starches tested showed the increases in indigenous bifidobacteria in mice fed although 40 % unmodified HACS showed the highest numbers. High amylose corn starch increased Lactobacillus numbers in the mice colon and acetylated HACS significantly reduced the population of coliforms.

The RS may be used to selectively modify gut function and that increasing butyrate availability may improve colonic health. HACS and the various commercial ingredients derived from them have been the focus of most of the studies regarding prebiotic and symbiotic effects. Given the current interest in developing new sources of commercial RS, there is great potential to increase the RS intake as well through consumption of many different types of processed foods.

Food Application of Prebiotics

Prebiotics can be used for either their nutritional advantages or technological properties, but they are often applied to offer a double benefit viz. an improved organoleptic quality and a better balanced nutritional composition [3]. Food applications are illustrated in Table 4.1. The use of inulin and nondigestible oligosaccharides as fiber ingredients is straight forward and often leads to improved taste and texture. When used in bakery products and breakfast cereals, this represents a major progress in comparison to classical dietary fibers. They give more crispiness and expansion to extruded snacks and cereals and increased bowl life [45].

They also help keep breads and cakes moist and fresh for longer. Their solubility further allows fiber incorporation in watery systems such as drinks, dairy products, and table spreads. They are also used more and more in functional foods as prebiotic ingredients, which stimulate the growth of health promoting gut bacteria and offer additional health benefits (Table 4.1).

Table 4.1: Food application of prebiotics

Application	Functionality
Dairy products (yoghurt, cheese, desserts, drinks)	Fat or sugar replacement, body and mouthfeel, foam stabilization, fiber, and prebiotic
Frozen desserts	Fat or sugar replacement, texture and mouthfeel, melting behavior
Fruit preparations	Sugar replacement, synergy with intense sweeteners, body and mouthfeel, fiber and prebiotic
Breakfast cereals	Sugar replacement, crispiness and expansion, fiber, extrudedsnacks and prebiotic
Baked goods and breads	Sugar replacement, moisture retention, fiber, and prebiotics
Fillings	Fat or sugar replacement, texture, and mouthfeel
Tablets and confectionery	Sugar replacement, fiber, and prebiotic
Chocolate	Sugar replacement, heat resistance and fiber
Dietetic products and meal	Fat or sugar replacement, synergy with replacers intense sweeteners, body and mouthfeel, fiber, and prebiotic
Table spreads and butter	Fat replacement, texture and spreadability, stability, products fiber, and prebiotic
Salad dressings	Fat replacement, mouthfeel and body
Meat products	Fat replacement, texture, stability and fiber

Probiotic Functionality and Disease Prevention

The word probiotic comes from pro, meaning "for," and bios, meaning "life." So probiotic literally means "for life." The term probiotic is used to describe nutritional supplements and other products that contain live bacteria. The historical association of probiotics with fermented dairy products led to extensive research. Investigations during the past few decades have demonstrated numerous health supportive properties of probiotics on human [5,6, 37].

Structurally, the GI or digestive tract is a hollow tube that runs from the mouth to the anus. Mastication (the chewing of food), peristalsis (the movement of food), enzymes, and stomach fluids break down the food into small, absorbable molecules. In the lining of the small intestines, specialized cells act as a barrier, separating needed nutrients from the molecules. By the time food leaves the small intestine and enters the colon, all of the nutrients in food will have entered the bloodstream [4]. In addition to digestion, the GI tract also provides several important immune response activities. Large numbers of white blood cells (lymphocytes) reside under the tonsils and in the appendix. Clumps of lymphocytes and lymphatic tissues make up Payer's patches, immune system structures located within the small intestine. Within the walls of the large intestine, huge numbers of immune system modulators and regulators reside [46, 47]. Evidence has demonstrated that the health status of the tonsils, appendix, and the small and large intestine have an impact on the health of the immune system. Numerous strains of gut microflora reside in significant numbers in the small intestine (10^6-10^8/g of small intestinal contents) and even greater numbers in the colon (10^{11}-10^{12}/g of colon contents) or large intestine. Microflora of the large intestine perform several activities beneficial to human health, including supporting healthy digestion through fermentation, promoting healthy bacterial and yeast balance, and stimulating certain immune system components [46]. The effectiveness of all probiotics is dependent on the ability of the organisms to reach the large intestine in a viable state and adhere to the intestinal wall. Only then can colonization of the microflora succeed. Researchers have discovered recently that certain broad spectrum probiotic combinations are able to function well within altered GI environments [2]. To date, the probiotic combination of *L. acidophilus, L. rhamnosus, B. bifidum, B. breve, B. longum,* and *B. lactis* shows great promise in supporting long term colon care, providing deep intestinal support and in helping the body respond during times of increased physical, emotional, or mental stress [48]. The probiotic supplement should be proven to function *in vivo*, tolerate harsh intestinal environments, and successfully adhere to the intestinal wall. Table 4.2 gives the list of the strains useful in probiotics formulations, and Table 4.3 describes their applications in various disease conditions [1].

Table 4.2: Beneficial effects of selected probiotics documented in clinical trials

Probiotic strains	Health benefits
Lactobacillus rhamnosus GG (Valio)	Effective in reducing the incidence of diarrheal diseases (rotavirus diarrhea, antibiotic-associated diarrhea, C. difficile diarrhea and traveler's diarrhea); Reduced risk of gastrointestinal disorders suchas inflammatory bowel diseases (pouchitis and Crohn's disease); Reduced incidence of H. pyloriinfections; Reduced incidence of atopic diseases; Prevention of atopic diseases; Inhibition of generationof carcinogenic products by reducing the activity of microbial enzymes; Normalization of intestinalpermeability; Reduced risk of respiratory infections ; Reduced risk of dental caries.
Lactobacillus casei Shirota (Yakult)	Effective in treatment of rotavirus diarrhea; Reduced risk of gastrointestinal disorders; Reduced activity of procarcinogenic enzymes; Induces production of INF.
Lactobacillus casei Defensis DN114 001 (Danone)	Shortens the diarrheal phase in children with rotavirus infection; effective in treatment and prevention of gastrointestinal infections; Maintains constant urease activity; Stimulation of immune system.
Lactobacillus acidophilus NCFM (Rhodia)	Improved lactose metabolism; Reduced risk of colon cancer by limiting DNA damage in colon cells, reducing the activity of procarcinogenic enzymes and binding mutagens; Reduced serum cholesterol level; Prevention of urogenital infections.
Lactobacillus johnsonii (Danone)	Reduced incidence of *H. pylori* infections; Reduction of inflammation; Stimulation of immune system.
Lactobacillus reuterii (Stonyfield, Biogaia)	Reduced incidence of rotaviral diarrhea in children; Shortens the duration of acute gastroenteritis.
Bifidobacterium bifidum	Reduces the incidence of rotavirus infections in children and stimulates rotavirus-specific antibodyresponse; Improved lactose digestion.
Saccharomyces boulardii	Reduced incidence of C. difficile diarrhea; Shortens the duration of acute gastroenteritis.

Table 4.3: Selected examples of diseases in which prevention and treatment probiotics may be applied

Reduced disease risk	Probiotics
Diarrhea in children, mainly rotavirus-associated diarrhea	*L. rhamnosus GG* *L. reuteri* *L. acidophilus* *B. bifidum + B. infantis* *S. boulardii*
Antibiotic-associated C. difficile diarrhea	*L. rhamnosus GG* *S. boulardii*
Traveler's diarrhea	*L. rhamnosus GG* *L. fermentum KLD* *L. acidophilus (unspecified strain)* *L. acidophilus + L. bulgaricus* *L. bulgaricus + B. bifidum + S. thermophilus* *S. boulardii*
Bacterial diarrhea	*L. rhamnosus GG* *L. acidophilus* *L. plantarum* *B. bifidum*
Bacterial gastroenteritis	*L. rhamnosus GG* *L. reuteri* *E. faecium SF68* *S. boulardii*
Inflammatory bowel diseases, e.g. pouchitis and Crohn's disease	*L. rhamnosus GG* *VSL#3 (containing four strains of lactobacilli, three strains of bifidobacteria and one strain of S salivarius subsp. thermophilus)* *L. reuterii* *L. salivarius UCC118* *B. longum infantis UCC35624* *S. boulardii*
Irritable bowel syndrome	*L. plantarum 299 V*
Hypercholesterolemia	*L. acidophilus (unspecified strain)* *L. plantarum*
Food allergies and atopic diseases	*L. rhamnosus GG* *L. paracasei F19* *B. lactis Bb-12*
Lactose intolerance	*Different LAB starters*
Colon cancer	*L. acidophilus* *L. casei Shirota* *L. rhamnosus GG* *B. longum* *Propionibacterium* sp.

In Europe, the probiotic food and beverage market has developed with outstanding success within the functional food industry. This has happened since the consolidation of this market around the mid 1980s. Although the application of bacterial strains claiming to improve consumers' health dates back to the 1920s, the first serious attempts to commercialize probiotics were performed in Japan in the 1950s with the foundation of Yakult Honsha Co. Ltd. At present, the application of new technologies such as microencapsulation has encouraged numerous food producers to integrate probiotic strains in their products [1, 49]. Some of these commercial probiotics are shown in Table 4.4 [50]. It should be noted that there are about 52 major companies involved in the European food and beverage probiotics market. Commercially, yogurt drinks are the most popular products within the probiotic market as they have a great acceptance among consumers. Table 4.5 gives examples of probitic foods available commercially in different countries [51].

Table 4.4: Popular commercial strains used in the probiotics market

Genera	Species	Strain	Company
Lactobacillus	*acidophilus*	La5	Chr HansenA/S
Lactobacillus	*acidophilus*	NCFM (HOWARU™ Dophilus)	DaniscoA/S
	acidophilus	LAFTI L10	DSM
	casei	Shirota	Yakult Honsha Co. Ltd
	coryne	Hereditum	Puleva Biotech SA
	delbrueckii	subsp. Bulgaricus	Centro Sperimentale del
	fortis		Nestle SA
	paracasei	F-19	MedipharmAB
	plantarum	299v	ProbiAB (& Institut Rosell)
	reuteri	ATCC 55730	BioGaiaAB
	reuteri	RC-14	Chr HansenA/S/Urex Biotech
	rhamnosus	HN001 (HOWARU™ Rhamnosus)	DaniscoA/S
	rhamnosus	ATCC 53103 (Gefilus)	Valio
Bifidobacterium	*bifidum*	Actif essensis	Danone (proprietary)
	bifidum	BB12	Chr HansenA/S
	lactis	HN019 (HOWARU™ Bifido)	DaniscoA/S
	lactis	B94	DSM
	regularis	(Activia-proprietary)	Danone
Saccharomyces	*boulardii*		Institut Rosell
	boulardii	(ULTRA-LEVURE/Florastor)	Biocodex
Streptococcus	*salivarius*	K12	BLISTechnologies

Table 4.5: Commercial examples of probiotics

Brand	Description	Producer
Actimel	Probiotic drinks, yogurt with *L. casei* Imunitass®cultures	Danone, France
Activia	Creamy yogurt containing Bifidus ActiRegularis®	Danone, France
Gefilus	A wide range of LGG products	Valio, Finland
Hellus	Dairy products containing *L. fermentum* ME-3	Tallinna PiimatoööstuseAS, Estonia
Jovita Probiotisch	Blend of cereals, fruit, and probiotic yogurt	H&J Bruggen, Germany
Pohadka	Yogurt milk with probiotic cultures	Valašské Mezirící Dairy, Czech Republic
ProViva	Refreshing natural fruit drink and yogurt in many different flavors containing *L. plantarum*	Skåne mejerier, Sweden
Rela	Yogurts, cultured milks and juices with *L. reuteri*	Ingman Foods, Finland
Revital Active	Yogurt and drink yogurt with probiotics	Olma, Czech Republic
Snack Fibra	Snacks and bars with natural fibers and extra minerals and vitamins	Celigüeta, Spain
SOYosa	Range of products based on soy and oats and includes a refreshing drink and a probiotic yogurt-like soy–oat product	Bioferme, Finland
Soytreat	Kefir-type product with six probiotics	Lifeway, USA
Yakult	Milk drink containing *L. casei* Shirota	Yakult, Japan
Yosa	Yogurt-like oat product flavored with natural fruits and berries containing probiotic bacteria (*L. acidophilus, B. lactis*)	Bioferme, Finland
Vitality	[illegible] ω-3	Müller, Germany
Vifit	Drink yogurts with LGG, vitamins, and minerals	Campina, The Netherlands

Summary

With the increasing consumer awareness regarding linkage of diet and health, research in probiotics seems a highly fascinating challenge. Even though the idea that probiotics are a panacea for a multitude of diseases as yet seems farfetched, use of these agents as an adjunct to other established therapies has definitely shown potential benefits. It is a well-documented fact that probiotics can provide a stabilizing influence on the human ecosystem. Although the precise mechanism of action needs to be illustrated to provide a scientific rationale for their use, the information regarding therapeutic effectiveness can help in designing large, double-blind, controlled clinical trials. The importance of specifying the probiotic strain used during each clinical study must be emphasized within the scientific community as it may have important implications for assessing the studies and for planning future studies. Looking at the immense clinical/research data indicating the wide therapeutic applications of probiotics, it seems important to develop these agents as pharmaceuticals. A pharmaceutical scientist can help in proposing the systems, which can result in improved adhesion of probiotics in the GIT. The future of probiotics will rely on better elucidation of their mechanism of action and on maximization of their therapeutic potential, a burning challenge for zealous scientists.

References

1 Fuller, R., *History and development of probiotics*, in *Probiotics* 1992, Springer. p. 1-8.
2. Schrezenmeir, J. and M. de Vrese, *Probiotics, prebiotics, and synbioticsa approaching a definition.* The American journal of clinical nutrition, 2001. **73**(2): 361s-364s.
3. Crittenden, R. and M.J. Playne, *Prebiotics.* Handbook of Probiotics and Prebiotics, 1999: p. 533-581.
4. Holzapfel, W.H. and U. Schillinger, *Introduction to pre-and probiotics.* Food Research International, 2002. **35**(2): 109-116.
5. Hammes, W.P. and C. Hertel, *Research approaches for pre-and probiotics: challenges and outlook.* Food Research International, 2002. **35**(2): 165-170.
6. Ziemer, C.J. and G.R. Gibson, *An overview of probiotics, prebiotics and synbiotics in the functional food concept: perspectives and future strategies.* International Dairy Journal, 1998. **8**(5-6): 473-479.
7. Van Loo, J., *Inulin-type fructans as prebiotics.* Prebiotics: Development & Application, 2006: 57-100.
8. Roberfroid, M.B., *Introducing inulin-type fructans.* British Journal of Nutrition, 2005. **93**(1): 13-26.
9. Roberfroid, M.B., *Inulin-type fructans: functional food ingredients.* The Journal of Nutrition, 2007. **137**(11): 2493S-2502S.
10. Cummings, J.H. and G.T. Macfarlane, *Gastrointestinal effects of prebiotics.* British Journal of Nutrition, 2002. **87**(S2): S145-S151.
11. Puupponen-Pimiä, R., *et al.*, *Development of functional ingredients for gut health.* Trends in Food Science & Technology, 2002. **13**(1): 3-11.

12. Simmering, R. and M. Blaut, *Pro-and prebioticsa the tasty guardian angels?* Applied Microbiology and Biotechnology, 2001. **55**(1): 19-28.
13. Tuohy, K.M., *et al., Using probiotics and prebiotics to improve gut health.* Drug Discovery today, 2003. **8**(15): 692-700.
14. Fooks, L.J., R. Fuller, and G.R. Gibson, *Prebiotics, probiotics and human gut microbiology.* International Dairy Journal, 1999. **9**(1): 53-61.
15. Salminen, S., *et al., Functional food science and gastrointestinal physiology and function.* British Journal of Nutrition, 1998. **80**(S1): S147-S171.
16. Lee, Y.K. and S. Salminen, *Handbook of probiotics and prebiotics* 2009: John Wiley & Sons.ISBN: 978-0-470-13544-0
17. Mattila-Sandholm, T., *et al., Technological challenges for future probiotic foods.* International Dairy Journal, 2002. **12**(2): 173-182.
18. Loo, J.V., *et al., Functional food properties of non-digestible oligosaccharides: a consensus report from the ENDO project (DGXII AIRII-CT94-1095).* British Journal of Nutrition, 1999. **81**(02): 121-132.
19. Bosscher, D., *Fructan prebiotics derived from inulin,* in *Prebiotics and Probiotics Science and Technology* 2009, Springer. 163-205.
20. Torres, D.P.M., *et al., Galacto Oligosaccharides: Production, Properties, Applications, and Significance as Prebiotics.* Comprehensive Reviews in Food Science and Food Safety, 2010. **9**(5): 438-454.
21. Sako, T., K. Matsumoto, and R. Tanaka, *Recent progress on research and applications of non-digestible galacto-oligosaccharides.* International Dairy Journal, 1999. **9**(1): 69-80.
22. Splechtna, B., *et al., Production of prebiotic galacto-oligosaccharides from lactose using β-galactosidases from Lactobacillus reuteri.* Journal of Agricultural and Food Chemistry, 2006. **54**(14): 4999-5006.
23. Macfarlane, G.T., H. Steed, and S. Macfarlane, *Bacterial metabolism and health related effects of galactoa oligosaccharides and other prebiotics.* Journal of Applied Microbiology, 2008. **104**(2): 305-344.
24. Rabiu, B.A., *et al., Synthesis and Fermentation Properties of Novel Galacto-Oligosaccharides by β-Galactosidases from Bifidobacterium Species.* Applied and Environmental Microbiology, 2001. **67**(6): 2526-2530.
25. Tzortzis, G., *et al., A novel galactooligosaccharide mixture increases the bifidobacterial population numbers in a continuous in vitro fermentation system and in the proximal colonic contents of pigs in vivo.* The Journal of Nutrition, 2005. **135**(7): 1726-1731.
26. Rycroft, C., *et al., A comparative in vitro evaluation of the fermentation properties of prebiotic oligosaccharides.* Journal of Applied Microbiology, 2001. **91**(5): 878-887.
27. Palframan, R.J., G.R. Gibson, and R.A. Rastall, *Effect of pH and Dose on the Growth of Gut Bacteria on Prebiotic Carbohydrates.* Anaerobe, 2002. **8**(5): 287-292.
28. Terada, A., *et al., Effect of lactulose on the composition and metabolic activity of the human faecal flora.* Microbial Ecology in Health and Disease, 1992. **5**(1): 43-50.
29. Tuohy, K.M., *et al., A human volunteer study to determine the prebiotic effects of lactulose powder on human colonic microbiota.* Microbial Ecology in Health and Disease, 2002. **14**(3): 165-173.
30. Bouhnik, Y., *et al., Lactulose ingestion increases faecal bifidobacterial counts: a randomised double-blind study in healthy humans.* European Journal of Clinical Nutrition, 2004. **58**(3): 462-466.
31. Crittenden, R.G. and M.J. Payne, *Production, properties and applications of food-grade oligosaccharides.* Trends in Food Science & Technology, 1996. **7**(11): 353-361.
32. Rycroft, C.E., *et al., A comparative in vitro evaluation of the fermentation properties of prebiotic oligosaccharides.* Journal of Applied Microbiology, 2001. **91**(5): 878-887.

33. Rossi, M., *et al.*, *Fermentation of fructooligosaccharides and inulin by bifidobacteria: a comparative study of pure and fecal cultures.* Applied and Environmental Microbiology, 2005. **71**(10): 6150-6158.
34. Hayakawa, K., *et al.*, *Effects of soybean oligosaccharides on human faecal flora.* Microbial Ecology in Health and Disease, 1990. **3**(6): 293-303.
35. Saito, Y., T. Takano, and I. Rowland, *Effects of soybean oligosaccharides on the human gut microflora in in vitro culture.* Microbial Ecology in Health and Disease, 1992. **5**(2): 105-110.
36. Sikorski, Z.E., *Chemical and functional properties of food components*2006: CRC Press.
37. Bansal, T. and S. Garg, *Probiotics: from functional foods to pharmaceutical products.* Current pharmaceutical biotechnology, 2008. **9**(4): 267-287.
38. Rastall, R.A. and V. Maitin, *Prebiotics and synbiotics: towards the next generation.* Current Opinion in Biotechnology, 2002. **13**(5): 490-496.
39. Brown, I.L., *Applications and uses of resistant starch.* Journal of AOAC International, 2004. **87**(3): 727-732.
40. Fuentes-Zaragoza, E., *et al.*, *Resistant starch as functional ingredient: A review.* Food Research International, 2010. **43**(4): 931-942.
41. Nugent, A.P., *Health properties of resistant starch.* Nutrition Bulletin, 2005. **30**(1): 27-54.
42. Schmiedl, D., *et al.*, *Production of heat-stable, butyrogenic resistant starch.* Carbohydrate Polymers, 2000. **43**(2): 183-193.
43. Jacobasch, G., *et al.*, *Hydrothermal treatment of Novelose 330 results in high yield of resistant starch type 3 with beneficial prebiotic properties and decreased secondary bile acid formation in rats.* British Journal of Nutrition, 2006. **95**(06): 1063-1074.
44. Wang, J., Z. Jin, and X. Yuan, *Preparation of resistant starch from starchaguar gum extrudates and their properties.* Food Chemistry, 2007. **101**(1): 20-25.
45. Roberfroid, M., *Prebiotics: the concept revisited.* The Journal of Nutrition, 2007. **137**(3): 830-837.
46. Goossens, D., *et al.*, *Probiotics in gastroenterology: indications and future perspectives.* Scandinavian Journal of Gastroenterology-Supplements, 2003. **38**(239): 15-16.
47. Guyton, A. and J. Hall, *Secretory functions of the alimentary tract.* Textbook of Medical Physiology, 1991. **9**: 815-32.
48. Collado, M., J. Meriluoto, and S. Salminen, *Development of new probiotics by strain combinations: is it possible to improve the adhesion to intestinal mucus?* Journal of Dairy Science, 2007. **90**(6): 2710-2716.
49. Douglas, L.C. and M.E. Sanders, *Probiotics and prebiotics in dietetics practice.* Journal of the American Dietetic Association, 2008. **108**(3): 510-521.
50. Frost-Sullivam. Strategic analysis of the European food and beverage products markets (#B956-88), London, Frost and Sullivan Ltd., 2007.
51. Sirb I, Appetite, 2008, 51: 456-467.

5

Seaweed/Microbial Source

Introduction

Whilst food has long been used to improve health, our knowledge of the relationship between food components and health is now being used to improve food. Although most foods can be considered "functional", in the context of this review, the term is reserved for foods and food components that have been demonstrated to provide specific health benefits beyond basic nutrition. Widespread interest in select foods that might promote health has resulted in the use of the term "functional foods". Seaweeds are known for their richness in polysaccharides, minerals and certain vitamins [1], but they also contain bioactive substances like polysaccharides, proteins, lipids and polyphenols, with antibacterial, antiviral and antifungal, as well as many such other properties [2]. This gives seaweed great potential as a supplement in functional food or for the extraction of compounds. Physiologically active substances in marine algae are classified into two types based on the difference in the mechanisms viz. (i) non-absorbed high-molecular materials like dietary fibres, and (ii) low molecular materials, which are absorbed and which affect the maintenance of human homeostasis directly. Seaweeds are among the first marine organisms chemically analyzed, with more than 3,600 articles published describing 3,300 secondary metabolites from marine plants and algae, and they still remain an almost endless source of new bioactive compounds [3].

Since 1975, three areas of research in marine natural products have emerged viz. bioactive metabolites, biotoxins and chemical ecology. More than 15,000 novel compounds have been isolated from marine organisms. Due to the wide range of competitive environments they survive in, marine organisms have developed unique defense strategies and bioactive compounds that, in some cases, are unparalleled by their terrestrial counterparts. The exploration of this chemical diversity for pharmaceutical purposes has revealed important chemical prototypes for the discovery of new agents, stimulating the use of sophisticated spectroscopic methods and development of new synthetic methodology. The main marine ingredients with poteintial food application and functionality are shown in Fig. 5.1.

Algae are divided in two general categories viz. (i) macroalgae such as red, green and brown algae, and (ii) microalgae such as blue-green algae. The aquaculture sector produces large amounts of seaweeds, such as Laminaria, Porphyra, and Gracilaria, and microalgae, including Dunaliella and Spirulina. There is great potential in the use of microalgae for production of food ingredients, as they are photoautrophic microorganisms that can grow on a very simple culture medium containing sea water, nitrate, phosphate, trace amounts of certain metals, and carbon-di-oxide. Excluding the bioactive secondary metabolites that target the pharmaceutical market, the most common compounds from algae having ready commercial applications besides phycocolloids include polyunsaturated fatty acids, steroids, carotenoids and lectins. Table 5.1 gives examples of typical marine microorganisms.

Table 5.1: Examples of typical marine microorganisms

Organism	Phylum	Examples
Bacteria	Actinobacteria	*Rhodococcus marinonascens* *Streptomyces griseus*
	Cyanobacteria	*Pochlorococcus* *Synechococcus*
	Proteobacteria	*Pseudoalteromonas* spp. *Shewanella* spp.
	Firmicutes	*Bacillus* spp.
Fungi	Ascomycota	*Halosphaeriales* spp. *Cladosporium* spp. *Leptosphaeria* spp.
	Basidiomycota	*Nia vibrissa Digitatispora marina*
	Zygomycota	*Mortierella* spp. *Mucor* spp. *Lichina pygmaea*
Microalgae and diatom	Heterokontophyta	*Nannochloropsis* spp. *Phaeodactylum tricornutum* *Ditylum brightwellii* *Thraustochytrium*
	Dinoflagellata	*Amphdinium*
Yeast	Saccharomycetaceae	*Candida, Torula*

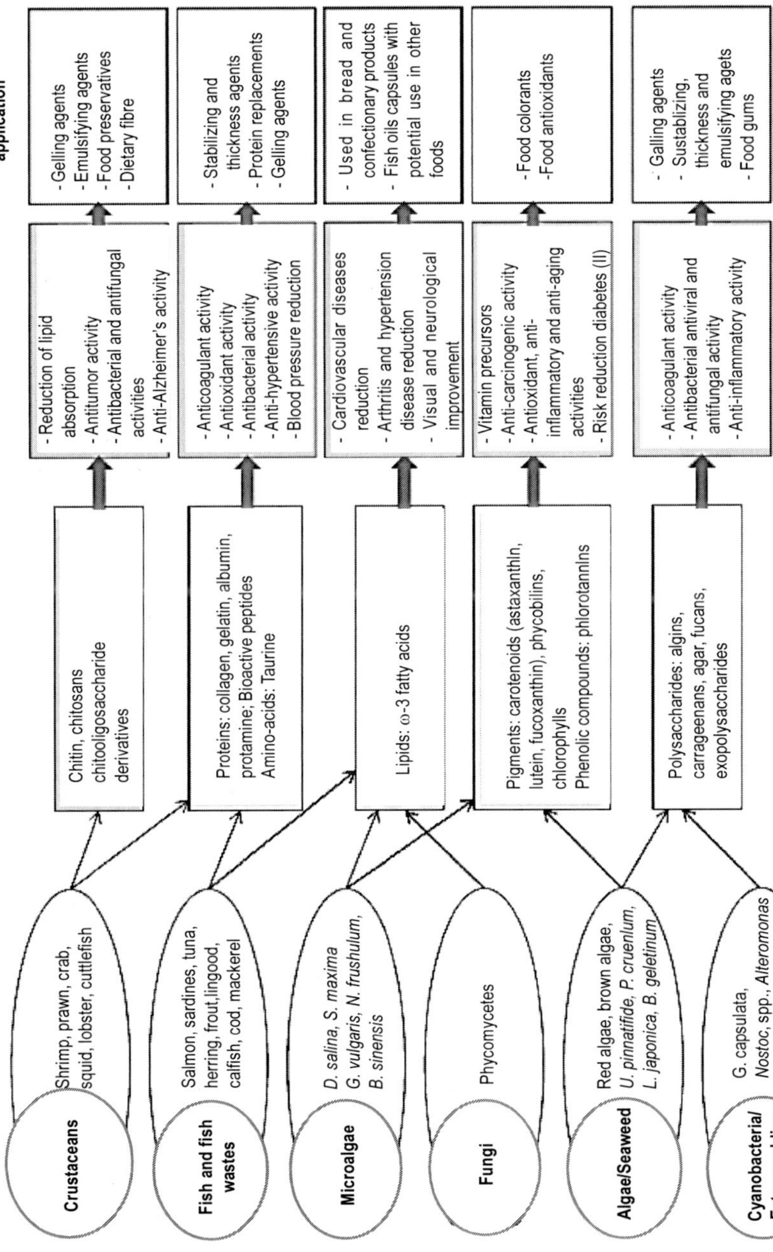

Fig. 5.1: Main marine ingredients with potential food application and functionality

Ingredients Functionality Derived from Seaweeds and Microbial Source

Polysaccharide

Marine algae contain large amounts of polysaccharides, notably cell wall structural, but also mycopolysaccharides and storage polysaccharides [2]. Polysaccharides are polymers of simple sugars (monosaccharides) linked together by glycosidic bonds, and they have numerous commercial applications in products such as stabilisers, thickeners, emulsifiers, food, feed, beverages, etc. The total polysaccharide concentrations in the seaweed species of interest range from 4% to 76% of dry weight.

The highest contents are found in species such as *Ascophyllum, Porphyra and Palmaria*; green seaweed species such as Ulvaalso have a high content, up to 65% of dry weight. Seaweeds are low in calories from a nutritional perspective. The lipid content is low, and even though the carbohydrate content is high, most of this is dietary fibres and not taken up by the human body. However, dietary fibres are good for human health as they make an excellent intestinal environment. The cell wall polysaccharides mainly consist of cellulose and hemicelluloses, neutral polysaccharides, and are thought to physically support the thallus in water.

The building blocks needed to support the thalli of seaweed in water are less rigid/strong compared to terrestrial plants and trees. The cellulose and hemicellulose content of the seaweed species of interest in this review is 2–10% and 9% dry weight, respectively. Examples of marine microorganisms producing extracellular polysaccharides are shown in Table 5.2.

Green algae contain sulphuric acid polysaccharides, sulphated galactans and xylans, brown algae alginic acid, fucoidan (sulphated fucose), laminarin (β-1,3 glucan) and sargassan and red algae agars, carrageenans, xylans, floridean starch (amylopectin-like glucan), water soluble sulphated galactan, as well as porphyran as mucopolysaccharides located in the intercellular spaces. Contents of both total and species specific polysaccharides show seasonal variations. The mannitol content varied markedly in the fronds of Saccharina and Laminaria species, with maximum amounts found during summer and autumn, from June to November. Several other polysaccharides are present in and utilised from seaweed, e.g. furcellaran, funoran, ascophyllan and sargassan; however, these are not described in this review. Seaweed polysaccharides are separated into dietary fibres, hydrocolloids, etc. in the following paragraphs, even though the polysaccharides belong to more than just one of the functional groups.

Table 5.2: Extracellular polysaccharides producing marine microorganisms

Microorganism	Properties
Aspergillus versicolor (Coral-associated fungus)	Molecular weight: 7 kDa Principal components: α-D-glucopyranose α-D mannopyranose. Yield: 0.2 g/l
Cyanothece sp.	Principal components: α-D-1,6-homoglucan (Cyanobacterium form sea water) Yield: 22.34 g/l
Keissleriella sp. YS 4108 (Marine sediment fungus)	Molecular weight: 0.13 MDa Principal components: galactose, glucose, rhamnose, mannose, glucuronic acid
Hahella chejuensis	Principal components: Glucose and galactose (Marine sediment bacteria) Yield 9.23 g/l
Geobacillus thermodenitrificans (Marine volcanic vent)	Molecular weight: 0.4 MDa. Principal component: Mannose, glucose
Pseudoalteromonas sp. CAM025 (Sea ice bacterium)	Molecular weight: 5.7 MDa. Principal component: Mannose, glucose, galactouronic acid
Weissella cibaria (LAB form fish intestine)	Yield: 14 g/l
Gyrodinium impudicum (Marine microalgae)	Molecular weight: 18.7 MDa. Principal component: Galactose with uronic acid, Yield: 134.7 mg/l
Porphyridium sp. (Red microalgae)	Molecular weight: 2.3 MDa. Principal components: Highly sulfated acidic heteropolymers

Carotenoids

Carotenoids are natural pigments derived from five-carbon isoprene units which are polymerized enzymatically to form regular, highly conjugated 40-carbon structures. β-Carotene is mainly found in plant tissues and exhibits the highest provitamin A activity. The concentration of β-carotene among marine plants and algae is species specific and is affected by stress factors, such as high light intensity, limited nitrates, and high salt concentrations. One of the major marine producers of β-carotene is the halophilic microalga *Dunaliella salina*, which is the highest β-carotene-enriched eukaryotic organism known.

β-Carotene derived from *Dunaliella* has been marketed in several forms, such as extract in edible oils (1.5–30% β-carotene) and dried Dunaliella powder in capsules or tablets (approximately 5% β-carotene). Astaxanthin is a red pigment common to several aquatic organisms including microalgae, sea grasses, shrimps, lobsters and fish, such as salmon and trout. Crustaceans are unable to synthesize carotenoids de novo and require astaxanthin or its precursors in their diet in order to acquire the desired color for the seafood market.

Astaxanthin, with an antioxidant activity 10 times stronger than other carotenoids (e.g. β-carotene, canthaxanthin and lutein), provides protection against cancer, inflammation, and UV light [4]. The beneficial properties of astaxanthin, in combination with the strong coloring ability, make it an important ingredient in the nutraceutical, cosmetics, food and animal feed industries.

The antioxidant properties of the algal carotenoids have also been shown to play a role in preventing man pathologies linked to oxidative stress [5]. Furthermore, experimental studies strongly suggest that β-carotene could prevent the onset of cancers, especially lung cancer [6]. However, other studies suggest that β-carotene could induce lung cancer in smokers. Recent studies have shown the correlation between a diet rich in carotenoids and a diminishing risk of cardiovascular disease, cancers (β-carotene, lycopene), as well as opthalmological diseases. Fucoxanthin is a xanthophyll and has a unique structure, including an unusual allenic bond and 5,6-monoepoxide in its molecule. Fucoxanthin is one of the most abundant carotenoids in nature. The content in seaweed varies during season and life cycle. It is quite stable in the presence of organic ingredients, apart from surviving the drying process and storage at ambient temperature. In pure form, fucoxanthin is vulnerable to oxidation [6]. The total carotenoid content of *F. serratus* was found to be aproximately 0.08% of the dried extracted cells, and fucoxanthin comprises about 70% of the total carotenoid.

Waste parts of cultivated *L. japonica* (kombu) are a good biosource for fucoxanthin extraction. Discards from processing (stipe, holdfast and blade material) can be used as a source for fucoxanthin extraction. These discards have similar content of fucoxanthin (178–196 mg per kg fresh weight). The recovery ratio of fucoxanthin has been shown to reach 82%, and a total of 1,490 g of fucoxanthin was obtained from 10 tonne of kombu waste. Fucoxanthin from fresh Undariais mostly found as the geometrical isomer all-trans (88 %). Trans forms are more stable; however, cis forms of fucoxanthin were found to exert a higher inhibitory effect compared to their trans counterparts on human leukaemia (HL-60) cells and colon cancer (Caco-2) cells. Uptake and incorporation of the trans form of fucoxanthin into cellular lipids were faster compared to cis counterparts. Fucoxanthin is easily converted to fucoxanthinol in human intestinal cells and in mice.

Experiments have shown that fucoxanthin from brown seaweed significantly reduces the viability of human prostate cancer cells and significantly reduces the percentage of tumour bearing mice and the average number of tumours per mouse when given in the drinking water. Other studies demonstrated anticancer effects, including inhibition of the proliferation of human leukaemia cell line (HL-60), and induced their apoptosis.

Sterols

Sterols are among the most important chemical constituents of micro and macro algae and a major nutritional component in the diet of aquacultured organisms. Microalgae are important components in the diet of many hydrobionts, especially bivalves. A mixture of several algal species typically improves larval development, presumably by avoiding deficiency resulting from a unialgal diet. The qualitative and quantitative sterol composition of microalgae used in hatcheries has direct implications in the phytosterol and cholesterol composition of bivalve larvae and can affect their growth performance.

Fatty acids

Algae are believed to be the primary producers of ω-3 polyunsaturated fatty acids (PUFA) in the marine food chain and the only plant source for eicosapentaenoic acid (EPA, 6) and docosahexaenoic acid (DHA) (Table 5.3). PUFA are essential for normal cell function, and after elucidation of their biological role in certain clinical conditions, such cardiovascular diseases, they entered the biomedical and nutraceutical markets. PUFAs are important in cellular and tissue metabolism, including the regulation of membrane fluidity electron and oxygen transport, as well as thermal adaptation. EPA and DHA, found at various levels in marine micro and macro algal species, have shown relatively high oxidative stability compared to fish oils. Recently, public preference for healthy food has increased the market visibility of PUFA and in particular the interest for the ω-3 family of EPA, DHA and α-linolenic acid.

Table 5.3: Examples of PUFA producing marine microbes

Organism		Total fatty acids (% of biomass)	PUFA (% of total) DHA	EPA
Aurantiochytrium limacinum	Protista	65	30–40	–
Shewanella violacea	Bacteria	50	–	11–17
Shewanella putrefaciens	Bacteria	50	–	24–40
Crypthecodinium cohnii	Dinoflagellate microalgae	20	30–35	–
Schizochytrium sp.	Unicellular protists	55	35–40	–
Pavlova salina	Microalgae	12	50	
Thraustochytrium aureum	Protists	20	51	–
Thraustochytrium roseum	Protists	25	50	–

Phycobiliproteins

Unlike chlorophylls and carotenoids, phycobiliproteins are water soluble and form particles (phycobilisomes) on the surface of the thylakoids rather than being embedded in the membranes. Phycobiliproteins consist of pigmented phycobilins, which are linear tetrapyrroles. Different combinations of the two principal phycobilins i.e. phycoerythrobilin (red) and phycocyanobilin (blue), which absorb different regions of wavelengths give different absorption spectra. Besides the phycoerythrin and phycocyanin, phycobiliproteins can also be divided into allophycocyanin and phycoerythrocyanin. Within phycobilisomes, phycobiliproteins play an important role in the photosynthetic process of the three types of algae viz. *Rhodophyta, Cyanophyta and Cyptophyta.* The additional photosynthetic pigments make light harvesting possible in deep waters as surface light wavelengths for some colours are almost completely absorbed below 10 m. Phycobiliproteins, in particular phycoerythrin, can constitute a major proportion of the red algal cell proteins, with levels of 1.2% and 0.5% of dry weight reported for *P. palmata* and *Gracilaria tikvahiae,* respectively [7]. Phycobiliproteins are currently being used as natural colourants for food and cosmetic applications and are used in foods such as chewing gum and dairy products in addition to lipsticks and eyeliners [8]. Niohan Siber Hegner Ltd. and Dainippon Ink & Chemical Inc. are two companies in Tokyo that market the phycobiliprotein phycocyanin as a natural food colourant [9]. Different phycobiliproteins have been shown to exhibit antioxidant, anti-inflammatory, neuroprotective, hypocholesterolemic, hepatoprotective, antiviral, anti-tumour, liverprotecting, atherosclerosis treatment, serum lipid-reducing and lipase inhibition activities.

Phenols and other phlorotannins

Phenols, sometimes called phenolics, are a class of chemical compounds consisting of a hydroxyl group (–OH) bonded directly to an aromatic hydrocarbon group. The simplest of the class is phenol, the parent compound used as a disinfectant and for chemical synthesis. Polyphenols from terrestrial plants are derived from gallic and ellagic acid, whereas the algal polyphenols are derived from polymerised phloroglucinol units (1,3,5-trihydroxybenzene). Phlorotannins are a group of phenolic compounds. Phlorotannins constitute an extremely heterogeneous group of molecules (structure and polymerisation degree heterogeneity) providing a wide range of potential biological activity. Phlorotannins are localised in physodes which are membrane bound cytoplasmic vesicles, and the fusion of physodes with cell membranes results in a secretion of phlorotannins.

Green and red seaweed have low concentrations of phenols compared to brown seaweed species, which have high concentrations of the phenol group phlorotannin. Phenol content varies from $<1\%$ to 14% of dry seaweed biomass, with *Ascophyllum* and *Fucusas* the species with the highest content (14%

and 12%). Phlorotannins would be potential candidates for the development of unique natural antioxidants for further industrial applications as functional foods, cosmetics and pharmaceuticals [10]. Besides phloroglucinol, there are many other flavonoids and their glycosides present in green, brown and red algae, such as catechin and epicatechin (upto 4.77 mg/g dry weight) or the flavonol glycoside rutin (upto 11.4 mg /g dry weight) in Porphyra. The flavone glycoside hespiridinis present in green, brown and red seaweeds and is especially high in certain green algae at 117 mg /g dry weight.

Halogenated compounds

Halogenated compounds are produced naturally, mainly by red and brown seaweed, dispelling the widespread notion that these chemicals are only of man made origin. Halogenated compounds are dispersed in several different classes of primary and secondary metabolites, including indoles, terpenes, acetogenins, phenols, fatty acids and volatile halogenated hydrocarbons (e.g. bromoform, chloroform, dibromomethane). In many cases, they possess biological activities of pharmacological interest, including antibacterial [11] and anti-tumoural. The most notable producers of halogenated compounds in the marine environment belong to the genus *Laurencia.* The compounds are predominantly derivates of esquiterpenes diterpenes, triterpenes, acetogenins, fatty acids and brominated indoles [12, 13]. In addition to antimicrobial and cytotoxic properties, *Laurencia* compounds may also play multifunctional ecological roles such as acting as a feeding defendant. Polyhalogenated monoterpenes from red algae of the genera *Plocamium* and *Chondrococcus* exhibit a wide range of pharmacological activities, including antimicrobial, antitubercular and anticancer activities [14]. The presence of halogen atoms (F, Cl, Br or I) in the fatty alkyl chain causes significant changes in their physicochemical characteristics, increasing their reactivity and changing the conformation of biological membranes.

Bioactive Metabolites from Macroalgae

Pharmaceutical compounds constitute one of the most important potential markets for algal products. Prior to 1950, the use of seaweed extracts and microalgae as drugs or drug sources was restricted to folk medicine. Still, up to date, there has been little commercial development of algal products as pharmaceutical agents. The vermifuge Q-kainic acid from the red alga Digenea simplex was marketed in the past, but has now been withdrawn in western countries. However, there is a tremendous potential for the development of algae as sources of pharmaceutical compounds, since in the recent years researchers have described a wide range of biological activities for metabolites produced by algae. Isolation of pharmacologically active compounds from marine algae has been the subject of many intensive investigations and comprehensive account of such studies in the field is given in numerous reviews.

In-vitro screens for the detection of a wide spectrum of biological activities for the initial evaluation of algal extracts and subsequent bioassay guided fractionations for the isolation of the active compounds constitute one of the most common methods of investigation. More than 3,280 structures from algae have been reported in the literature, amounting to approximately 30% of the total number of natural products isolated from marine organisms. Among macroalgae, significantly more rich in secondary metabolites appear the brown and red algae, with the latter being the top producers of halogenated metabolites. Algal metabolites have exhibited a wide spectrum of activity, with the majority of them having been evaluated for their cytotoxic and antibiotic properties. The development of high-throughput screening systems and the enhanced sensitivity of pharmacological assays allow now extensive biological studies on compounds that are isolated in sub-milligram quantities.

Bioactive Compounds from Marine Microbes

Marine microorganisms have received mounting attention in biodiscoveries due to their exclusive physicochemical characteristics which have been acquired as an adaptation to prevailing extreme conditions in the marine environment. It has been noticed that those unique marine microbes and their biologically active metabolites are potential sources to be used as sustainable food and pharmaceutical ingredients. Even though dozens of research articles demonstrate the immense potentials of marine microbial metabolites as lead compounds in drugs and pharmaceutical developments, their role as food ingredients is poorly addressed. However, recent advances in food technology have opened up a number of novel avenues to develop natural substances as food or food ingredients. In this context, this section aims at revealing and discussing prospective applications of marine microorganisms and their metabolites in modern nutraceuticals and functional foods.

Carbohydrates

Over the past few decades, various terrestrial and aquatic sources have been explored for bioactive polysaccharides, and a wide range of biological activities including anticoagulant, anti-inflammatory, antiviral and antitumor activities were reported [15]. Interestingly, among the identified sources microbial polysaccharides received great attention due to ease of extraction in pure form. Microbial polysaccharides are generally divided into three groups based on its present location in the microbial cell; cell wall polysaccharides, intercellular polysaccharides and exocellular polysaccharides (EPS) [16]. The latter, EPS, are commercially valuable products of debris. It is widely accepted that microbial EPS offer many advantages over plant polysaccharides due to their novel functions and stable chemical and physical properties [17]. In the extreme environmental conditions of oceans, marine microorganisms secrete structurally and functionally diverse EPS to defend against biotic as well as abiotic pressure in the ocean. EPS form a layer surrounding the marine microbial cells which helps them to withstand or resist adverse and extreme conditions [17, 18]. Nichols and his research team [19] reported that EPS extracted from six closely related marine *Pseudoalteromon* has showed a great structural variation despite the close phylogenetic relationships of the strains (Table 5.4). This result indicated the magnitude of structural and functional diversity of marine microbial EPS. With great biodiversity of the marine microbial community, it is clear that the marine environment provides a magnificent opportunity to explore novel microorganisms since they are constantly diffused into the culture medium, easy to isolate from the culture media and free from protein and cell biologically active EPS compared to its terrestrial counterpart.

Table 5.4: Representative marine metabolites with different therapeutic activity

Compound	Type	Source
Antifungal activity		
Aurantoside B	Polyketide	*Siliquariaspongia japonica* sponge
Phorboxazole A	Macrolide	*Phorbas* sp. sponge
Halishigamide A	Macrolide	*Halichondria* sp. sponge
Halichondramide	Macrolide	*Halichondria* sp. sponge
Bengazole A	Oxazole-containing fatty-acid ester	*Jaspis* sp. sponge
Ptilomycalin A	Polycyclic	*Ptilocaulis spiculifer* guanidine alkaloid sponge
Antituberculosis activity		
Alkaloid (+)-8-hydroxymanzamine	Heterocyclic ring system attached to α-carboline moiety	*Pachypellina* sp. sponge and *Petrosiidae* genus
Axisonitrile-3	Cyanosesquiterpene	*Acanthella klethra* sponge
Litosterol	C-19 hydroxysteroid	Okinawan soft coral *Litophyton viridis*
Puupehenones	Shikimate-sesquiterpene	Sponges
Antiviral activity		
Papuamides A	Cyclic depsipeptides	*Theonella mirabilis* *T. swinhoei* sponges
Avarone	Sesquiterpene hydroquinone	*Dysidea avara* sponge
Gymnochrome D	Brominated phenanthropery lenequinone	Fossil crinoid *Gymnocrinus richeri* pigments
Microspinosamide	Cyclic depsipeptide	*Sidonops microspinosa* sponge
Solenolide A	Diterpene lactone	Gorgonian of the genus *Solenopodium*

Compound	Type	Source
Hennoxazole A	Oxazole-containing alkaloid	*Polyfibrospongia* sp. sponge
Thyrsiferol	Triterpene	Red alga *Laurencia venusta*
Spongiadiol	Tetracyclic furanoditerpene	Deep-water *Spongia* sp.
Antibacterial activity		
Squalamine	Aminosterol	Dogfish shark *Squalus acanthias*
Cribrostatin 3	Isoquinolines	*Cribrochalina* sp. sponge
Bromosphaerone	Bromoditerpenes	*Sphaerococcus coronopifolius* red alga
Pestalone	Chlorinated benzophenone	*Rosenvingea* sp. brown alga
Jorumycin	Dimeric isoquinoline alkaloid	*Jorunna funebris*
Antiprotozoal activity		
Cyclic peroxides	Peroxide	*Plakortis* aff *angulospiculatus* sponge
Sigmosceptrellin-B	Alkaloid	*Diacarnus erythraeanus* red sea sponge
Plakortide	Cyclic peroxylactone	*Plakinastrella onkodes* sponge
Ascochyta salicorniae	Fungus	*Ulva* sp. green alga
Anthelmintic activity		
Dihydroxytetrahy-drofuran	Tetrahydrofuran	*Notheia anomala* south Australian marine brown alga *Amphimedon* sp. sponge
Amphilactams		
Geodin A magnesium salt lactam tetramic acid	Macrocyclic polyketide	*Geodia* sp. sponge

In addition to microbial polysaccharide, marine microbes have been explored for bioactive monosaccharides. Due to several adaptations to hard marine environment, marine derived novel microbial species have been reported as high yield producers of both glucosamine and chitosan compared to those of terrestrial species. These strains have a potential to be used as alternative source for production of vegetarian glucosamine which has growing demand in modern nutraceuticals. However, further studies are necessary for validation and scale up of the glucosamine production using marine microbial strains.

Polyunsaturated Fatty Acids (PUFA)

Due to well documented healing properties against chronic diseases, long chain PUFA has earned considerably high market share as a dietary supplement or nutraceutical [20]. Further, in 2004 the U.S. Food and Drug Administration declared "qualified health claim" status to ω–3 fatty acids eicosapentaenoic acid (EPA) and docosahexaenoic acid (DHA). Therefore, both EPA and DHA have been extensively used as nutraceuticals to prevent some chronic diseases. Among various PUFA sources, marine sources such as herring, mackerel, sardine and salmon are considered as the prime sources of ω-3 fatty acids, in particular EPA and DHA. However, growing demand on marine-derived ω–3 fatty acids has become a threat for these fish species due to over fishing for oils, and hence, there is an urgent need of alternative sustainable sources of high quality PUFA. On the other hand, sea food derived ω-3 fatty acids also raise some quality problems such as heavy metal contamination and unpleasant smell and taste [21]. In the search of alternative sources of PUFA, marine microorganisms are under the spot light since they are the primary producers of PUFA in the marine environment [22].

Numerous marine microorganisms, including microalgae, bacteria, yeast, protists, dinoflagellates and filamentous fungi have been reported as producers of long chain PUFA in marine environment. Among them, only oleaginous organisms (microorganisms which produce more than 25% lipid on a dry cell weight basis) can be tailored to produce high amounts of ω-3 fatty acids, particularly EPA and DHA [23]. Marine protists and dinoflagellates have recently received much attention as potent producers of bioactive PUFA. Thraustochytrids, which belong to marine protists Labyrinthulomycetes containing higher level of DHA in fatty acid profile, has been identified as a potential strain. A recent comprehensive article [24] on Thraustochytridsreveals various plus points of these organisms to be developed as alternative producer of ω-3 fatty acids. Moreover, Schizochytrium spp., heterotrophic marine *Thraustochytrids* that have the ability to produce around 35% – 40% of their total fatty acid as DHA, is currently cultured for commercial production of PUFA [25].

In addition, marine microalgae have also been identified as a rich source of long chain PUFA where fatty acid content accounts for 10–20% of total biomass in some species. PUFA content in marine microalgae is comparatively higher than that of fresh water microalgal species as marine microalgae produce higher amounts of PUFA to survive in marine environments. *Nannochloropsis* spp., *Porphyridium cruentum, Phaeodactylum tricornutum* and *Chaetoceros calcitrans* have been reported as the best EPA producing microalgae, while *Isochrysis galbana* and *Cryptecodinium* spp. are more suitable for extraction of DHA [6]. However, many algal species that have the ability to produce high amount of PUFA are obligate phototrophs and, in many cases, applying technologically advanced bioreactors for culturing those species make it cost prohibitive for commercial production. In this regard, heterotrophic algal species offer added advantage as they can be grown in conventional fermenters to achieve high biomass in the absence of light.

Pigments

From the simplest organisms like cyanobacteria up to higher plants and animals are producing natural pigments for various purposes. Some of these natural pigments possess promising therapeutic potentials including, prevention of chronic diseases. In this context, various plant, animal and microbial species have been explored as sources of natural pigments, particularly with the aim of identifying therapeutic pigments since bioactive natural pigments have received much interest in the production of modern nutraceutical and functional foods (Table 5.5) [5]. Owing to the advantages offered by fermentative production and comparative simplicity of extraction, microbial pigments received much interest in natural pigment explorations than plant based pigments; some microbial pigments are already being produced commercially. Despite some toxicological barriers for microbial pigments in certain countries, red koji from fungus *Monascus* [26], β-carotene from fungus *Blakeslea* [27], β-carotene from microalgae. *Dunaliella* and *astaxanthin* from Haematococcus [28] are some of the successful stories of microorganism derived food grade pigments. Exploring marine microorganisms for natural pigment is not a novel approach; particularly marine microalgae have been studied for years as a source of natural pigment. Interest on marine pigment resurged with novel trends in bioactive natural pigments. Many reports have revealed that a number of carotenoids such as β-carotene, astaxanthin, cantaxanthin, lutein and violaxanthin, which bear promising pharmaceutical potentials, are abundantly produced by marine microalgae (Table 5.5).

Table 5.5: Marine microorganism-derived bioactive pigments

Pigment	Producing organisms	Chemical structure	Health claim
β-Carotene (Orange red)	*Dunaliella salina,* *Synechococcus* spp. *Nannochloropsis gaditan*	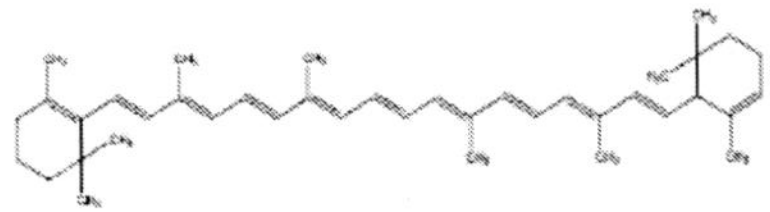	Antioxidant Anticancer Source of provitamin A
Zeaxanthin (Yellow)	*Paracoccus* sp. *Zeaxanthinibacter enoshimensis*	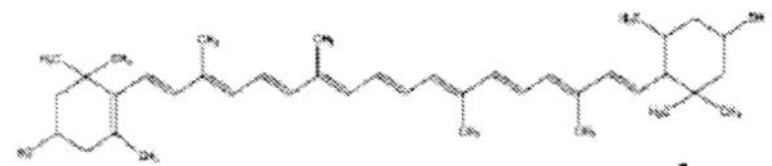	Improve vision Antioxidant
Violaxanthin (Orange)	*Dunaliella tertiolecta*		Antiproliferative
Fucoxanthin (Green, orange)	*Phaeodactylum tricornutum*	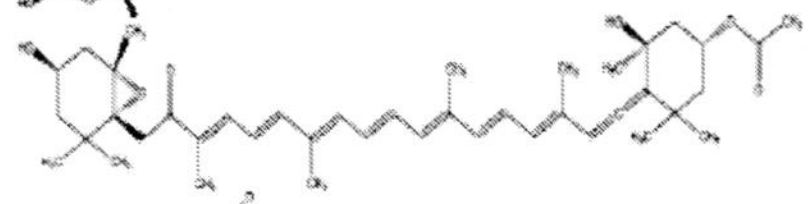	Reduce cholesterol Anticancer Antidiabetics
Violacein (Violet)	*Pseudoalteromonas tunicate*	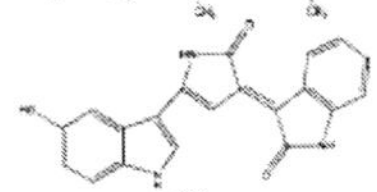	Antitumoural Antimicrobial Antiparasitary
Prodiginines (Red)	*Pseudoalteromonas rubra* *Pseudoalteromonas denitrificans*	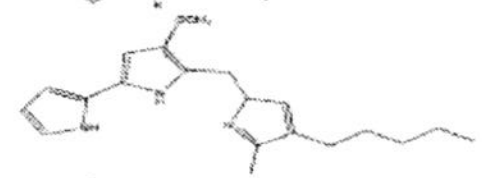	Antimicrobial Immunosuppressive
Glaukothalin (Blue)	*Proteobacteria* spp.	NA[a]	Antitumoral activity
Eumelanin (Brown-black)	*Marinomonas mediterranea*	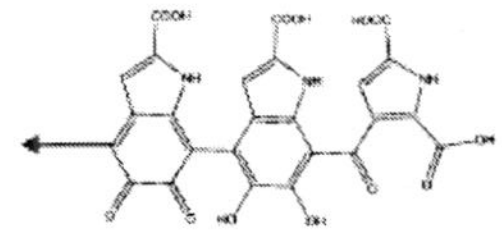	Anticardiovascular disease Photoprotectants Antioxidant

NA-Not available

Table 5.6: Marine microorganism-derived nutracuticals and functional foods currently available in market

Product	Microorganism	Brand name	Company
ω-3 oil	*Schizochytrium* sp.	Algae Omega	Nordic Naturals, Inc.
ω-3 DHA supplement	*Schizochytrium* sp.	Baby's DHA	Nordic Naturals, Inc
Astaxanthin	*Spirulina platensis*	BioAstin Hawaiian Astaxanthin	Nutrex Hawaii
Multi-vitamin	*Spirulina platensis*	Spirulina Pacifica®	Nutrex Hawaii
β-carotene	*Dunaliella salina*	Nature's Life Mixed Carotenoids	NutraMarks, Inc.
Phytoplankton powder	Marine phytoplankton	Sea Trinity	Pure Healing Foods
Assorted algae pastes (Functional food for aquaculture)	*Tetraselmis*, *Nannochloropsis*, *Isochrysis*	Instant Algae	Brine Shrimp Direct, Inc.
Concentrated marine phytoplankton	A blend of marine phytoplankton	UMAC CORE	UMAClife

Protein and Bioactive Peptide

Bioactive proteins and protein hydrolysates are two of most common terms in modern nutritional supplements. However, in nutritional point of view, all of the high protein sources cannot be used to develop supplements without considering their biological value which is the amount, or percentage of protein that the body is able to absorb [29, 30]. Likewise, there are several drawbacks of meat, dairy and soy protein which are used as the most common sources of protein supplements. Besides high biological value, animal-derived proteins have faced growing challenges in modern nutritional market due to spreadable animal diseases and novel trend in vegetarian diet [30].

In this regard, with high biological value of microbial proteins, marine microorganism derived proteins could be used as an alternative source of protein for food and pharmaceutical products [30]. Among the alternative protein sources, marine microalgae are considered as potential source of high quality proteins. The protein content of some microalgae species, including *Chlorella, Spirulina, Scenedesmus, Dunaliella, Micractinium, Oscillatoria, Chlamydomonas,* and *Euglena*, accounts for more than 50% of the dry weight and these proteins bear high biological value [31]. In the case of *S. platensis*, approximately 80% of the protein nitrogen could be extracted from hexane defatted biomass. Therefore, *S. platensisis* considered as a prime source of bioactive proteins in marine environment. Moreover, species such as Spirulina and Chlorella with biological values of 77.6% and 71.6%, respectively have not only been recognized as sources of high quality of protein, but also as potential sources of bioactive proteins. Microalgal proteins comprise essential amino acids, specifically lysine and branched chain amino acid (BCAA) lucine, isoleucine and valine which account for 35% of the essential amino acids in human muscle proteins. Compositional analysis of microalgae proteins clearly indicates that this high quality protein can be effectively used as direct supplements or could be used for formulation of other health products such as nutraceuticals [32, 33]. In addition, potentials of microalgae proteins for farm animal nutrition have also been evaluated.

Protein based therapeutic is another leading section in biotechnology since protein is highly specific and sensitive molecule which has less number of side effects. In the development of therapeutic proteins, marine microalgae offer the potential to produce high yields of recombinant proteins rapidly and at much lower cost than traditional cell culture [33, 34]. Despite the wide use of bacteria and yeast based bioreactors in production of therapeutic proteins, marine microalgae received considerable attention as a bioreactor with the ability of post-transcriptional and post translational modifications of biosynthesized proteins. Indeed, microalgae are an ideal candidate of protein bioreactor as it combines the ease of cultivation inherent to microorganisms and protein modification ability of plant cell. Recently, light has been shed on *Chlamydomonas reinhardtii*, a single-celled eukaryotic marine microalga for production of recombinant therapeutic proteins. For decades, *C. reinhardtiihas* been serving as a model organism for studying chloroplast and its functions, but immense biotechnological potentials of this well-known unicellular flagellates were discovered recently [35].

In addition to the therapeutic proteins, peptides also play a vital role in metabolic regulation and modulation after releasing from the parent protein. Peptides with therapeutic potentials are referred as bioactive peptides and these peptides have potential to be used as nutraceuticals and functional food ingredients for health promotion and disease risk reduction. Bioactive peptides are 2–20 amino acids long protein fragments; generally remain latent within the parent protein molecule.

Marine Probiotics

The gastrointestinal tract harbors an array of microbial species which is known as gut flora and in human intestine, it exceeds 100 trillion microorganisms. The most prominent intestinal microbial species, including *Bacteroides, Prevotella, Eubacterium, Clostridium, Bifidobacterium, Staphylococcus, Enterococcus, Streptococcus, and Enterobacter* exert many beneficial actions to the host, particularly in digestion of unutilized energy sources. However, there are many strains in a healthy intestine which may act as pathogenic strains at certain conditions [36]. Thus, it has been noticed that reducing the amount of potentially harmful or pathogenic species while promoting a favorable environment for beneficial microbiota are vital for well-being. In this regard, there are many proposed ways to enhance the beneficial functions of intestinal microflora. Live microorganisms which are beneficial to gut microflora and have the ability to suppress the action of disease causing bacteria are referred as a probiotic strains, and this stands as one of the best available strategies of promoting useful microbes in the gut [37]. On the other hand, prebiotics allow specific changes, both in the composition and/or activity in the gastro-intestinal microflora that confers benefits upon host well being and health [38]. Live strains from the genera *Lactobacillus* and *Bifidobacterium* are commonly used as probiotics. Fermented dairy products, especially yogurts, drinks and capsules with freeze dried bacteria, are the most popular vehicles for delivering these organisms into the gastro-intestinal tract [39]. Recent studies have revealed that marine-derived *Lactobacillus* species.

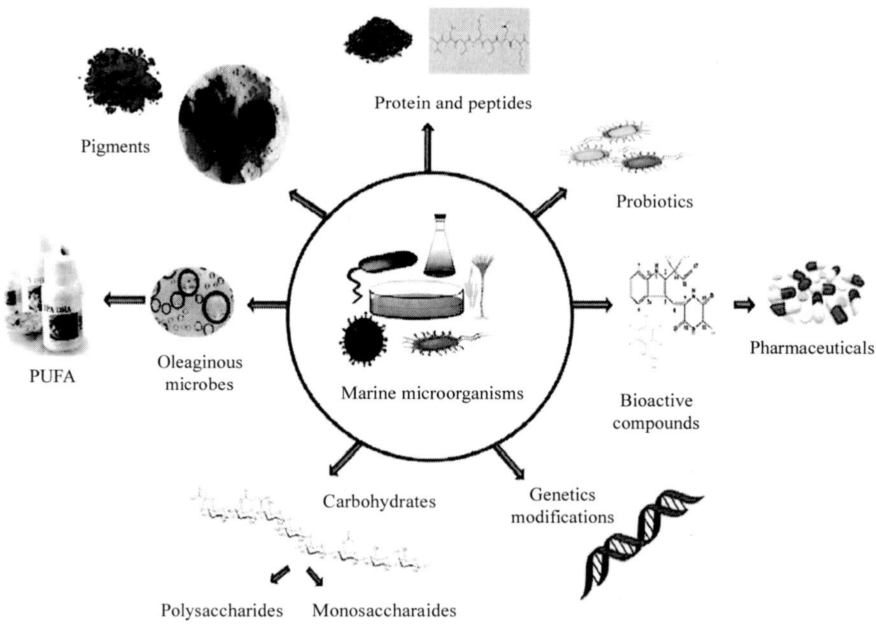

Fig. 5.2: Potential application of marine microorganism in health product development

Summary

Undoubtedly marine environment offer unique and diverse bioactive compounds with an array of health promoting abilities, even today, several marine-derived biologically active compounds are used as food and food supplements. Moreover, it is clear that marine environment is an ideal source for exploring novel functional materials. In the search of novel biologically active metabolites, marine microorganism derived functional ingredients has exhibited that they are unlike those found in terrestrial counterpart. In order to develop advanced nutraceuticals with pinpointed health claims, these microbial metabolites are good candidates. Applications of active metabolites of marine microorganisms for food and pharmaceutical products have just started the journey. The health claims of novel metabolites that have been mentioned in this chapter are acquired only through *in-vitro* and *in-vivo* studies and thus, comprehensive studies on mode of action, biological consequences and possible side effects have to be conducted in order to use those materials as food ingredients. On the other hand, there are a number of legal barriers to be addressed when developing marine microorganism derived ingredients as a food supplement. Especially, microbial metabolites should come from strains that have been recognized as safe, and hence, application of marine microbial products in food and food supplements would be a challenge as marine microbiology is not a matured field. The use of modern technologies and knowledge in the field of microbiology is becoming very important since safety levels and standards for marine microbial strains remain largely to be defined. Indeed, many novel as well as conventional techniques is applicable to reveal the biological and environmental role of marine microbes and it would facilitate effective isolation and characterization of food-grade marine microbial species. Further, improved methodologies in fermentation technologies are necessary for commercial production of marine microbial metabolites cost effectively. Particularly, culturing the strains that have been identified from extreme environments still remains a challenge. Recent advances of biotechnology in the field of microbiology and marine science would overcome these hurdles soon. The application of marine microorganism derived novel metabolites in modern nutraceutical and functional food with advanced health claims is not so far.

References

1. Hong, D.D., H.M. Hien, and P.N. Son, *Seaweeds from Vietnam used for functional food, medicine and biofertilizer.* Journal of applied phycology, 2007. **19**(6): 817-826.
2. Lahaye, M. and A. Robic, *Structure and functional properties of ulvan, a polysaccharide from green seaweeds.* Biomacromolecules, 2007. **8**(6): 1765-1774.

3. Holdt, S.L.V. and S. Kraan, *Bioactive compounds in seaweed: functional food applications and legislation.* Journal of Applied Phycology, 2011. **23**(3): 543-597.
4. Guerin, M., M.E. Huntley, and M. Olaizola, *Haematococcus astaxanthin: applications for human health and nutrition.* Trends in Biotechnology, 2003. **21**(5): 210-216.
5. Grung, M., *et al.*, *Algal carotenoids 51. Secondary Carotenoids 2. Haematococcus pluvialis aplanospores as a source of -astaxanthin esters.* Journal of applied phycology, 1992. **4**(2): 165-171.
6. Haugan, J.A. and S. Liaaen-Jensen, *Algal carotenoids 54. Carotenoids of brown algae (Phaeophyceae).* Biochemical Systematics and Ecology, 1994. **22**(1): 31-41.
7. Chronakis, I.S., *et al.*, *The behaviour of protein preparations from blue-green algae (< Spirulina platensis strain Pacifica) at the air/water interface.* Colloids and Surfaces A: Physicochemical and Engineering Aspects, 2000. **173**(1): 181-192.
8. Sekar, S. and M. Chandramohan, *Phycobiliproteins as a commodity: trends in applied research, patents and commercialization.* Journal of Applied Phycology, 2008. **20**(2): 113-136.
9. Dufossé, L., *et al.*, *Microorganisms and microalgae as sources of pigments for food use: a scientific oddity or an industrial reality?* Trends in Food Science & Technology, 2005. **16**(9): 389-406.
10. Nagayama, K., *et al.*, *Bactericidal activity of phlorotannins from the brown alga Ecklonia kurome.* Journal of Antimicrobial Chemotherapy, 2002. **50**(6): 889-893.
11. Vairappan, C.S., *et al.*, *Antibacterial halogenated metabolites from the Malaysian< i> Laurencia species.* Phytochemistry, 2001. **58**(2): 291-297.
12. Gschwend, P.M., J.K. Macfarlane, and K.A. Newman, *Volatile halogenated organic compounds released to seawater from temperate marine macroalgae.* Science, 1985. **227**(4690): 1033-1035.
13. Vairappan, C.S., *Potent antibacterial activity of halogenated metabolites from Malaysian red algae,Laurencia majuscula Rhodomelaceae,Ceramiales).* Biomolecular engineering, 2003. **20**(4): 255-259.
14. Kladi, M., C. Vagias, and V. Roussis, *Volatile halogenated metabolites from marine red algae.* Phytochemistry Reviews, 2004. **3**(3): 337-366.
15. Bergé, J.-P. and G. Barnathan, *Fatty acids from lipids of marine organisms: molecular biodiversity, roles as biomarkers, biologically active compounds, and economical aspects*, in *Marine biotechnology* 2005, Springer. p. 49-125.
16. Sutherland, I.W., *Novel and established applications of microbial polysaccharides.* Trends in Biotechnology, 1998. **16**(1): 41-46.
17. Baird, J., P. Sandford, and I. Cottrell, *Industrial applications of some new microbial polysaccharides.* Nature biotechnology, 1983. **1**(9): 778-783.
18. Poli, A., G. Anzelmo, and B. Nicolaus, *Bacterial exopolysaccharides from extreme marine habitats: production, characterization and biological activities.* Marine drugs, 2010. **8**(6): 1779-1802.
19. Nichols, C.M., J. Guezennec, and J. Bowman, *Bacterial exopolysaccharides from extreme marine environments with special consideration of the southern ocean, sea ice, and deep-sea hydrothermal vents: a review.* Marine biotechnology, 2005. **7**(4): 253-271.
20. Tapiero, H., *et al.*, *Polyunsaturated fatty acids (PUFA) and eicosanoids in human health and pathologies.* Biomedicine & pharmacotherapy, 2002. **56**(5): 215-222.
21. Venegas-Calerón, M., O. Sayanova, and J.A. Napier, *An alternative to fish oils: metabolic engineering of oil-seed crops to produce omega-3 long chain polyunsaturated fatty acids.* Progress in lipid research, 2010. **49**(2): 108-119.
22. Skerratt, J.H., J.P. Bowman, and P.D. Nichols, *Shewanella olleyana sp. nov., a marine species isolated from a temperate estuary which produces high levels of polyunsaturated*

fatty acids. International journal of systematic and evolutionary microbiology, 2002. **52**(6): 2101-2106.

23. Kris-Etherton, P., *et al., Polyunsaturated fatty acids in the food chain in the United States.* The American Journal of Clinical Nutrition, 2000. **71**(1): 179-188.
24. Gupta, A., C.J. Barrow, and M. Puri, *Omega-3 biotechnology: thraustochytrids as a novel source of omega-3 oils.* Biotechnology Advances, 2012. **30**(6): 1733-1745.
25. Ganuza, E., A. Anderson, and C. Ratledge, *High-cell-density cultivation of Schizochytrium sp. in an ammonium/pH-auxostat fed-batch system.* Biotechnology Letters, 2008. **30**(9): 1559-1564.
26. Dewapriya, P. and S.K. Kim, *Marine microorganisms: An emerging avenue in modern nutraceuticals and functional foods.* Food Research International, 2014. **56**: 115-125.
27. Li, D., *et al., Antibacterial bisabolane-type sesquiterpenoids from the sponge-derived fungus Aspergillus sp.* Marine Drugs, 2012. **10**(1): 234-241.
28. Lamers, P.P., *et al., Carotenoid and fatty acid metabolism in nitrogen-starved Dunaliella salina, a unicellular green microalga.* Journal of biotechnology, 2012. **162**(1): 21-27.
29. Kim, S.K. and I. Wijesekara, *Marine-derived Peptides: Development and Health Prospects.* Marine Proteins and Peptides: Biological Activities and Applications, 2013: p. 1-3.
30. Ngo, D.H., *et al., Biological activities and potential health benefits of bioactive peptides derived from marine organisms.* International Journal of Biological Macromolecules, 2012. **51**(4): 378-383.
31. McCarren, J., *et al., Microbial community transcriptomes reveal microbes and metabolic pathways associated with dissolved organic matter turnover in the sea.* Proceedings of the National Academy of Sciences, 2010. **107**(38): 16420-16427.
32. Nasri, R. and M. Nasri, *Marine-derived bioactive peptides as new anticoagulant agents: a review.* Current Protein and Peptide Science, 2013. **14**(3): 199-204.
33. Harnedy, P.A. and R.J. Fitzgerald, *Bioactive Proteins and Peptides from Macroalgae, Fish, Shellfish and Marine Processing Waste.* Marine Proteins and Peptides: Biological Activities and Applications, 2013: p. 5-39.
34. Specht, E., S. Miyake-Stoner, and S. Mayfield, *Micro-algae come of age as a platform for recombinant protein production.* Biotechnology letters, 2010. **32**(10): 1373-1383.
35. Kim, M.J., P. Dewapriya, and S.K. Kim, *Floridoside suppresses pro-inflammatory responses by blocking MAPK signaling in activated microglia.* Biochemistry and Molecurar Biology Reports, 2013. **46**(8): 398-403.
36. Cani, P.D., *et al., Role of gut microflora in the development of obesity and insulin resistance following high-fat diet feeding.* Pathologie Biologie, 2008. **56**(5): 305-309.
37. Correia, M.I.T., J.C. Liboredo, and M.L. Consoli, *The role of probiotics in gastrointestinal surgery.* Nutrition, 2012. **28**(3): 230-234.
38. Floch, M.H., *et al., Recommendations for probiotic use—2011 update.* Journal of Clinical Gastroenterology, 2011. **45**: S168-S171.
39. Kolida, S. and G.R. Gibson, *Prebiotic capacity of inulin-type fructans.* The Journal of Nutrition, 2007. **137** (11): 2503S-2506S.

6

Spices and Herbs

Introduction

Spices and aromatic herbs have been used since antiquity as preservatives, colorants, and flavour enhancers. Spices, which have long been the basis of traditional medicine in many countries, have also been the subject of study, particularly by the chemical, pharmaceutical, and food industries, because of their potential use for improving health. Both *in-vitro* and *in-vivo* studies have demonstrated how these substances act as antioxidants, digestive stimulants, and hypolipidemics and show antibacterial, anti-inflammatory, antiviral, and anti-carcenogenic activities. These beneficial physiological effects may also have possible preventative applications in a variety of pathologies. The chapter presents an overview of the potential of spices and aromatic herbs as functional foods ingredients.

Recent years have seen an increased interest on the part of consumers, researchers, and the food industry into how food products can help maintain the health of the organism. Industry should take into account that the use of functional ingredients is related to the health of consumers. The "spice health claims" are very expensive and need a lot of investment. The specifications for spice health claims are still relatively unclear and so industry could claim that there is a certain amount of inherited cultural knowledge and little scientific evidence. Moreover, it is important to note that some rejections related to spice healthy aspects are not a rejection of the ingredient or product, but just that the scientific evidence submitted is incomplete. Human clinical trials are also important and the value of such trials for the food industry is undeniable. But all too often nutrients are pulled out of context following the same methodology as used for the testing of drugs. Hence, in many cultures the use of spices to cure pathologies is important and may well serve as a base on which to rest scientific evidence in support of health benefits. Spices as a whole food do not normally form part of the usual food chain, and randomized clinical trials are the best because such compounds can be tested and retested successfully.

Safety of Spice Consumption

Extensive animal studies carried out to evaluate the safety aspect of spices have indicated that even at much higher dietary levels (upto 100 times the normal intake), red pepper, black pepper and turmeric have no adverse effects on growth, organ weights, feed efficiency ratio, nitrogen balance, and blood constituents. In the past two to three decades, many more beneficial physiological effects of spices have been experimentally documented (Table 6.1), which suggest that the use of these food adjuncts extend beyond taste and flavour. Among the health problems that affect mankind, diabetes, cardiovascular disease, and inflammatory disorders including arthritis, and cancer have received considerable attention [1].

Table 6.1: Experimentally documented beneficial health effects of spices

Beneficial health effect	Spices observed to exert
Lowering of blood cholesterol	Garlic, Onion, Fenugreek, Turmeric/Curcumin, Red pepper/Capsaicin
Prevention and dissolution of chole-sterol gallstones	Curcumin, Capsaicin
Protection of erythrocyte integrity in hypercholesterolemic condition	Curcumin, Capsaicin, Garlic
Hypoglycemic potential	Fenugreek, Garlic, Onion, Turmeric, Cumin
Amelioration of diabetic nephropathy	Curcumin, Onion
Antioxidant effect	Turmeric/Curcumin, Capsaicin, Eugenol
Anti-inflammatory and anti-arthritic	Turmeric/Curcumin, Capsaicin, Eugenol
Antimutagenic/cancer preventive	Turmeric/Curcumin, Garlic, Ginger/Gingerol, Mustard
Digestive stimulant action	Curcumin, Capsaicin, Piperine, Ginger, Cumin, Ajowan, Fennel, Coriander, Onion, Mint
Antimicrobial	Turmeric/Curcumin, Garlic, Asafetida

Chemical Composition of Spices

The first distinction to be made is that between a culinary herb and a spice. In general, the leaves of a plant used in cooking are denominated as culinary herbs, while any other part of the plant is known as a spice. Spices can be leaf (e.g. bay leaf), buds (clove), bark (cinnamon), root (ginger), berries (grains of pepper), seeds (cumin), or even the stigma of the flower (saffron). Both spices and herbs can be used fresh, dried, whole, chopped, or ground and due to their color, aroma and/or flavour characteristics, are used in the preparation of foods and drinks. In their composition can be found proteins, fiber, sugars, essential oils, minerals, and pigments, besides bioactive compounds such as phenolic acids, flavonoids, sterols, and cumarins [2].

Many of the functional properties presented by spices are associated with the presence, type, and concentration of phenolic compounds, although the exact composition will depend on several factors, such as the part of the plant used, its vegetative state, environmental conditions, harvesting technique, etc. [3]. Table 6.2 [4] shows the main compounds of a phenolic nature in several spices. Other compounds present in spices are the essential oils (EOs), which, due to their content in terpenes, monoterpenes, and sesquiterpenes (as hydrocarbons, alcohols, ketones, etc., which may be acyclic, monocyclic, bicyclics, tricyclics), are responsible for many of their functional properties. These, in general terms, are composed of more than seventy components [5], some of which may represent more than 85% of the total content, while others may only be present in trace amounts. However, the role played by these minor compounds is very important since evidence suggests that they may contribute significantly to the functional properties of EOs, in which they sometimes act synergically. As with the phenolic compounds, the composition, type, and concentration of the EOs is influenced by climate, harvesting time, part of the plant used, extraction method, etc. [6, 7]. For example, the four major components of *Origanum vulgare subsp. Hirtum* harvested in autumn are carvacrol (1.7–69.6%), thymol (0.2–42.8%), pcymene (17.3–51.3%), and α-terpinene (0.6–3.6%) while Veres et al., (2003) analyzing the same variety harvested at another time of the year and in a different geographical area found the major components to be carvacrol (76.4%), α-terpinene (6.6%), p-cymene (4.7%), and thymol (0.23%) [8].

Antioxidant Properties

In recent years many ingredients and by-products of the agro-food industry have been studied as possible sources of natural antioxidants (Table 6.2). Among such compounds are spices, which have shown their ability to slow down the process of lipid oxidation. This process is one of the principal causes of food spoilage and may occur during the storage of raw materials, processing, or the storage of the final product [9]. Such oxidation leads to a significant loss of a food's nutritional value since it involves a loss of vitamins and essential fatty acids. It also affects the food's sensorial quality changes in color, texture, and taste which shortens its shelf life and can result in rejection on the part of consumers. To avoid such spoilage the food industry has resorted to synthetic antioxidant formulations, some of the most commonly used being butylated hydroxytoluene (BHT), butylated hydroxyanisole (BHA), and propyl gallate (PG) [10]. However, because of their synthetic origin, their safety and efficacy are frequently questioned. The result has been a growing interest in substances

Table 6.2: Common herbs, active ingredients, and purported function [8]

Name	Scientific name	Active compounds	Purported function	Body system
Ginkgo biloba	*Ginkgo biloba*	Ginkolides: flavonoid glycosides, and diterpene lactones	Memory, cognition	Brain, circulatory system
St. John's wort	*Hypericum perforatum*	Hypericin, pseudohypericin, hyperforins, terpenes, catechin-type tannins, proanthocyanidins	Antidepressant for mild to moderate depression	Brain
Kava kava	*Piper methysticum*	Kavalactones	Antianxiety, relaxation	Brain
Valerian	*Valeriana officalis*	Valeptriates, valerenic acid, sesquiterpenes	Sleep inducement	Brain
Hawthorn berry	*Crataegus oxycantha*	Amygdalin, crategolic acid, oligomeric proanthocyanadins	Cardiac insufficiency	Circulatory system
Echinacea	*Echinacea purpurea* *E. Epallida* *E. angustifolia*	Echinosides, caffeic, ferulic acids and cichoric acid, inulin and fructans	Immunomodulators	Innate immunity
Peppermint oil	*Mentha x piperita*	Menthol	Gastric intestinal distress	GI tract
Ginger	*Zingiber officinale*	Gingerols, gingerdiols	Antiemetic, nausea	GI tract
Licorice	*Glycyrrhiza glabra*	Glycyrrhetic acid, triterpene, saponins	Expectorant	Respiratory system
Saw palmetto berry	*Serenoa repens*	β-sitosterol, sterols, fatty acids	α reductase inhibitor	Prostate gland
Feverfew	*Tancetum parthenuum*	Sesquiterpene lactones	Migraine prophylactic	Brain

of a vegetal origin that show antioxidant potential for use as natural additives to replace the synthetic antioxidants, whose use is increasingly restricted due to the secondary effects they may produce [11].

The antioxidant activity of spices, essential oils, and their components has been the subject of many studies [12-14]. Many of the assays related with these studies have been conducted *in-vitro*. The action mechanism set in motion by the antioxidant activity of these compounds is still not clearly understood. Among possible mechanisms are the scavenging of free radicals, hydrogen donation, the chelating of metallic ions, or the capacity to act as substrate of radicals such as superoxide or hydroxyl [15]. These bioactive compounds with their antioxidant properties may also interfere with propagation reactions [16], or inhibit the enzymatic systems involved in initiation reactions [17, 18]. Differences in the antioxidant activities of flavonoids depend on their different molecular structures, especially as regards the degree of hydroxylation and methylation of the compounds [19]. The determination of the antioxidant capacity of spices and their derivatives in foods is being given greater importance by researchers and those involved in the agro-food industry. Table 6.3 provides a list of studies into the antioxidant capacity of several spices and their derivates [20]. An overview of studies testing the antioxidant activity of some spices and essential oils in food is presented in Table 6.4 [20].

Table 6.3: Antioxidant properties of various spices

Common name	Scientific name	Major components	Concentration, (%)
Sage	*Salvia officinalis*	Camphor	38
		1,8-cineole	24
Rosemary	*Rosmarinus officinalis*	α–pinene	2–46
		1,8-cineole	3–89
Cumin	*Cuminum cyminum*	γ–terpinene	15–30
		Cuminal	20–36
		P-cymene	18–20.5
Oregano	*Origanum vulgare*	Carvacrol	Trace-88
		Thymol	Trace-43
		P-cymene	Trace-52
Cinnamon	Cinnamomum zeylanicum	Cinnamaldehyde	77.1
		β–caryophyllene	6
		α–terpineol	4

Table 6.4: Overview of studies testing the antioxidant activity of some spices and essential oils in food

Food Group	Food	Spices	Concentration (%)	Combination with another preservative / process
Meat	Cured pork fat	Extract of rosemary	0.3	
	Meat balls turkey mince	Sage	1	
	Dehydrated chicken meat	Extract of rosemary	1	
	Beef mince	Extract of rosemary	0.25	Irradiation (1–4 kGy) wrapped in PVC
	Patties from muscle of cattle	Extract of rosemary	0.2–0.25	Wrapped in MAP
	Meat balls poultry mince	Extract of rosemary	1–1.5	Frozen storage
	Beef steaks	Oleoresin rosemary	0.05 and 0.1	Sodium tripolyphosphate
	Fresh, minced chicken	Rosemary	0.10	
	Pork sausage	Rosemary extract	2.50	Refrigerated
	Cooked chicken patties	Tea catechins	0.01	
	Beef, skinless lean pork, and cured pork back fat	Rosemary powder	0.10	
	Pork patties	Rosemary	0.10	
	Beef patties	Oregano	0.5	Modified atmosphere packaging
	Pork patties	Catechins	0.25	
	Chicken fat	Sage, rosemary	0.50	
	Bologna sausage	Thyme essential oil	0.02	
	Bologna sausage	Oregano essential oil	0.02	
	Bologna sausage	Oregano essential oil	0.02	
	Pork patties	Sage	0.05	
	Chicken nuggets	Sage	0.10	Salt (0.75%)
	Raw beef steaks	Oleoresin rosemary	0.10	Sodium tripolyphosphate
	Buffalo Ghee meat	Methanolic extracts of black cumin	0.12	
	Lard	Black cumin extract	0.40	
	Goat meat for fermented sausage	Rosemary	0.05	
	Lamb meat	Potato peel extract	0.04	Irradiated

Food Group	Food	Spices	Concentration (%)	Combination with another preservative / process
	Minced beef	Oregano oil	0.5–1	
	Beef muscle	Oregano oil	1.00	
Fish	Raw fish	Clove, rosemary, mace, oregano, Thyme and sage	0.10	
	Mackerel oil	Dried oregano	1	
	Fish oil	Sage	5	
	Sardine oil	Rosemary	0.02	α-Tocopherol (0.05%)
	Anchovy lipids	Rosemary extract	0.10	Tocopherol (0.1%) + Vacuum packaging
	Crushed bonito meat	Rosemary	0.02	α-Tocopherol (0.05%)
	Catfish cooked	Clove, rosemary, oregano, sage and thyme	0.10	
Oil	Refined olive oil	Oregano	5	
	Rice bran oil	Rosemary, sage, Oregano, ginger and thyme	2	
	Refined olive oil	Rosemary	5	
	Maize oil	Black cumin extracted with Diethyl ether	0.4	
	Sunflower oil	Thyme, rosemary and sage	2	

Antibacterial Properties

Many organoleptic and sensorial properties of foods diminish with time. There are many factors involved (temperature, light, oxygen, etc.) but the most important is undoubtedly the action of microorganisms. Many techniques including refrigeration, freezing, water activity reduction and the restriction of nutrients, acidification, modified atmosphere packaging fermentation, or the addition of antimicrobial compounds have been used, some since antiquity, to preserve foods from microorganic attack. To these may be added new technologies such as high pressure, electric pulses, nanotechnology, and irradiation [21].

The addition of microbial agents to foods is a particularly effective method for controlling microbial contamination [22]. The agents used in food are compounds that are added or found naturally therein and which act by inhibiting the growth of the microorganisms [23]. As stated above, the food industry is under pressure from consumers to replace synthetic preservatives by natural ones and spices tend to be the main source of these [24]. Among these natural antimicrobials are spices and their derivates such as essential oils (EO). The main advantage of using EO and which makes their use in the food industry so widespread is that they are classified as GRAS (Generally Recognized As Safe) [25], although it must be borne in mind that their maximum efficacy must be determined and they should not provoke appreciable changes in the organoleptic properties of the foods to which they are added.

Antimicrobial compounds in food, whether or not processed, can increase the shelf life of the same by reducing the growth of microorganisms or by reducing their viability [26]. Originally, spices and the EO were added to foods in order to modify or potentiate a given taste, and only indirectly to increase the food's useful life by reducing the microbial load or inhibiting their growth and multiplication [27].

The antimicrobial properties of EO have been demonstrated against a large number of microbial strains, among them *Staphylococcus* spp., *Lactobacillus* spp., and *Enterobacteriaceae*, as well as certain molds (e.g. *Aspegillus* spp.). Generally, the composition, structure, and functional groups of EOs play an important role in determining their antimicrobial capacity. Compounds containing phenolic groups are responsible for these antimicrobial properties, although other compounds also present the same properties [28].

Possible action modes of the constituents of spices have been proposed [29, 30], although no specific mechanism has been confirmed. The effect of phenolic compounds and terpenes on the growth and production of toxins may be the result of their propensity to alter the permeability of the microbe cell wall, leading to the loss of macromolecules; they might also interact with the proteins present in the cell wall, deforming its structure and leading to a loss of functionality [31]. The composition, structure, and functional groups of EO play an important role in determining their antimicrobial capacity. It has been demonstrated that the concentration of an EO to be added to a foodstuff is greater (1–3%) than the quantity added to *in-vitro* assays to obtain the same level of antimicrobial activity, which is one of the reasons that their use in the food sector is limited, since high concentrations lead to organopleptic alterations reducing consumer acceptability [32].

The interaction between the phenolic groups of EO and proteins, lipids, and aldehydes may partially explain the reduction of antimicrobial effects. It has been suggested that the extracts derived from spices be used as antimicrobial agents forming part of a technological barrier system that involves the preservation of food by the multiple and simultaneous action of a series of compounds and/or actions [33].

The development of multicomponent antimicrobial systems for foods will depend on totally understanding the action mechanisms of specific agents to be able to seek potential effective combinations [34]. The joint use of essential oils, along with other factors, will facilitate their use as antimicrobial agents without affecting the organoleptic characteristics of the product concerned. For example, low pH values, low temperatures, or low oxygen levels will permit lower concentrations of EO to be used [35]. Several factors may increase microbial resistance to EO e.g. the greater availability of nutrients in foods than in the culture media used in in-vitro tests, which may permit bacteria to repair damaged cells more quickly [36].

Not only intrinsic characteristics of the food such as the proportion of fatty acids, water content, protein content, presence of antioxidants, pH salt, etc. may be relevant in this respect, but also extrinsic factors, such as temperature, vacuum packaging, gas, air, the particular characteristics of the mocroorganisms in question, may also influence bacterial sensitivity [37]. In general, the susceptibility of bacteria to the antimicrobial effect of EO increases as the pH of the food, the storage temperature, and quantity of oxygen in the packaging decrease [35]. At low pH, the hydrophobicity of Eos increases, facilitating their solubility in the membrane lipids and hence helping their antimicrobial action [38]. It is accepted that high levels of fat and/or protein in foods protect bacteria against the action of EO. If the EO dissolves in the lipid phase of the food there will be less EO available to act on the bacteria present in the aqueous phase [39], while carbohydrates do not appear to offer the bacteria as much protection against the action of EO [40]. On the other hand, a high concentration of water and/or salt facilitates the action of EO [35]. Table 6.5 lists how certain

Table 6.5: Antibacterial activity of some spices and essential oils in food [20]

Food group	Food	Spice	Concentration (%) applied	Bacterial species	Preservation combination
Meat	Raw chicken meat mince	Cinnamon extracts	0.80	*Aeromonas hydrophila*	
	Cooked chicken sausage	Mustard oil	0.10	*Escherichia coli*	
	Hotdogs	Clove oil	1	*Listeria monocytogenes*	
	Minced beef	Oregano oil	0.05–0.1	Natural flora	MAP
	Pork liver sausage	Rosemary oil	1	*Listeria monocytogenes*	
	Chicken frankfurters	Clove oil	1–2	*Listeria monocytogenes*	
	Chicken noodles	Sage oil	0.2–0.5	*Bacillus cereus; Staphylococus aureus*	
	Cooked pork	Coriander oil	1250 μg/cm^2	*Aeromonas hydrophila*	
	Ham	Cilantro oil	0.1–6	*Listeria monocytogenes*	Vacuum packaging
	Cooked chicken	Clove oil	2	*Aeromonas hydrophila* *Listeria monocytogenes*	
	Pâté	Mint oil	0.5–2	*Listeria monocytogenes*	
	Beef	Oregano oil or clove oil	1	*Listeria monocytogenes* Scott	4°C
	Cooked pork	Clove oil	500 μg/cm^2	*Aeromonas hydrophila*	
Fish	Cod fillets	Oregano oil	0.05	*Photobacterium phosphoreum*	Packaged in modified atmosphere
	Mackerel broth	Clove oil	0.50	*Enterobacter aerogenes*	2% NaCl
	Salmon fillets	Oregano oil	0.05	*Photobacterium phosphoreum*	
	Asian sea bass	Thyme oil	0.05	*Natural flora*	Refrigerated storage
	Fish roe salad	Mint oil	0.5–2.0	*Salmonella enteritidis*	

Food group	Food	Spice	Concentration (%) applied	Bacterial species	Preservation combination
Dairy	Yoghurt	Clove oil	0.005–0.5 in milk before fermentation	*Streptococcus thermophilus*	
	Mutton cheese	Cinnamon powder	6	*Listeria monocytogenes*	30°C
	Mozzarella cheese	Clove oil	0.5–1	*Listeria monocytogenes*	
	Soft cheese	Ground cloves	1	*Listeria monocytogenes*	
	Full fat cheese	Clove oil	1	*Listeria monocytogenes*	
	Low fat cheese	Bay, clove, cinnamon oil	1	*Salmonella enteritidis*	
Vegetables	Lettuce	Thyme oil	0.1–10 mL L^{-1}	*Scherichia coli* O157:H7	
	Aubergine salad	Oregano oil	1	*Escherichia coli* O157:H7	Low pH
	Lettuce	Thyme oil	1	*Shigella sonnei; Shigella flexneri*	EO added to washing water
Fruits	Apple juice	Cinnamon	0.10	*Listeria monocytogenes*	5°C
	Kiwifruit	Carvacrol	1 mM in dipping solution	Natural flora	
Others	Spaghetti sauce	Thyme	1	*Shigella* spp.	
	Spaghetti sauce	Basil	1	*Shigella* spp.	

spices and their derivates have been used as antimicrobial agents in different foods [20].

Anti-Inflammatory Activity

The topical application of spices and extracts is common practice for alleviating a variety of discomforts such as backache, rheumatism, skin rashes, and inflammatory processes in general [41]. The inflammatory process is triggered by several chemical and/or biological aspects that include pro-inflammatory enzymes and cytokines, low molecular weight compounds such as eicosanoids or the enzymatic degradation of tissues [42]. Several studies have related cyclooxigenase-2 (COX-2) with the inflammatory process [43]. This enzyme is an isoform of cyclooxigenase (COX) which is responsible for catalyzing araquidonic acid to prostaglandin. The other isoform is cyclooxigenase-1 (COX-1) which regulates homeostasis processes [42]. Many studies have demonstrated the anti-inflammatory capacity of spices [44]. This capacity is basically due to the presence of flavonoids that inhibit the development of inflammation provoked by a variety of agents [45]. Among these flavonoids, galangin, which is present in Lippia graveolens [46], is capable of inhibiting cyclooxigenase (COX) and lipooxygenase, weakening the action of polygalacturonase and reducing the expression of the inducible isoform of ciclooxigenasa (COX-2) [47]. The anti-inflammatory activity arises from the inhibition of eicosanoic acid and of the lipidic peroxidation of the cell membrane through the inhibition of cyclooxigenase and 5-lipooxigenase enzymatic system [48].

Antiviral Properties

Another functional property of spices and herbs is their antiviral activity. It has been shown that the carnosol and carnosic acid, the two main components of rosemary, show inhibitory activity over the human immunodeficiency virus (HIV) [49]. Similarly, it has been claimed that apigenin, a flavonoid characteristic of several spices (thyme, sage, oregano, rosemary), inhibited HIV-1 in latent infection models through a mechanism that probably included inhibition of viral transcription [50].

Several studies have described the antiviral activity of curcumin [51, 52]. In earlier studies, curcumin was seen to inhibit the activity of histone acetyltransferase, thus impeding the genic promotion of the type 1 herpes simple virus (HSV-1) [53]. It was shown that curcumin acts against the replication of coxsackievirus by disrupting the ubiquitin-proteasome system (UPS) [54]. The antiviral effect of essential oils, fundamental components of spices, has also been demonstrated. For example, it was shown in earlier studies that both *in-vitro* and in a mouse model, that eugenol, the principal component of clove bud (*Eugenia caryophyllus*) essential oil presented antiviral activity against the types 1 and 2 herpes simple viruses (HSV-1 and HSV-2) [55]. In recent times, one of the researchers analyzed the *in-vitro* effect of the EO of melissa (*Melissa officinalis*) against HSV-1 and HSV-2 and found that it showed strong antiviral activity as long as it was added after the virus had penetrated the cell [55]. In an analysis of the effect of the EO of several herbs and spices against HSV-2, it was found that hyssop (*Hyssopus officinalis*), thyme (*Thymus vulgaris* L.), anis (*Pimpinella anisum*), ginger (*Zingiber officinale*), camomile (*Chamaemelum nobile*), and sandalwood (*Santalum × album*) all showed antiviral activity, possibly through their interaction with the development of the virus [56].

Anticancerous Properties

Many of the non-nutritive components of herbs and spices are known to possess potential activity as chemoprotective agents against cancer. Among the action mechanisms proposed for these compounds are [57]:

- Inhibiting of the phase I enzymes or blocking carcinogen formation
- Induction of phase II (detoxification) enzymes
- Scavenging DNA reactive agents
- Modulation of homeostatic hormones
- Suppression of hyper-cells proliferation induced by carcinogen
- Induction of apoptosis
- Depression of tumor angiogenesis.
- Inhibition of phenotypic expressions of preneoplastic and neoplastic cells.

Many studies have discussed the anticarcinogenic properties of aromatic herbs and spices [58-61]. Particularly mentioned in this respect has been rosemary and its components ursolic acid, carnosol, etc.; turmeric and its principal component curcumin, and pepper (principal component capsaicin). Tsai *et al.* (2007) demonstrated that 500µg/mL of rosemary suppressed the production of NO, which is responsible for oxidative stress in many diseases, including cancer. In studies using hamsters, it was shown that the application of 30µg/ mL of a rosemary extract protected DNA from oxidative damage, the action mechanism being related with the capacity of rosemary to sequester OH radicals and singlet oxygen [62]. A component of rosemary was shown to have anticarcinogenic activity against several cancerigenic lines, the authors suggesting that cell apoptosis would cause the loss of DNA, externalization of the phosphatidylserine cell membrane, and depolarization of the mitochondrial membrane [63]. Prior to this, studies demonastrated that the application of carnosol prevented the damage to DNA caused by 7,12-dimethylbenz anthracene (DMBA) which provoked the formation of tumors in rat mammary glands [64]. A similar study indicated that an extract from rosemary leaves lengthened the latency period of tumors and reduced the incidence of malign tumors induced by DMBA in rats [65]. Another component of rosemary, ursolic acid, has also been to show anticarcinogenic activity. Bonaccorsi *et al.* (2008) indicated that this acid inhibited reverse transcriptase activity of tumoral cells, which is related to the control of neoplastic cell proliferation and differentiation. Ramos *et al.* (2008) indicated that ursolic acid acted as an anticarcinogenic agent through cell mediated mechanisms

Antiulcerous Properties

Another functional property of both herbs and spices is the protection they offer against ulcers [66, 67], a capacity due fundamentally to the presence of phenolic compounds, especially flavonoids [68, 69]. The way in which these compounds act seems to vary greatly. It is shown that flavonoids provoke an increase in prostaglandins, which have an inhibitory effect on acidic secretions and thus prevent the formation of ulcers [70]. Speroni and Ferri (1993) suggested that flavonoids increase the prostaglandin content of the gastric mucosa, protecting it from the formation of ulcers [71]. Other authors suggest that ulcerous lesions are related with oxygen reactive species. Flavonoids inhibit lipidic peroxidation by substantially increasing the activity of glutathione peroxidase [72]. Some studies point to the anti-ulcerous capacity of flavonoids like kaempherol and quercetin, compounds that are present in several spices [73]. Another theory proposed is that the anti-ulcerous capacity is due to the joint action of flavonoids with other substances, such as terpenes, sterols, tannins, carbohydrates, glycosides, saponins, and traces of alkaloids, which, together with flavonoids, are to be found in herbs and spices [74].

Antidiabetic Properties

Diabetes is the most common metabolic disease in the world and is still increasing. The International Diabetes Federation mentioned that 194 million people had diabetes in 2003, which will increase to 333 million by 2025 [75]. According to the World Heath Organization, it is the third most important disease after cardiovascular and oenological disorders. One of the ways to control diabetes mellitus is through the diet and it is here that herbs and spices can play a part. Indeed numerous studies have described their anti-diabetic activity. For example, cinnamon (*Cinnamomum zeylanicum*) is by far the most valued spice for improving diabetes [76]. *In-vitro* experiments, past research have shown that it has a potentiating effect on insulin due to the high concentrations of chrome, a known insulin potentiator [77]. According to one study, the ingestion of 1–6 g of cinnamon in the diet decreases serum glucose levels after 40 days [78]. In past work it could be seen that cinnamon extracts increase levels of insulin beta and transporter 4 glucose in diabetic rats thereby decreasing the serum levels of glucose [79]. One of the research indicated that cinnamaldehyde, a component of cinnamon, has a hypoglucemic effect on diabetic rats [80]. Extracts of cinnamon activate glucogen synthase, increase glucose uptake, inhibit glucogen synthase kinase-3, and activate insulin receptor kinase and inhibit dephosphorylation of the insulin receptor, leading to maximal phosphorylation of the insulin receptor [81]. All of these effects would lead to increased insulin sensitivity. Another species showing potential anti-diabetic activity is turmeric (*Curcuma longa*). Srinivasan (2005) stated that the daily intake of curcumin (coloring principle of turmeric) not only reduced the fasting blood sugar level, but also lowered the dosage of insulin needed for normoglycaemia [82]. It has been reported that the curcumin reduced blood glucose in alloxan induced diabetic rats and suggested that curcumin had a better hypoglycemic effect than turmeric although the mechanism of the anti-diabetic action was left unclear [83]. Few research work have indicated that the anti-diabetic effects of curcumin are partly due to a reduction in hepatic glucose production caused by activation of AMP kinase and inhibition of hepatic glucose-6-phosphatase activity and phosphoenolpyruvate carboxykinase activity [84]. The anti-diabetic effect of turmeric is due to the presence of curcuminoids and sesquiterpenoids, which also have a synergic effect. Fenugreek, too, shows anti-diabetic activity [85]. In one of the past work, it was analyzed that the effect of oral administering of fenugreek extracts on diabetic mice, resulted in reduction in serum levels of glucose [86].

Antilithogenic Effect

The inhibitory effect of a curcuma mixture (*Temoe Lawak Singer*) on lithogenesis in rabbits has been reported [87]. Studies on experimental induction of cholesterol gallstones in mice and hamsters by feeding a lithogenic diet have revealed that the incidence of gall stones is 40%–50% lower when the animals are maintained on 0.5% curcumin or 5 mg% capsaicin containing diet [88]. Animal studies have also revealed significant regression of preformed cholesterol gallstones by these spice principles in a 10-week feeding trial [89]. The possibility of such a beneficial prevention and regression of cholesterol gallstones by other known hypocholesterolemic spices remains to be examined. The antilithogenicity of curcumin and capsaicin is considered to be due to lowering of cholesterol concentration and enhancing the bile acid concentration, both of which contribute to lowering of the cholesterol saturation index and, hence, reduce crystallization . In addition to their ability to lower the cholesterol saturation index, the antilithogenecity of these spice principles may also be due to their influence on biliary proteins (Fig. 6.1) [90].

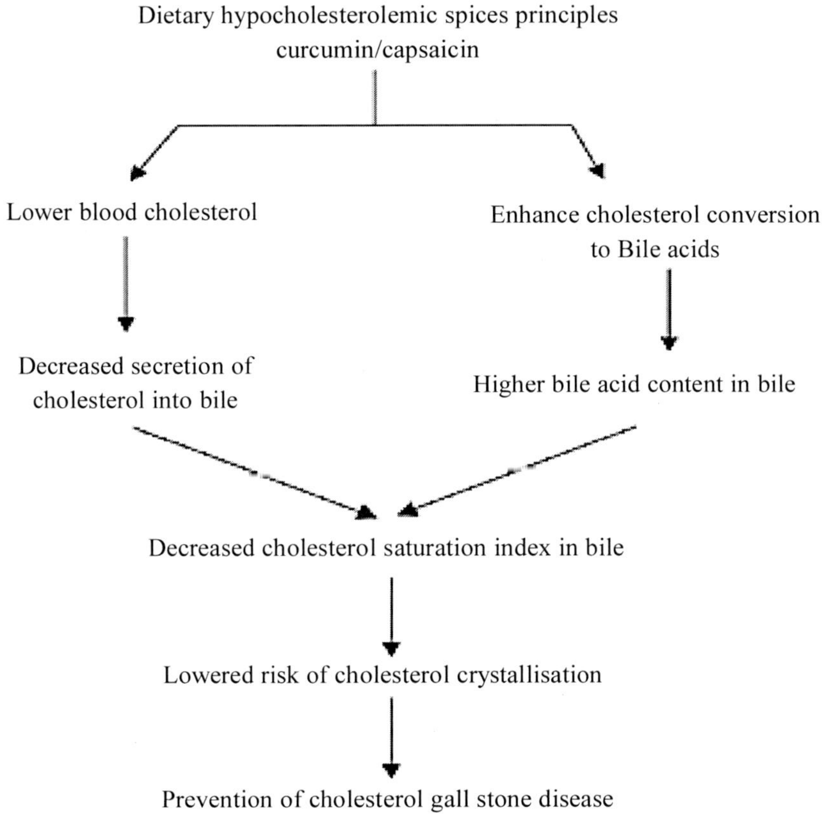

Fig. 6.1: Dietary hypocholesterolemic spices principles

Summary

Besides the role that they play in imparting taste and flavor to the food we eat, the long list of beneficial physiological effects that spices and herbs have on health suggests that they might well be considered as essential (and natural) components of our diet. Every spice has its own potentially beneficial property and there is the possibility, too, that they might have synergic effects, increasing their culinary and health related potential. However, despite this multiplicity of potentially beneficial activity that herbs and spices possess, more in-depth information is required concerning how exactly exposure to their components achieves a specific molecular response. Only when such information is available will it be possible to define strategies of intervention to obtain maximum benefits from herbs and spices without exposure to undesirable side effects.

References

1. Srinivasan, K., *Spices as influencers of body metabolism: an overview of three decades of research.* Food Research International, 2005. **38**(1): 77-86.
2. Pradeep, K.U., P. Geervani, and B. Eggum, *Common Indian spices: nutrient composition, consumption and contribution to dietary value.* Plant Foods for Human Nutrition, 1993. **44**(2): 137-148.
3. Cosentino, S., *et al., In-vitro antimicrobial activity and chemical composition of Sardinian thymus essential oils.* Letters in Applied Microbiology, 1999. **29**(2): 130-135.
4. Wildman, R.E., *Handbook of nutraceuticals and functional foods*2006: CRC press.
5. Russo, M., *et al., Essential Oil Chemical Composition of Wild Populations of Italian Oregano Spice (Origanum v ulgare ssp. h irtum (Link) Ietswaart): A Preliminary Evaluation of Their Use in Chemotaxonomy by Cluster Analysis. 1. Inflorescences.* Journal of Agricultural and Food Chemistry, 1998. **46**(9): 3741-3746.
6. Kawase, K.Y.F., *et al., Changes in essential oil of origanum vulgare* International Journal of Research and Reviews in Applied Sciences 2013. **14**.(2): 238-247.
7. Azizi, A., F.Yan, and B. Honermeier, *Herbage yield, essential oil content and composition of three oregano (Origanum vulgare L.) populations as affected by soil moisture regimes and nitrogen supply.* Industrial Crops and Products, 2009. **29**(2): 554-561.
8. Kokkini, S., *et al., Autumn essential oils of Greek oregano.* Phytochemistry, 1997. **44**(5): 883-886.
9. Tepe, B., *et al., Antimicrobial and antioxidant activities of the essential oil and various extracts of Salvia tomentosa Miller (Lamiaceae).* Food Chemistry, 2005. **90**(3): 333-340.
10. Vergara-Valencia, N., *et al., Fibre concentrate from mango fruit: Characterization, associated antioxidant capacity and application as a bakery product ingredient.* LWT-Food Science and Technology, 2007. **40**(4): 722-729.
11. Zheng, W. and S.Y. Wang, *Antioxidant activity and phenolic compounds in selected herbs.* Journal of Agricultural and Food Chemistry, 2001. **49**(11): 5165-5170.
12. Mata, A., *et al., Antioxidant and antiacetylcholinesterase activities of five plants used as Portuguese food spices.* Food Chemistry, 2007. **103**(3): 778-786.
13. Milan, K., *et al., Enhancement of digestive enzymatic activity by cumin (Cuminum cyminum L) and role of spent cumin as a bionutrient.* Food Chemistry, 2008. **110**(3): 678-683.
14. Bozin, B., *et al., Phenolics as antioxidants in garlic (Allium sativum L., Alliaceae).* Food Chemistry, 2008. **111**(4): 925-929.
15. Van Acker, S.A., *et al., Structural aspects of antioxidant activity of flavonoids.* Free Radical Biology and Medicine, 1996. **20**(3): 331-342.
16. Cotelle, N., *et al., Antioxidant properties of hydroxy-flavones.* Free Radical Biology and Medicine, 1996. **20**(1): 35-43.
17. You, K.M., H.-G. Jong, and H.P. Kim, *Inhibition of cyclooxygenase/lipoxygenase from human platelets by polyhydroxylated/methoxylated flavonoids isolated from medicinal plants.* Archives of Pharmacal Research, 1999. **22**(1): 18-24.
18. Hoult, J., M.A. Moroney, and M. Payá, *Actions of flavonoids and coumarins on lipoxygenase and cyclooxygenase.* Methods in Enzymology, 1993. **234**: 443-454.
19. Meyer, A.S., *et al., Fruit hydroxycinnamic acids inhibit human low-density lipoprotein oxidation in-vitro.* Journal of Agricultural and Food Chemistry, 1998. **46**(5): 1783-1787.

20. Viuda-Martos, M., *et al.*, *Spices as functional foods.* Critical reviews in Food Science and Nutrition, 2010. **51**(1): 13-28.
21. Martos, M.V., *et al.*, *Antimicrobianos naturales: uso de los aceites esenciales de especias.* Alimentación, equipos y tecnología, 2008. **27**(233): 44-47.
22. Kabuki, T., *et al.*, *Characterization of novel antimicrobial compounds from mango (Mangifera indica L.) kernel seeds.* Food Chemistry, 2000. **71**(1): 61-66.
23. Smid, E.J. and L.G. Gorris, *Natural antimicrobials for food preservation.* Food Science and technology , New York ,Marcek Dekker, 1999: p. 285-308.
24. Deba, F., *et al.*, *Chemical composition and antioxidant, antibacterial and antifungal activities of the essential oils from Bidens pilosa Linn. var. Radiata.* Food control, 2008. **19**(4): 346-352.
25. Kabara, J., *Phenols and chelators.* Food Preservatives, 1991: p. 200-214.
26. Beuchat, L.R. and D.A. Golden, *Antimicrobials occurring naturally in foods.* Food technology, USA, 1989.
27. Kim, J., M.R. Marshall, and C.-i. Wei, *Antibacterial activity of some essential oil components against five foodborne pathogens.* Journal of Agricultural and Food Chemistry, 1995. **43**(11): 2839-2845.
28. Dorman, H. and S. Deans, *Antimicrobial agents from plants: antibacterial activity of plant volatile oils.* Journal of Applied Microbiology, 2000. **88**(2): 308-316.
29. Sofos, J.N., *et al.*, *Naturally occurring antimicrobials in food.* Regulatory Toxicology and Pharmacology, 1998. **28**(2): 71-72.
30. Davidson, P. and A. Naidu, *Phyto-phenols.* Natural food antimicrobial systems, 2000: 265-294.
31. Fung, S. Taylor, and J. Kahan, *Effects of butylated hydroxyanisole (BHA) and butylated hydroxytoluene (BHT) on growth and aflatoxin production of Aspergillus flavus.* Journal of Food Safety, 1977. **1**(1): 39-51.
32. Lis Balchin, M. and S. Deans, *Bioactivity of selected plant essential oils against Listeria monocytogenes.* Journal of Applied Microbiology, 1997. **82**(6): 759-762.
33. Nychas, G., *et al.*, *Antimicrobials from herbs and spices.* Natural antimicrobials for the minimal processing of foods, 2003: p. 176-200.
34. Gill, A.O. and R.A. Holley, *Mechanisms of bactericidal action of cinnamaldehyde against Listeria monocytogenes and of eugenol against L. monocytogenes and Lactobacillus sakei.* Applied and Environmental Microbiology, 2004. **70**(10): 5750-5755.
35. Skandamis, P.N. and G.-J.E. Nychas, *Development and evaluation of a model predicting the survival of Escherichia coli O157: H7 NCTC 12900 in homemade eggplant salad at various temperatures, pHs, and oregano essential oil concentrations.* Applied and Environmental Microbiology, 2000. **66**(4): 1646-1653.
36. Burt, S., *Essential oils: their antibacterial properties and potential applications in foods —a review.* International Journal of Food Microbiology, 2004. **94**(3): 223-253.
37. Tassou, C., E. Drosinos, and G. Nychas, *Effects of essential oil from mint (Mentha piperita) on Salmonella enteritidis and Listeria monocytogenes in model food systems at 4 and 10 C.* Journal of Applied Bacteriology, 1995. **78**(6): 593-600.
38. Juven, B., *et al.*, *Factors that interact with the antibacterial action of thyme essential oil and its active constituents.* Journal of Applied Bacteriology, 1994. **76**(6): 626-631.
39. Mejlholm, O. and P. Dalgaard, *Antimicrobial effect of essential oils on the seafood spoilage micro-organism Photobacterium phosphoreum in liquid media and fish products.* Letters in Applied Microbiology, 2002. **34**(1): 27-31.
40. Shelef, L., E. Jyothi, and M. Bulgarellii, *Growth of Enteropathogenic and Spoilage Bacteria in Sage Containing Broth and Foods.* Journal of Food Science, 1984. **49**(3): 737-740.

41. Ramadan, M.F., *Nutritional value, functional properties and nutraceutical applications of black cumin (Nigella sativa L.): an overview.* International Journal of Food Science & Technology, 2007. **42**(10): 1208-1218.
42. Dao, T.T., *et al., Synthesis and inhibitory activity against COX-2 catalyzed prostaglandin production of chrysin derivatives.* Bioorganic & Medicinal Chemistry Letters, 2004. **14**(5): 1165-1167.
43. Cho, H., *et al., Modulation of the activity of pro-inflammatory enzymes, COX-2 and iNOS, by chrysin derivatives.* Pharmacological research, 2004. **49**(1): 37-43.
44. Fang, F., *et al., Isolation and identification of cytotoxic compounds from Bay leaf (Laurus nobilis).* Food Chemistry, 2005. **93**(3): 497-501.
45. Mani, F., *et al., Propolis: effect of different concentrations, extracts and intake period on seric biochemical variables.* Journal of Ethnopharmacology, 2006. **105**(1): 95-98.
46. Lin, L.-Z., *et al., Identification and quantification of flavonoids of Mexican oregano (Lippia graveolens) by LC-DAD-ESI/MS analysis.* Journal of Food Composition and Analysis, 2007. **20**(5): 361-369.
47. Raso, G.M., *et al., Inhibition of inducible nitric oxide synthase and cyclooxygenase-2 expression by flavonoids in macrophage J774A. 1.* Life sciences, 2001. **68**(8): 921-931.
48. Houghton, P.J., *et al., Fixed oil of Nigella sativa and derived thymoquinone inhibit eicosanoid generation in leukocytes and membrane lipid peroxidation.* Planta Medica, 1995. **61**(01): 33-36.
49. Aruoma, O., *et al., An evaluation of the antioxidant and antiviral action of extracts of rosemary and Provencal herbs.* Food and Chemical Toxicology, 1996. **34**(5): 449-456.
50. Critchfield, J.W., S.T. Butera, and T.M. Folks, *Inhibition of HIV activation in latently infected cells by flavonoid compounds.* AIDS research and human retroviruses, 1996. **12**(1): 39-46.
51. Joe, B., M. Vijaykumar, and B. Lokesh, *Biological properties of curcumin-cellular and molecular mechanisms of action.* Critical Rviews in Food Science and Nutrition, 2004. **44**(2): 97-111.
52. Lin, L. and K. H. Lee, *Structure-activity relationships of curcumin and its analogs with different biological activities.* Studies in Natural Products Chemistry, 2006. **33**:
53. 785-812.Kutluay, S.B., *et al., Transcriptional coactivators are not required for herpes simplex virus type 1 immediate-early gene expression in-vitro.* Journal of virology, 2009. **83**(8): 3436-3449.
54. Si, X., *et al., Dysregulation of the ubiquitin-proteasome system by curcumin suppresses coxsackievirus B3 replication.* Journal of Virology, 2007. **81**(7): 3142-3150.
55. Benencia, F. and M. Courreges, *In-vitro and in-vivo activity of eugenol on human herpesvirus.* Phytotherapy Research, 2000. **14**(7): 495-500.
56. Schnitzler, P., *et al., Melissa officinalis oil affects infectivity of enveloped herpesviruses.* Phytomedicine, 2008. **15**(9): 734-740.
57. Tanaka, T. and S. Sugie, *Inhibition of colon carcinogenesis by dietary non-nutritive compounds.* Journal of Toxicologic pathology, 2007. **20**(4): 215-235.
58. Aggarwal, B.B., *et al., Curcumin: the Indian solid gold*, in *The molecular targets and therapeutic uses of curcumin in health and disease*2007, Springer. p. 1-75.
59. Bonaccorsi, I., *et al., Endogenous reverse transcriptase as a mediator of ursolic acid's anti-proliferative and differentiating effects in human cancer cell lines.* Cancer letters, 2008. **263**(1): 130-139.
60. Ramos, A.A., *et al., Antigenotoxic effects of quercetin, rutin and ursolic acid on HepG2 cells: evaluation by the comet assay.* Toxicology letters, 2008. **177**(1): 66-73.
61. Tsai, P.-J., *et al., Comparison of NO-scavenging and NO-suppressing activities of different herbal teas with those of green tea.* Food Chemistry, 2007. **103**(1): 181-187.

62. Slameòová, D., *et al.*, *Rosemary-stimulated reduction of DNA strand breaks and FPG-sensitive sites in mammalian cells treated with visible light-excited Methylene Blue.* Cancer letters, 2002. **177**(2): 145-153.
63. Dörrie, J., K. Sapala, and S.J. Zunino, *Carnosol-induced apoptosis and downregulation of Bcl-2 in B-lineage leukemia cells.* Cancer letters, 2001. **170**(1): 33-39.
64. Singletary, K., C. MacDonald, and M. Wallig, *Inhibition by rosemary and carnosol of 7, 12-dimethylbenz anthracene (DMBA)-induced rat mammary tumorigenesis and in-vivo DMBA-DNA adduct formation.* Cancer letters, 1996. **104**(1): 43-48.
65. Sancheti, G. and P. Goyal, *Effect of rosmarinus officinalis in modulating 2 dimethylbenz (a) anthracene induced skin tumorigenesis in mice.* Phytotherapy Research, 2006. **20**(11): 981-986.
66. Borrelli, F. and A.A. Izzo, *The plant kingdom as a source of anti-ulcer remedies.* Phytotherapy Research, 2000. **14**(8): 581-591.
67. Kakub, G. and M. Gulfraz, *Cytoprotective effects of Bergenia ciliata Sternb, extract on gastric ulcer in rats.* Phytotherapy Research, 2007. **21**(12): 1217-1220.
68. Batista, L.M., *et al.*, *Gastric antiulcer activity of Syngonanthus arthrotrichus Silveira.* Biological and Pharmaceutical Bulletin, 2004. **27**(3): 328-332.
69. Hiruma-Lima, C.A., *et al.*, *Antiulcerogenic activity of Alchornea castaneaefolia: Effects on somatostatin, gastrin and prostaglandin.* Journal of Ethnopharmacology, 2006. **104**(1): 215-224.
70. Vilegas, W., *et al.*, *Isolation and structure elucidation of two new flavonoid glycosides from the infusion of Maytenus aquifolium leaves. Evaluation of the antiulcer activity of the infusion.* Journal of Agricultural and Food Chemistry, 1999. **47**(2): 403-406.
71. Speroni, E. and S. Ferri, *Gastroprotective effects in the rat of a new flavonoid derivative.* Acta Horticulturae, 1993: p. 249-249.
72. Martin, M., *et al.*, *Anti-oxidant mechanisms involved in gastroprotective effects of quercetin.* Zeitschrift fur Naturforschung C-Journal of Biosciences, 1998. **53**(1): 82-88.
73. Leite, J.P.V., *et al.*, *Isolation and HPLC quantitative analysis of flavonoid glycosides from Brazilian beverages (Maytenus ilicifolia and M. aquifolium).* Journal of Agricultural and Food Chemistry, 2001. **49**(8): 3796-3801.
74. Osadebe, P. and F. Okoye, *Anti-inflammatory effects of crude methanolic extract and fractions of Alchornea cordifolia leaves.* Journal of Ethnopharmacology, 2003. **89**(1): 19-24.
75. Shaw, J.E., R.A. Sicree, and P.Z. Zimmet, *Global estimates of the prevalence of diabetes for 2010 and 2030.* Diabetes Research and Clinical Practice, 2010. **87**(1): 4-14.
76. Khan, A., *et al.*, *Insulin potentiating factor and chromium content of selected foods and spices.* Biological Trace Element Research, 1990. **24**(2-3): 183-188.
77. Anderson, R.A., *et al.*, *Isolation and characterization of polyphenol type-A polymers from cinnamon with insulin-like biological activity.* Journal of Agricultural and Food Chemistry, 2004. **52**(1): 65-70.
78. Khan, A., *et al.*, *Cinnamon improves glucose and lipids of people with type 2 diabetes.* Diabetes care, 2003. **26**(12): 3215-3218.
79. Cao, H., M.M. Polansky, and R.A. Anderson, *Cinnamon extract and polyphenols affect the expression of tristetraprolin, insulin receptor, and glucose transporter 4 in mouse 3T3-L1 adipocytes.* Archives of Biochemistry and Biophysics, 2007. **459**(2): 214-222.
80. Subash Babu, P., S. Prabuseenivasan, and S. Ignacimuthu, *Cinnamaldehyde—a potential antidiabetic agent.* Phytomedicine, 2007. **14**(1): 15-22.
81. Jarvill-Taylor, K.J., R.A. Anderson, and D.J. Graves, *A hydroxychalcone derived from cinnamon functions as a mimetic for insulin in 3T3-L1 adipocytes.* Journal of the American College of Nutrition, 2001. **20**(4): 327-336.

82. Srinivasan, K., *Plant foods in the management of diabetes mellitus: spices as beneficial antidiabetic food adjuncts.* International journal of Food Sciences and Nutrition, 2005. **56**(6): 399-414.
83. Arun, N. and N. Nalini, *Efficacy of turmeric on blood sugar and polyol pathway in diabetic albino rats.* Plant Foods for Human Nutrition, 2002. **57**(1): 41-52.
84. Fujiwara, H., *et al., Curcumin inhibits glucose production in isolated mice hepatocytes.* Diabetes Research and Clinical Practice, 2008. **80**(2): 185-191.
85. Nishiyama, T., *et al., Curcuminoids and sesquiterpenoids in turmeric (Curcuma longa L.) suppress an increase in blood glucose level in type 2 diabetic KK-Ay mice.* Journal of Agricultural and Food Chemistry, 2005. **53**(4): 959-963.
86. Eidi, A., M. Eidi, and M. Sokhteh, *Effect of fenugreek (Trigonella foenum-graecum L) seeds on serum parameters in normal and streptozotocin-induced diabetic rats.* Nutrition Research, 2007. **27**(11): 728-733.
87. Beynen, A., J. Visser, and J. Schouten, *Inhibitory effect of lithogenesis by ingestion of a Curcuma mixture (Temoe lawak Singer).* Journal of Food Science and Technology, 1987. **24**(5): 253-256.
88. Hussain, M. and N. Chandrasekhara, *Effect on curcumin on cholesterol gall-stone induction in mice.* The Indian Journal of Medical Research, 1992. **96**: 288-291.
89. Hussain, M. and N. Chandrasekhara, *Effect of curcumin and capsaicin on the regression of preestablished cholesterol gallstones in mice.* Nutrition Research, 1994. **14**(10): 1561-1574.
90. Hussain, M. and N. Chandrasekhara, *Biliary proteins from hepatic bile of rats fed curcumin or capsaicin inhibit cholesterol crystal nucleation in supersaturated model bile.* Indian Journal of Biochemistry & Biophysics, 1994. **31**(5): 407-412.

Section III: Functional Food Products: Formulation and Processing Technology

7 Microencapsulation

Introduction

Encapsulation may be defined as a process to entrap one substance (active agent) within another substance (wall material). The encapsulated substance, except active agent, can be called the core, fill, active, internal or payload phase. The substance that is encapsulating is often called the coating, membrane, shell, capsule, carrier material, external phase, or matrix. In the food industry, encapsulation process can be applied for a variety of reasons. Encapsulation is a useful tool to improve delivery of bioactive molecules (e.g. antioxidants, minerals, vitamins, phytosterols, lutein, fatty acids, lycopene) and living cells (e.g. probiotics) into foods [1]. In most cases, encapsulation refers to a technology in which the bioactive components are completely enveloped, covered and protected by a physical barrier, without any protrusion of the bioactive components. Also, encapsulation has been defined as a technology of packaging solids, liquids, or gaseous materials in small capsules that release their contents at controlled rates over prolonged periods and under specific conditions. Produced particles usually have diameters of a few nm to a few mm.

Encapsulation was originally introduced in the area of biotechnology to make production processes more efficient as the matrix around the cells allows for rapid and efficient separation of the producer cells and the metabolites. Such technologies developed approximately 60 years ago, are of significant interest to the pharmaceutical sector especially for drug and vaccine delivery, but also have relevance for the food industry. In recent years, the food industry requires the addition of functional compounds in products. These compounds are usually highly susceptible to environmental, processing and/or gastro-intestinal conditions and, therefore, encapsulation has imposed an approach for effective protection of those. Functional compounds are used to control flavour, colour, texture or preservation properties. Bioactive compounds with various potential health benefits are included, too. There is a multitude of possible benefits of encapsulated ingredients in the food industry. Encapsulation aims to preserve stability of the bioactive compounds during processing and storage and to prevent undesirable interactions with food matrix. Mainly bioactive food compounds are characterised by rapid inactivation. These compounds would profit from an encapsulation procedure, since it slows down the degradation processes (e.g., oxidation or hydrolysis) or prevents degradation until the product is delivered at the desired sites.

Thus, the bioactive component would be kept as fully functional. Also, this technology may provide barriers between sensitive bioactive materials and the environment, and thus, to allow taste and aroma differentiation, mask bad tasting or smelling, stabilize food ingredients or increase their bioavailability. In addition to the above, encapsulation can be applied for modification of physical characteristics of the original material in order to (a) allow easier handling, (b) to help separate the components of the mixture that would otherwise react with one another, and (c) to provide an adequate concentration and uniform dispersion of an active agent.

Material Used for Encapsulation

A lot of substances may be used to coat or encapsulate solids liquids, or gases of different types and properties. However, regulations for food additives are more rigid than for pharmaceuticals. Different compounds, widely accepted for drug encapsulation, have not been approved for use in the food industry, because many of these substances have not been certified for food applications as generally recognized as safe (GRAS) materials. Infact, the whole food process should be designed in order to meet the safety requirements of governmental agencies such as the European Food Safety Authority (EFSA) or Food and Drug Administration (FDA) in the USA, Food Safety & Standards Authority (FSSAI) in India [2].

The most important criteria for selection of an encapsulation material are functionality that encapsulate should provide to the final product, potential restrictions for the coating material, concentration of encapsulates, type of release, stability requirements and cost constrains. Materials used for design of protective shell of encapsulates must be food-grade, biodegradable and able to form a barrier between the internal phase and its surroundings. The majority of materials used for encapsulation in the food sector are biomolecules. Except to be natural, materials have to provide maximal protection of the active material against environmental conditions, to hold actives within capsules structure during processing or storage under various conditions, not to react with the encapsulated material, to have good rheological characteristics at high concentration if it is needed and to have easy work ability during the encapsulation.

Among all materials, the most widely used for encapsulation in food applications are polysaccharides. Starch and their derivates such as amylose, amylopectin, dextrins, maltodextrins, polydextrose, syrups and celluloses and modified celluloses are commonly used. Plant exudates and extracts viz. gum arabic, gum tragacanth, gum karaya, mesquite gum, galactomannans, pectins and soluble soybean polysaccharides are employed. Marine extracts such as carrageenans and alginate are also present in foods. Microbial and animal polysaccharides like dextran, chitosan, xanthan and gellan are also exploited. Apart from natural and modified polysaccharides, proteins and lipids are also appropriate for encapsulation.

Examples of the most common milk and whey proteins are caseins, gelatine and gluten. Among lipid materials suitable for food applications there are fatty acids and fatty alcohols, waxes (beeswax, carnauba wax, and candellia wax), glycerides and phospholipids. In addition to the above, other materials are employed such as PVP, paraffin, shellac, inorganic materials [1]. It is impossible to number all criteria to select a proper material for encapsulation. For sure, the type of an active, and its characteristics, and an application where the encapsulates are going to be used for are first on the list. Except this, cost constraint stays a key factor for choosing the most appropriate materials. No matter what is the material in question, the conversion of the physico-chemical characteristics of the materials will be the pre-condition for successful food product development. So, it is pre-requisite to study and analyse all the properties of potential wall material in order to conclude and predict its behaviour under conditions present in food formulations.

Encapsulation Techniques

There are number of techniques available for encapsulation of food compounds. Since encapsulating compounds are very often in a liquid form, many technologies are based on drying. Different techniques like spray drying, spray-bed-drying, fluid-bed coating, spray-chilling, spray-cooling or melt injection are available to encapsulate active agents [3, 4].

Spray drying is one of the oldest and the most widely used encapsulation technique in the food industrial sector. It is a flexible, continuous, but more important an economical operation. It produces particles of good quality, which size is less than 40 micron. This feature is desired from the standpoint of sensorial and textural characteristics of final products. Although spray-dryers are widespread in the food industry, there are several disadvantages of this technique such as complexity of the equipment, non-uniform conditions in the drying chamber and it is not always easy to control particle size. About 80–90% of encapsulates are spray dried ones, rest of them are mostly prepared by spray chilling, freeze drying, melt extrusion and melt injection [5].

Extrusion methods consist of dropping droplets of an aqueous solution of polymer (most often this is 0.6 -3 wt% sodium alginate) and active into a gelling bath (in case of alginate, gelling bath is 0.05-1.5 M calcium chloride solution). The dripping tool can be simply a pipette, a syringe, a vibrating nozzle, a spraying nozzle, jet cutter or atomizing disk. In comparison to other extrusion techniques, jet cutter was found to be the best technology for large-scale/industrial applications [6]. Electrostatic extrusion is especially effective for production of very small particles, down to 50 micron. An alternative extrusion technology is co-extrusion. It might be utilized to prepare spherical microbeads with a hydrophobic core and a hydrophilic or hydrophobic shell.

Another frequently used technique is emulsification. It is utilised in case of water soluble food active agents and there are two combinations of emulsions viz. water/oil emulsions or oil/water emulsions and water/oil/water double emulsions [7]. An oil-in-water emulsion can be dried by different drying methods such as spray or freeze drying, and thus to produce a powder. Such dried emulsions might be encapsulates or an instant formulation for numerous food products. Liquid emulsions may also be used as delivery systems in foods [8]. Oil-in-water emulsions are suitable for the delivery of lipids and lipid soluble bioactives. Kinetically stable oil-in-water emulsions are made by homogenising a mixture of either an oil or an oil containing a lipid-soluble bioactive, with an aqueous solution containing the encapsulating material. Spontaneously formed, thermodynamically stable microemulsions may also be loaded with nutraceuticals and used as delivery systems. Garti and Amar (2008) have discussed the importance of understanding the nature of the microstructures and phase transitions in micro and nanoemulsions for the effective delivery of nutraceuticals [9]. Ways of improving the release characteristics of conventional primary emulsions for controlled or triggered release delivery systems of bioactives, by developing multilayered emulsion formulations was studied in past [10]. Preparation of water-in-oil-in-water (w/o/w) emulsions by membrane filtration was explored earlier to encapsulate a model hydrophilic bioactive, with a view to protecting functional food ingredients for controlled release application [11].

Spray chilling or spray cooling are technologies to produce lipid coated active agents. The difference between these two techniques is the melting point of lipids. In case of spray chilling it is in range of 34–42 °C and for spray cooling temperature is higher. The agent could be dissolved in lipids, present as dry particles or present as aqueous emulsions. The spray cooling is a technique with possibility to achieve high yields and it can be run in both continuous and batch processing modes. In case of spray chilling, the particles are kept at a low temperature in a set-up similar to the fluidized bed spray granulation [4, 12]. Fluid bed coating is an encapsulation technique where a coating is applied onto powder particles in a batch processor or a continuous set-up. The powder particles are suspended by an air stream at a specific temperature and sprayed with an atomized, coating material. The coating material might be an aqueous solution of cellulose or starch derivatives, proteins and gums.

Vacuum and freeze drying are very similar drying processes, but the first one is faster and cheaper, because it operates at a temperature above the freezing point of the solvent. The major disadvantages of freeze drying are the high energy input and long processing time. In addition, during processing a barrier with an open porous structure between the active agent and its surroundings is formed; this high porous wall offers poor protection when prolonged release of an active is required. Of the different chemical micro-encapsulation processes available, only gelation and coacervation are widely used in the food industry. All current chemical methods are batch processes, although there is significant effort going into the development of continuous processes. Biopolymer structures as complexes and coacervates may be formed under conditions where the two biopolymers carry opposite charges. These structures may have potential for controlled release and delivery of bioactives in foods [13, 14]. The formation of native whey protein isolate low methoxy pectin complexes by electrostatic interaction has potential for entrapment of water-soluble ingredients in acidic foods. Also, molecular inclusion in cyclodextrins and liposomal vesicles provide some specific features to bioactives; these techniques are more expensive, and therefore, are less exploited. Cyclodextrins have a lipophilic inner pocket of about 5-8 Å, in which an active molecule with the right size can be reversibly entrapped in an aqueous environment. However, the small size of the ring forming hole limits its loading capacity. Liposomes are particles with size ranges from 30 nm to several microns. The mechanism for the creation of liposomes is basically the hydrophilic–hydrophobic interactions between phospholipids and water molecules. A brief list of food ingredients that have potential for microencapsulation are presented in Table 7.1.

Table 7.1: Type of ingredient that have been microencapsulated

Types of ingredients
Flavouring agents (including sweeteners, seasonings and spices)
Acids, bases and buffers (e.g. citric acid, lactic acid and sodium bicarbonate)
Lipids (e.g. fish oils, milk fat and vegetable oils)
Enzymes (e.g. proteases) and microorganisms (e.g. probiotic bacteria)
Amino acids and peptides
Vitamins and minerals
Antioxidants
Polyphenols
Phytonutrients
Soluble fibres

Need for Microencapsulation in Functional Food

Many bioactives (e.g. ω-3 oils, carotenes and polyphenols) need to be protected against degradation. For example, ω-3 oils are very susceptible to oxidation, leading to the development of off-flavours and off-odours. Microencapsulation protects the sensitive oils from exposure to oxygen, light and metal ions during processing and storage [15]. Protecting the core from degradation and from interactions with other food components can extend the shelf stability of the ingredient itself, as well as that of the final food product to which it is added. Microencapsulation can facilitate or improve handling of ingredients during production processes used in the final applications.

The conversion of a liquid ingredient into a powder offers significant convenience, as it is much easier to store, weigh and add a powdered ingredient, compared to its liquid version. Microencapsulation can aid in the addition and more uniform blending of bioactive ingredients into a food formulation. Bioactive ingredients in their pure or very concentrated forms are usually added in very low amounts (sometimes at ppm levels). Addition of a few milligrams or grams of ingredients into hundreds of kilos or tonnes of products can lead to uneven distribution within the food matrix, especially when ingredients are dry blended. Microencapsulated forms with much lower payload can be used in these applications to facilitate a more homogeneous blending of these highly potent bioactive ingredients into food because the lower payloads provide a larger amount of the microencapsulated ingredient to be added to the mix into which it has to be blended.

Bioactive ingredients may require microencapsulation to improve or modify their functionality and release characteristics. Understanding the core, the final application, and the mechanism required to release the core is essential for effective design of the microcapsule's release characteristics. Different release characteristics can be achieved depending on the requirement, e.g. controlled, sustained or delayed release. Bioactive ingredients are often known to possess undesirable tastes and/or odours that require masking before they can be used successfully in food formulations. A significant challenge associated with nutraceutical ingredients is the need to mask bitterness and aftertaste. With new developments in understanding the science of taste, the introduction of new bitterness blockers and sweetness potentiators in food formulations can be combined within a microencapsulation system to allow controlled, delayed or sustained release of bioactives. During the addition of bioactive ingredients into food, it is essential that both the bioactivity and the bioavailability are maintained to ensure that the bioactives achieve the desired function in the body. When direct addition of the bioactive could compromise its bioavailability, it needs to be protected by microencapsulation. The protection of the core from the acidic pH of the stomach during transit through the gastrointestinal (GI) tract may potentially enable more efficient delivery of the bioactive to the target site in the body and may also reduce the dosage required to achieve the heath benefits. Advances in the development of microencapsulation technology for food applications have been driven by the need for (i) the core to be encapsulated, (ii) new and alternative materials that are cost-effective encapsulants, and (iii) materials which will withstand the processes widely used in the food industry. More recent developments in microencapsulation technologies for food applications have focused on applying the technology to more cost-effective food-grade encapsulant materials and processes available in the food industry.

The need for controlled release and delivery of bioactive food ingredients to target sites in the body continues to drive other new developments. Converting stable microcapsule formulations (emulsions, dispersions, suspensions, coacervates) into powders is still the preferred option for production of microencapsulated bioactive ingredients, as it offers more convenience and flexibility. An understanding of how these formulations will behave during the drying process and on reconstitution is critical to the success of powdered preparations.

With the primary reason or purpose of microencapsulation being clearly identified, other important factors need to be seriously considered to ensure proper selection of encapsulant materials and processes that are cost effective, practical and scalable. Important considerations during the development of microencapsulated products include (i) core properties, e.g. chemical structure, solubility and stability, (ii) product format,e.g. liquid or powder format depending on final application, (iii) physical properties of the microencapsulated ingredient e.g. particle size, bulk density and colour, (iv) payload i.e. amount of bioactive loading in the microcapsule, (v) release trigger mechanism e.g.

dissolution, pressure, heat and shear, (vi) storage conditions and shelf-life requirements e.g. refrigerated or ambient storage, and (vii) legal and regulatory requirements for addition into food in the country of its application. From a commercial perspective, there is the additional factor of material and production costs and whether the final food product can bear the additional cost of using a microencapsulated ingredient.

Examples of Microencapsulates in Functional Food

There are several technical challenges in developing functional ingredients for incorporation into foods. They must satisfy the sensory demands of the consumers and ensure that the bioactive can be delivered to specific sites in the GI tract to exert the desired health benefit. Microencapsulation has been applied to a number of food ingredients to develop them into tailor made bioactive ingredients [15].

The increasing number of microencapsulated food ingredient launches have been the result of more creative translation and adaptation of microencapsulation techniques originally developed in the pharmaceutical industries. New encapsulant materials and more cost effective formulations and processes have enabled the food industry to develop these new ingredients with added value and functionality. In recent years, the addition of microencapsulated ingredients into a wider range of food products ensures that it does not significantly affect the cost of the final food product. This is a significant issue as food has very low profit margins compared to pharmaceutical.

Vitamins and minerals

Fortification with vitamins and minerals is often challenging due to their susceptibility to degrade during processing and storage and to react with other components in the food system. Vitamins and minerals are generally sensitive to temperature, moisture, light and pH, and their potency is often compromised by their reaction with other ingredients or premature release. Vitamins and minerals are added to a range of food products to (i) replace those that are lost during processing and storage, (ii) meet special nutritional needs, e.g. for infants and elderly persons, and (iii) prevent disease in specific consumer or at-risk groups. Traditionally, higher levels than that are required in the end product have been added to overcome losses during processing and storage. These high overages may be avoided by using microencapsulated forms. For water soluble vitamins (vitamins B and C) and minerals (iron and calcium), spray drying, spray chilling, fluidised bed coating and spinning disk coating have been used to manufacture dry powder microcapsules. Where liquid microcapsule formats are preferred, microencapsulation in liposomal delivery systems can be used.

There is also the possibility of entrapping water soluble vitamins in double emulsions.

It is demonstrated that vitamin B_{12} in the inner phase of an oil/water/oil emulsion stabilised by caseinate–dextran conjugates, instead of pure protein, reduced the release of the vitamin under acidic conditions [16]. For lipid soluble vitamins (vitamins A, D, E and K) and provitamin A (β-carotene), table emulsion formulations or spray dried emulsions are commonly used as delivery systems. Emulsion based systems are often used for delivery of lipid soluble bioactives [17]. However, where there are specific interactions between hydrophobic bioactives and a protein, an aqueous based system may be exploited. Earlier, scientists have demonstrated that casein micelles were useful for delivery of vitamin D_2 [18].

Microencapsulation has benefits when used for delivery of iron and calcium in foods. Direct addition of iron into foods may reduce its bioavailability through interaction with tannins, phytates and polyphenols. Free iron is also known to catalyse the oxidation of fats, vitamins and amino acids. These interactions can affect the sensory characteristics of the final food formulation, as well as decrease the nutritional value of the food due to iron induced catalysis of deteriorative reactions. Many of these limitations of direct addition of iron may be overcome by microencapsulation.

Other microencapsulation technologies used for encapsulation of iron include liposomal delivery systems and application of lipid coats by fluidised bed coating [19]. Molecular inclusion of iron using cyclodextrins may also be used in its delivery. The interaction of calcium with proteins can cause unwanted coagulation or precipitation of the protein, especially in calcium-fortified protein beverages. Calcium is naturally present in dairy products, but there is interest in fortifying other protein products with calcium, such as soy protein beverages. Calcium fortification of protein based beverages may be achieved with the addition of calcium chelating agents; this may result in an undesirable taste when high levels of calcium fortification are desired. Microencapsulation of calcium can prevent its negative interaction with other food components (e.g. soy proteins) in the food environment.

Functional fatty acids

Functional fatty acids, particularly docosahexaenoic acid, eicosapentaenoic acid, linolenic acid and conjugated linoleic acid, have attracted significant attention due to their potential health benefits [20]. Emulsion-based technologies and spray-drying are currently the most common approaches employed for microencapsulation and delivery of functional fatty acids into food [17].

Omega-3 fatty acids are highly susceptible to oxidation and have an inherent fishy taste and odour. Therefore, most food applications of ω-3 fatty acids require microencapsulation for protection from oxidation and to mask the fishy taste and odour. Significant research has been carried out on microencapsulation of ω-3 fatty acids. An increasing number of food companies are developing new functional food products containing ω-3 fatty acids. This increase in the number of food products launched containing ω-3 fatty acids has also been driven by the qualified health claims that were allowed by Food and Drug Administration (FDA) in 2004. Technologies that have been successfully used to encapsulate ω-3 oils include emulsification and spray drying, coacervation, cyclodextrin complexation and liposomal preparations.

Probiotics

Probiotics present two sets of problems when considering their size (typically between 1 and 5 mm diameter), which immediately excludes nanotechnologies, and the fact that they must be kept alive. This latter aspect has been crucial in selecting the appropriate microencapsulation (ME) technology. At least five ME methods have been applied to probiotics viz. spray coating, spray drying, extrusion, emulsion and gelparticle technologies (which include spray-chilling). As bioactive or encapsulating compounds can be in liquid form, many techniques are based on spraying. In spray chilling, a molten matrix with low melting point (32–42 °C) containing the bioactive compound is atomized through a nozzle into a vessel.

This process is similar to spray-drying with respect to the production of fine droplets; however, it is based on the injection of cold air into the vessel to enable solidification of the gel particle, rather than on hot air which dries the droplet into a fine powder particle (Fig 7.1). The liquid droplet thus solidifies and entraps the bioactive product. In spray-coating, the core material needs to be in a solid form and is kept in motion in a specially designed vessel, either by injection of air at the bottom or by a rotary action. A liquid coating material is sprayed over the core material which solidifies to form a layer at the surface. The coating material can be injected from many angles, and this influences the properties of the coating. Of the various technologies mentioned, spray coating and gel particle technologies are most often used (Fig 7.1).

The production of gel particles by spray chilling is considered the least expensive encapsulation technology. Unfortunately, this approach has rarely been reported for probiotics, presumably because the current methods are much more easily carried out under laboratory conditions. It is our view that spray chilling deserves greater attention for the ME of probiotics, because it could expand the range of matrices used. Moreover, this technology could be used to generate smaller beads which, as mentioned previously, might be desirable in food processing.

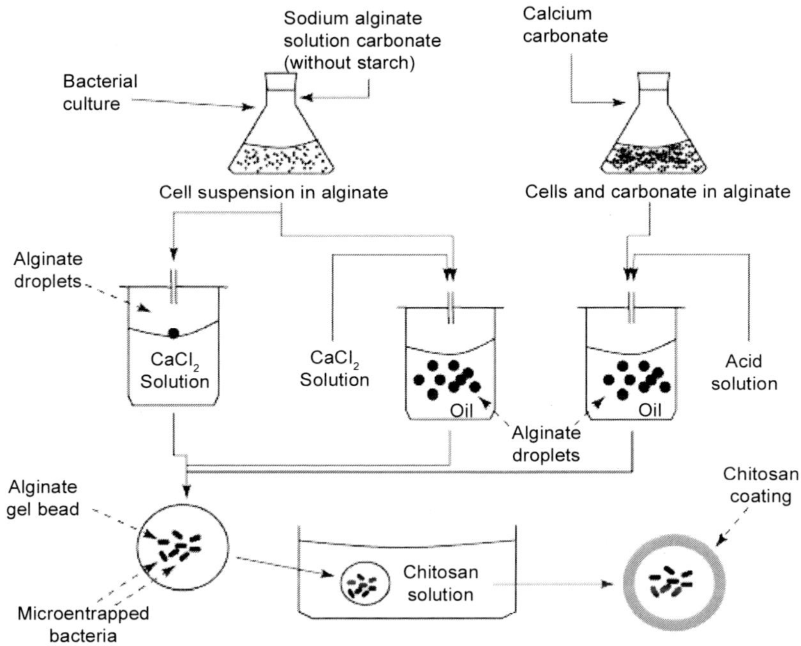

Fig. 7.1: Gel particle technologies for the microencapsulation of probiotics

Probiotics are live microorganisms that must remain alive during processing, storage and gastric transit to fulfil their desired function in the body [21]. Much clinical data have been accumulated to support the role of probiotics in human health by benefiting the immune system, strengthening the mucosal barrier and suppressing intestinal infection [22]. This has driven interest in adding probiotics to a wider range of food products, other than traditional fermented dairy products such as yoghurt. As probiotics are sensitive to heat and moisture, keeping them alive during food processing and storage is not easy. Even in fermented dairy product applications, the survival of probiotics during storage still remains a challenge for the industry (Table 7.2).

Important issues to consider for successful delivery of microencapsulated bioactives into commercial food products are listed in Table 7.3.

Tables 7.2: Comparison of three technologies used for the microencapsulation (ME) of probiotics

Technology	Characteristics	Benefits	Problems
Gel particles	Cells entrapped in a gel matrix: polymer-based system (e.g. alginate–pectin)	Can be used in a fermenter to produce biomass	Lower yields in biomass production
		Particles can be air dried	Particles can affect food texture; need small size (e.g. <50 μm)
	Gel particle coatings can be applied (e.g. chitosan)	Many methodologies can be used	Particles might dissolve in environments that contain chelating compounds (e.g. phosphates, citrates)
		Spray-chilling is the cheapest ME technology	
	High cell load (>10^{11} cfu/g) in pectin or alginate beads, but lower for starch granules	Cell release in the intestine Carrier can have bioactive properties	
Spray coating	True coating; cells are in a core powder upon which a coat (generally lipid based) is sprayed	High protection to short exposure periods (a few days) to acids, humidity and oxygen;	Phase separation in beverages if coating is lipid based
	High cell load (>10^{11} cfu/g)	Multiple layers can be added for controlled release or density adjustments	Slow cell release at low temperatures
	Most widely used commercially		
Spray drying	A liquid suspension of the bioactive product is sprayed at the top of a large vessel, creating a mist of fine droplets, in conjunction with hot air	Many possibilities for coating materials	High temperatures used in the process could kill many strains; might require a pre-heating adaptation step
	Cells 'coated' individually with the ingredients in the drying medium	Rapid cell release if ingredients dissolve rapidly	Short-term protection (a few hours) linked to particle dissolution
	Medium cell load (10^{10}–10^{11} cfu/g powder)	Less effects on food /beverage texture if particles dissolve	

Table 7.3: Important issues to consider for successful delivery of microencapsulated bioactives into commercial food products

Important issues	Action required or questions to ask
Regulatory standard	Check the regulatory standards in each country for addition of bioactives. Can the bioactive be added to the chosen food? Are the ingredients used as encapsulants allowed in the chosen food? What levels are required if there is to be a health claim?
Food product application format	What is the format of the final product chosen? For a powder, blending applications require good control of particle size, moisture and bulk density. For a liquid, rehydration and redispersion behaviour of powdered encapsulated bioactives is important.
Protection and release characteristics	What processing stresses has the ingredient to survive during incorporation into the food? Under what conditions or in response to what trigger is bioactive released?
Stage or point of addition	Is the food manufacturing plant set-up automated? At what point during production will the ingredient be added? Will the ingredient be added using an automated process or manually?
Interaction with other ingredients	Is there a need to avoid interaction of the bioactive with other ingredients in the final food during processing and storage?
Final product characteristics	What are the storage conditions and shelf-life of the final food product? What flavour characteristics are present in the food chosen? Does the food product have a delicate flavour or strong flavour that can mask some undesirable taste and aroma.

Processes that have been used to encapsulate probiotics include spray coating, spray drying, extrusion, and emulsification and gel particle technologies (Table 7.4). Of these technologies, the technique most widely investigated by researchers involves the use of polysaccharides to form gelled particles [23]. However, the use of gelled particles for microencapsulation of probiotics has not been widely adopted by commercial companies, as it is a batch process. The use of alginate chitosan microcapsules has also been explored to improve the mechanical strength of the capsules to survive *in-vitro* digestion [24]. The application of a lipid coating by a fluid-bed technique has also been used for probiotic encapsulation [25]. Probiotics encapsulated in lipid-based materials are used, in a limited range of food products, with varying degrees of success. The application of high melting lipids and waxes allows protection of probiotics from high-moisture environments and thermal protection below the melting point of the coat. Starch based encapsulation was also explored, but results showed no effect on improving the viability of *Bifidobacterium longumstrains* [26].

Table 7.4: Beneficial effects of probiotic microencapsulation

Benefit	Product
Facilitates the production of oxygen sensitive cultures	Dried probiotic culture
Facilitates the recovery of centrifugation sensitive cultures	Dried probiotic culture
Facilitates the recovery of high EPS producing cultures	Dried probiotic culture
Less contamination problems	Dried probiotic culture
Cultures can be air dried	Dried probiotic culture
Improved survival on exposure to gastric solutions	Nutraceutical
Improved survival on exposure to bile solutions	Nutraceutical
Improved stability during storage in dried form	Nutraceutical
Improved acidification rate	Dried sausages
Improved survival on heating	Biscuits, powder
Improved survival on freezing	Ice cream, milk-based medium
Improved retention in the finished product	Cheese
Protection against bacteriophages	Fermented milks
Protection against yeast contaminants	Fermented milks
Improved survival during storage	Yoghurt, mayonnaise, milk

Spray drying has always been an attractive process for production of powdered food ingredients because it is a continuous, high volume and cost effective process. A number of researchers have explored spray drying for production of probiotic microcapsules with varying degrees of success [1, 27]. The most important step still remains the selection and formulation of an encapsulant that can protect the probiotics during drying. A novel microencapsulation technology using protein carbohydrate conjugate in the matrix provided significant protection to probiotic bacteria during spray drying, during exposure to acidic pH and during non-refrigerated storage at low to intermediate water activity. The use of appropriate materials and process conditions applied during microencapsulation has the potential to enable the addition of probiotics to a much wider range of food products with intermediate water activity which does not require refrigeration.

Phytochemicals

Phytochemicals are biologically active plant chemicals, with increasing evidence that they can reduce the risk of chronic diseases [28]. Ingredients claimed to be rich in phytochemicals are extracted from plant sources. Once isolated from their natural environment, these bioactive ingredients generally require microencapsulation to stabilise the active component and mask undesirable tastes, colours and odours. The phytochemicals of interest to the food industry include phytosterols, tocopherols, carotenoids, coenzyme Q10, curcumin, garlic extracts and polyphenols (e.g. resveratrol). Resveratrol is a naturally occurring non-flavonoid polyphenolic compound present in plants such as grapes, berries and peanuts, as well as in cocoa and chocolate [29]. Resveratrol is photosensitive and benefits from microencapsulation to maintain its stability when added to food products. It has been shown that the encapsulation of resveratrol in yeast cells can offer protection and enhance its stability as an ingredient [30]. The use of chitosan–alginate coacervates as an encapsulant has also exhibited potential for preparation of encapsulated powder ingredients from aqueous (water soluble) antioxidant plant extracts. The use of natural fruit fibres as encapsulating agents for the microencapsulation and spray drying of sticky bioactive extracts (*Hibiscus sabdariffa*) has been explored in past [31]. Extracts containing curcumin have been encapsulated using commercially available lecithin to form liposomes by homogenisation or microfluidisation [32]. The delivery of curcumin through oil-in-water nanoemulsions has been shown to enhance its anti-inflammatory activity in animal tests [33].

Proteins, amino acids, peptides and enzymes

Proteins have traditionally been encapsulated for pharmaceutical applications.The demand for more protein in food and beverages is on the rise [34]. Whey, casein and soy proteins are commonly used in high protein food formulations either in their native or hydrolysed forms. Protein derived peptides and amino acids are also being isolated from their source to enable addition at the correct dosage required for physiological health functions. The direct addition of these components into food and beverage formulations can result in an undesirable bitter taste and astringency. Encapsulation of casein hydrolysates in lipospheres has been found to reduce the bitterness [35-37]. Encapsulation may also be used to preserve the activity of enzymes. Components in garlic have been shown to offer beneficial health effects [38], and microencapsulation of garlic powder results in protection of alliinase activity, which improves the ratio of alliin to allicin conversion under *in-vitro* conditions.

Fibre

The trend of adding dietary fibres to food and beverage formulations that traditionally do not contain these fibres is increasing due to the increasing evidence of health benefits of high fibre diets [39, 40]. Examples of dietary fibres for which the FDA has allowed health claims are glucan from oats and psyllium fibre. β-glucan, a cholesterol lowering soluble fibre, shown to reduce the risk of heart disease was allowed an FDA health claim in 1997. Later, in 1998, the FDA extended the health claim for soluble fibre to psyllium fibre. Other dietary fibres added to food and beverage formulations include indigestible gums, polysaccharides, oligosaccharides and lignins [41]. High levels of fibre need to be added in the final food formulation in order to make a health claim.

The problems associated with the addition of high levels of dietary fibres to food and beverages are the unpalatability of the high fibre ingredients and the significant effects they have on the viscosity of the final product. This has resulted in the development of expensive, refined fibre ingredients, e.g. polydextrose. Microencapsulation can minimise palatability problems as well as minimise water absorption during formulation and processing. Much cheaper sources (e.g. indigestible gums) can also be added at a much higher levels if the fibre in food formulations is encapsulated with materials that can reduce hydration and water absorption during processing. Chito-oligosaccharide, as a functional ingredient, offers a range of health benefits; however, direct addition to milk can affect its flavour and colour.

Development of Microencapsulated Ingredients

During the development of functional foods using microencapsulated food ingredients, the selection of ingredients and processes was traditionally based on empirical approaches. Past researchers have suggested that an alternative concept is to use a retro-design approach that relies more on a fundamental understanding of the required performance of the ingredient in a complex food environment [42]. This approach encompasses an understanding of the effects of processing and the factors controlling the chemical and physical events that govern the stability and release properties of a microencapsulated product; however, the test of whether a microencapsulation system is suitably tailored for its end product application is its acceptance in the marketplace. The route from concept to acceptance of functional foods by consumers has many stages and requires input from scientists, technologists, nutritionists and an understanding of the regulatory processes. It is important to ensure that when microencapsulation is used to deliver active ingredients into foods, it provides a simple, efficient and cost effective solution compared to direct addition of bioactives.

Nano Encapsulation

An important step in the development of functional ingredients is the design of formulation procedures for the stabilisation, solubilisation and delivery of the active components in the food product to which they are added. During the last few years, a new concept to prepare nanoparticles using amphiphilic cyclodextrin (CD) molecules for such purposes was considered [4,5, 43]. CDs are cyclic oligomers of d-glucopyranose that can be produced through the transformation of starch by certain bacteria such as Bacillus macerans. CDs have the ability to alter physical, chemical and biological properties of guest molecules through the formation of inclusion complexes. The widely used natural, cyclodextrins consist of six, seven and eight d-glucopyranose residues, respectively. Each cyclodextrin has its own ability to form inclusion complexes with specific guests, which depends on a proper fit of the guest molecule into the hydrophobic cyclodextrin cavity.

Cyclodextrins have become so important that there is now what we may call a 'cyclodextrin science', involving the search for improved properties and commercial applications of these molecules. Not only the relevant aspects of CDs are explored, but also applications of nuclear magnetic resonance (NMR), spectroscopy, calorimetry, computational chemistry and X-ray crystallography among other techniques are further developed for the elucidation of the structures of these compounds and their complexes. In the food, cosmetics, toiletry and tobacco industries, CDs have been widely used either for the stabilisation of flavours, vitamins and natural colours or to eliminate off-flavours, microbiological contaminations and other undesired compounds [44, 45]. In the chemical industry, CDs are also used as catalysts to improve the selectivity of reactions, as well as for the separation and purification of industrial scale products. The most common applications of cyclodextrins are to enhance the solubility, stability and bioavailability of guest molecules. However, natural cyclodextrins have relatively low solubility, both in water and organic solvents, which thus limits their uses in formulations. Various kinds of cyclodextrin derivatives have been prepared to modify the physicochemical properties and inclusion capacity of natural cyclodextrins as novel drug carriers.

In an aqueous solution, the slightly polar CD cavity is occupied by water molecules and, therefore, they can be readily substituted by appropriate 'guest molecules', which are less polar than water. It is known that in order to form the complexes, water has to migrate from the interior of the cavity and to be replaced by the ligand. On the other side, once the complex is formed, the addition of water may cause the breakdown of the system. Thus, the interactions with water determine not only the ability to form the complexes, but also their stability. Under storage of the CD-encapsulated materials, water sorption properties are fundamental for defining the release of the guest molecule. It has been reported that the stability of the dried complexes is governed by the shape of the water sorption isotherms. Besides of polarity, the spatial structure of the ligand affects the degree of inclusion, its release and water adsorption.

The aim of this chapter is to analyse the influence of water adsorption of CDs, their derivatives and their complexes on the chemical and physical stability of encapsulated compounds. These data are useful for selecting storage conditions for encapsulated nutraceuticals or for predicting the shelf life of functional products formulated with nanoencapsulated compounds. In the last 30 years, there was a progressive increase in the use of CD-assisted molecular nanoencapsulation in the food industry and also in the number of publications and patents related to the production of CD. The widespread utilisation of CD is reflected in pharmaceutical, food, chemical and other industrial areas. Its use has increased annually around 20–30%; most of that in food products [46-48]. CDs and their complexes are usually colourless, non-hygroscopic powders and they are used in food industry with a variety of objectives

Summary

Microencapsulation technology holds promise for the successful delivery of bioactive ingredients into functional foods, and has the potential to enhance the functionality of bioactive ingredients, thus maximising the health benefits available to consumers from these foods. Microencapsulation can offer significant advantages for improved delivery and protection of bioactive ingredients in food, which would not have been possible by direct addition. New developments in a range of microencapsulation technologies continue to address different functionality challenges that occur when formulating bioactive ingredients into functional foods. Opportunities for use of microencapsulation in the food industry continue to grow as greater demands are being made on the integrity of the capsules to control the release and delivery of the core material at a specific time during digestion and to a specified site in the body. This often requires tailor-made microencapsulated ingredients that are fit for this purpose to be individually developed to take into account the final food application and format for delivery of the bioactive ingredients.

As pointed out earlier, microencapsulation (ME) could be useful in helping to deliver bioactive ingredients both to the food matrix itself and to the GI tract. To date, the literature shows that ME has primarily served for the delivery of bioactives into the matrix and, as yet, has not been fully explored for more efficient delivery in the GI tract. Thus, more research needs to be carried out to investigate the additional benefits of ME on the stability of bioactive ingredients in the gastric environment and on the release of bioactive ingredients into the GI tract. The ME of bioactive compounds for use in functional foods must increasingly be examined in a larger context, where functional benefits could extend from processing to improved assimilation in the GI tract.

Another area that is likely to see intense research activity in the future is the use of co-encapsulation. In this regard, many emulsion and spray coating technologies offer significant opportunities for the co-encapsulation of various hydrophobic and hydrophilic bioactives. Finally, the paradigm of not wanting added bioactive compounds to alter food sensory properties or appearance might well be reconsidered. There are already commercial products available (yoghurt and breakfast cereal) where particles containing probiotics are clearly seen in the foods and even advertised on the label. Therefore, ME might even be used to create particles that clearly show consumers that the bioactive ingredients are present in the functional foods, thus promoting marketing strategies for product differentiation.

References

1. Anal, A.K. and H. Singh, *Recent advances in microencapsulation of probiotics for industrial applications and targeted delivery.* Trends in Food Science & Technology, 2007. **18**(5): 240-251.
2. Gouin, S., *Microencapsulation: industrial appraisal of existing technologies and trends.* Trends in Food Science & Technology, 2004. **15**(7): 330-347.
3. Krasaekoopt, W., B. Bhandari, and H. Deeth, *Evaluation of encapsulation techniques of probiotics for yoghurt.* International Dairy Journal, 2003. **13**(1): 3-13.
4. F. Gibbs, S.K., Inteaz Alli, Catherine N. Mulligan, Bernard, *Encapsulation in the food industry: a review.* International Journal of Food Sciences and Nutrition, 1999. **50**(3): 213-224.
5. Desai, K.G.H. and H. Jin Park, *Recent developments in microencapsulation of food ingredients.* Drying technology, 2005. **23**(7): 1361-1394.
6. Risch, S.J., *Encapsulation: overview of uses and techniques.* FAO.1995.
7. McClements, D.J., *Food emulsions: principles, practice, and techniques*1998: CRC press.
8. Ezhilarasi, P., *et al., Nanoencapsulation techniques for food bioactive components: a review.* Food and Bioprocess Technology, 2013. **6**(3): 628-647.
9. Garti, N. and I. Yuli-Amar, *Micro-and nano-emulsions for delivery of functional food ingredients.* Delivery and controlled release of Bioactives in Foods and Nutraceuticals, 2008: 149-183.
10. Guzey, D. and D.J. McClements, *Formation, stability and properties of multilayer emulsions for application in the food industry.* Advances in Colloid and Interface Science, 2006. **128**: 227-248.
11. Shima, M., *et al., Protection of Lactobacillus acidophilus from the low pH of a model gastric juice by incorporation in a W/O/W emulsion.* Food Hydrocolloids, 2006. **20**(8): 1164-1169.
12. Soottitantawat, A., *et al., Microencapsulation of menthol by spray drying and its release characteristics.* Innovative Food Science & Emerging Technologies, 2005. **6**(2): 163-170.
13. Rodríguez Patino, J.M. and A.M. Pilosof, *Protein–polysaccharide interactions at fluid interfaces.* Food Hydrocolloids, 2011. **25**(8): 1925-1937.
14. Jones, O.G., E.A. Decker, and D.J. McClements, *Formation of biopolymer particles by thermal treatment of â-lactoglobulin–pectin complexes.* Food Hydrocolloids, 2009. **23**(5): 1312-1321.
15. Sanguansri, P. and M.A. Augustin, *Nanoscale materials development–a food industry perspective.* Trends in Food Science & Technology, 2006. **17**(10): 547-556.
16. Fechner, A., *et al., Stability and release properties of double-emulsions stabilised by caseinate–dextran conjugates.* Food Hydrocolloids, 2007. **21**(5): 943-952.
17. Coupland, J.N. and D.J. McClements, *Lipid oxidation in food emulsions.* Trends in Food Science & Technology, 1996. **7**(3): 83-91.
18. Semo, E., *et al., Casein micelle as a natural nano-capsular vehicle for nutraceuticals.* Food Hydrocolloids, 2007. **21**(5): 936-942.

19. Xia, S. and S. Xu, *Ferrous sulfate liposomes: preparation, stability and application in fluid milk.* Food Research International, 2005. **38**(3): 289-296.
20. Mandarino, J.M.G., A.C. Roessing, and V.d.T. Benassi, *Functional Fatty Acids* 2005: Embrapa Soja.
21. Lin, D.C., *Probiotics as functional foods.* Nutrition in Clinical Practice, 2003. **18**(6): 497-506.
22. Fuller, R., *History and development of probiotics*, in *Probiotics* 1992, Springer. 1-8.
23. Janes, K., P. Calvo, and M. Alonso, *Polysaccharide colloidal particles as delivery systems for macromolecules.* Advanced Drug Delivery Reviews, 2001. **47**(1): 83-97.
24. Reis, C.P., *et al., Review and current status of emulsion/dispersion technology using an internal gelation process for the design of alginate particles.* Journal of Microencapsulation, 2006. **23**(3): 245-257.
25. Lee, P.K. and P.H. Richardson, *Controlled release encapsulated bioactive substances*, 2004, Google Patents.
26. Lahtinen, S., *et al., Effect of starch and lipid based encapsulation on the culturability of two Bifidobacterium longum strains.* Letters in Applied Microbiology, 2007. **44**(5): 500-505.
27. Picot, A. and C. Lacroix, *Production of Multiphase Water Insoluble Microcapsules for Cell Microencapsulation Using an Emulsification/Spray drying Technology.* Journal of Food Science, 2003. **68**(9): 2693-2700.
28. Lakkis, J.M., *Encapsulation and controlled release technologies in food systems*2008: John Wiley & Sons.
29. Baur, J.A., *et al., Resveratrol improves health and survival of mice on a high-calorie diet.* Nature, 2006. **444**(7117): 337-342.
30. Shi, G., *et al., Stabilization and encapsulation of photosensitive resveratrol within yeast cell.* International Journal of Pharmaceutics, 2008. **349**(1): 83-93.
31. Chiou, D. and T. Langrish, *Development and characterisation of novel nutraceuticals with spray drying technology.* Journal of Food Engineering, 2007. **82**(1): 84-91.
32. Fang, Z. and B. Bhandari, *Encapsulation of polyphenols–a review.* Trends in Food Science & Technology, 2010. **21**(10): 510-523.
33. Wang, X., *et al., Enhancing anti-inflammation activity of curcumin through O/W nanoemulsions.* Food Chemistry, 2008. **108**(2): 419-424.
34. Bhatia, R.B., *et al., Aqueous sol-gel process for protein encapsulation.* Chemistry of Materials, 2000. **12**(8): 2434-2441.
35. Morais, H., *et al., Casein hydrolysates using papain: peptide profile and encapsulation in liposomes.* Acta Alimentaria, 2005. **34**(1): 59-69.
36. Mendanha, D.V., *et al., Microencapsulation of casein hydrolysate by complex coacervation with SPI/pectin.* Food Research International, 2009. **42**(8): 1099-1104.
37. Favaro-Trindade, C., *et al., The use of spray drying technology to reduce bitter taste of casein hydrolysate.* Food Hydrocolloids, 2010. **24**(4): 336-340.
38. Bekheet, S.A., *A synthetic seed method through encapsulation of in vitro proliferated bulblets of garlic (Allium sativum L.).* Journal of Biotechnology, 2006. **9**: 415-426.
39. Lanza, E. and R. Butrum, *A critical review of food fiber analysis and data.* Journal of the American Dietetic Association, 1986. **86**(6): 732-743.
40. Casterline, J.L., C.J. Oles, and Y. Ku, *In vitro fermentation of various food fiber fractions.* Journal of Agricultural and Food Chemistry, 1997. **45**(7): 2463-2467.
41. Cho, S.S. and P. Samuel, *Fiber ingredients: Food applications and health benefits* 2010: CRC Press. London
42. Ubbink, J. and J. Krüger, *Physical approaches for the delivery of active ingredients in foods.* Trends in Food Science & Technology, 2006. **17**(5): 244-254.

43. Fathi, M., M. Mozafari, and M. Mohebbi, *Nanoencapsulation of food ingredients using lipid based delivery systems.* Trends in Food Science & Technology, 2012. **23**(1): 13-27.
44. Szente, L. and J. Szejtli, *Cyclodextrins as food ingredients.* Trends in Food Science & Technology, 2004. **15**(3): 137-142.
45. Astray, G., *et al.*, *A review on the use of cyclodextrins in foods.* Food Hydrocolloids, 2009. **23**(7): 1631-1640.
46. Hedges, A.R., W.J. Shieh, and C.T. Sikorski, *Use of cyclodextrins for encapsulation in the use and treatment of food products.* 1995.
47. Cravotto, G., *et al.*, *Cyclodextrins as food additives and in food processing.* Current Nutrition & Food Science, 2006. **2**(4): 343-350.
48. Del Valle, E.M., *Cyclodextrins and their uses: a review.* Process Biochemistry, 2004. **39**(9): 1033-1046.

8

Supercritical Extraction

Introduction

In recent years, there has been a growing interest in the functional foods because they can provide physiological benefits additional to nutritional and energetic, as for instance, antihypertensive, antioxidant or anti-inflammatory [1]. A functional food can be defined as a food that produces a beneficial effect in one or more physiological functions, increases the welfare and/or decreases the risk of suffering a particular disease. Furthermore, new types of products, derived from food, called nutraceuticals have recently been developed. These products, usually employed as food supplements, are marketed as tablets and pills, and can provide important health benefits.

Among the different compounds with functional properties, antioxidants are the most widely studied [2, 3]. These compounds can play an important role in food technology because of their usefulness against lipid peroxidation. Usually, food production, processing and storage can generate important losses of endogenous antioxidants that limit their own protection against lipid oxidation. The traditional extraction methods used to obtain these type of products have several drawbacks; they are time consuming, laborious, have low selectivity and/or low extraction yields. Moreover, these traditional techniques employ large amounts of toxic solvents. At present, extraction methods able to overcome the above mentioned drawbacks are being studied. Supercritical fluid extraction (SFE) and subcritical water extraction (SWE) are among the more promising processes [4]. These extraction techniques provide higher selectivities, shorter extraction times and do not use toxic organic solvents. The inception of supercritical fluid extraction in food processing technology dates back the early 1970s. The initial application of SFE in food processing was in the decaffeination of coffee beans [5] followed by a number of other applications related to the extraction of hops, aroma compounds, decaffeination of tea, and many more.

By the 1980s, supercritical fluids were extensively used in many extraction and reaction schemes as well as in the chromatographic analysis of food products [6]. However, it was not until the late 1990s that supercritical fluid processing technology was embraced as a 'green' processing platform. Nutraceutical products touting labels such as 'naturally processed' or 'supercritical CO_2 treated' have gained greater appeal among consumers and government agencies that are cognizant of food safety issues.

Process Description

When a fluid is forced to a pressure and temperature above its critical point, it becomes a supercritical fluid (Fig 8.1). Under these conditions, various properties of the fluid are placed between those of a gas and those of a liquid. Although the density of a supercritical fluid is similar to a liquid and its viscosity is similar to a gas, its diffusivity is intermediate between the two states. Thus, the supercritical state of a fluid has been defined as a state in which liquid and gas are indistinguishable from each other, or as a state in which the fluid is compressible (i.e. similar behaviour to a gas) even though possessing a density similar of a liquid and, therefore, similar solvating power. Because of its different physicochemical properties, SFE provides several operational advantages over traditional extraction methods [7]. Due to their low viscosity and relatively high diffusivity, supercritical fluids have better transport properties than liquids, can diffuse easily through solid materials and can, therefore, give faster extraction yields. Range of values of important physicochemical properties viz. density, diffusivity and viscosity of gases, liquids and supercritical fluids is provided in Table 8.1.

One of the main characteristics of a supercritical fluid is the possibility of modifying the density of the fluid by changing its pressure and/or its temperature. Since density is directly related to solubility [8, 9], by altering the extraction pressure, the solvent strength of the fluid can be modified. Other advantages, compared to other extraction techniques, are the use of solvents generally recognized as safe (GRAS), the higher efficiency of the extraction process (in terms of increasing yields and lower extraction times), and the possibility of direct coupling with analytical chromatographic techniques such as gas chromatography (GC) or supercritical fluid chromatography (SFC).

As for the solvents, there is a wide range of compounds that can be used as supercritical fluids; carbon dioxide is the most commonly used because of its moderate critical temperature (31.3 °C) and pressure (72.9 atm) (Table 8.2). Carbon dioxide is a gas at room temperature, so once the extraction is completed, and the system decompressed, a substantial elimination of CO_2 is achieved without residues, yielding a solvent free extract. On an industrial scale, when carbon dioxide consumption is high, the operation can be controlled to recycle it.

However, supercritical CO_2, because of its low polarity, is less effective in extracting more polar compounds from natural matrices. To overcome this problem, modifiers (also called co-solvents) are commonly used. Modifiers are highly polar compounds that, added in small amounts, can produce substantial changes of the solvent properties of supercritical CO_2 [10]. According to the specific requirements, the design of a supercritical fluid extraction system can be relatively simple or highly complex. Basically, it is possible to differentiate between analytical instruments and preparative systems (pilot or industrial scale). The analytical systems are utilized in sample preparation prior to, for example, a chromatographic analysis, to obtain milligrammes to grammes of extracts. There are several configurations, depending on their degree of automation.

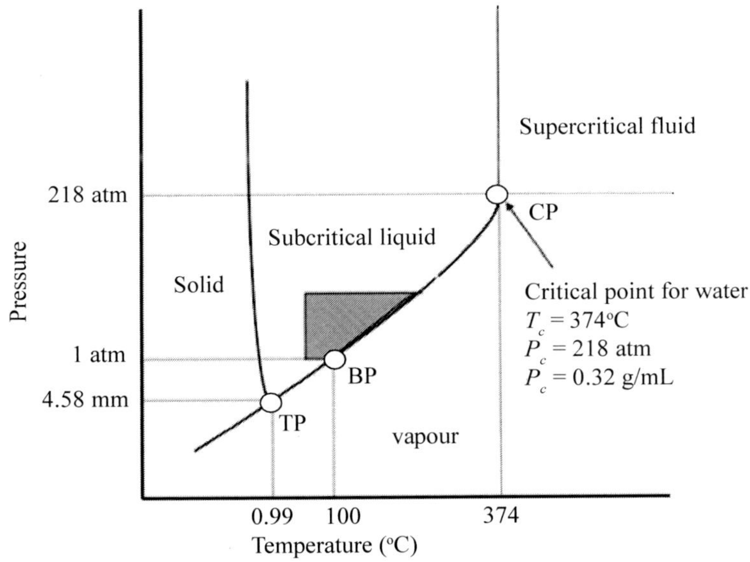

TP - Tripple point BP - Boiling point CP - Critical point

Fig. 8.1: Phase diagram of water

Table 8.1: Range values of several physicochemical properties of gases, liquids and supercritical fluids

State of fluid	Density (ρ, g/cm³)	Diffusivity (DAB, cm²/s)	Viscosity (μ, g s/cm)
Gas			
p = 1 atm; T = 21 °C	10-3	10^{-1}	10^{-4}
Liquid			
p = 1 atm; T = 15–30 °C	1	$<10^{-5}$	10^{-2}
Supercritical			
$p = P_C$; $T = T_C$	0.3–0.8	10^{-3}–10^{-4}	10^{-4}–10^{-3}

Table 8.2: Critical data of some pure components

Component	T_C (°C)	P_C (MPa)
Ethylene	9.4	5.04
Carbon dioxide	31.1	7.38
Ethane	32.3	4.87
Nitrous oxide	36.6	7.26
Propane	96.8	4.25
n-Hexane	234.5	3.01
Acetone	235.1	4.70
Methanol	239.6	8.09
Ethanol	240.9	6.14
Ethyl acetate	250.2	3.83
Water	374.1	22.06

T_C – Critical temperature, P_C – Critical pressure

The preparative systems are used to extract grammes of compounds when working on a pilot scale. In these preparative systems, two different configurations can be found for processing solid or liquid samples. Basically, a preparative system on a pilot scale plant consists of a solvent pump, that delivers the fluid throughout the system, a modifier pump if necessary, an extraction cell or extraction column, according to the system configuration (for solids or liquids), and one or more separators (also called solubility of different food components fractionation cells) in which the extract is collected and the solvent depressurized. Likewise, the extraction cell or column and the separators are commonly equipped with independent control of temperature and pressure, in such a way that fractionation of the extracted compounds can be carried out by a stepwise depressurization (Figure 8.2).

Therefore, different compounds can be obtained within each separator, depending on their differential solubility in the supercritical fluid. Additionally, it is possible to install a refrigerated system especially designed to trap the most volatile compounds, as well as a recycling system to recycle the fluid employed.

Comparison of $SC–CO_2$ and Solvent Extraction

The advantages of SFE over conventional processes such as extraction by solvents and separation by distillation are automation, the reduction in operational steps, safe operation due to the use of nonorganic solvents and the use of moderate temperature in the critical range favourable for heat labile foods. The main advantage of SFE is the excellent quality of the resulting product [9]. A comparison of SFE and solvent extraction is shown in Table 8.3 below.

Table 8.3: Comparison of $SC–CO_2$ extraction with solvent extraction

Solvent Extraction	Supercritical Extraction
Solvent presence is unavoidable, the residual level (generally ppm) of the solvent depends on the type of solvent used.	Is totally free of solvents and hence very pure.
Heavy metal content is also unavoidable and depends on the solvent, the method of solvent recycling, the source of the raw material, and the material used to construct the contact parts of the machinery.	Totally free of heavy metals since they are not extractable even if they are present in the raw material. No heavy metals are present in CO_2 or the equipment.
Inorganic salt content cannot be avoided, using the same concept as above.	Totally free of inorganic salts using the same explanation as above.
Polar substances get dissolved along with the lipophilic substances from the raw material due to poor selectivity of the solvent. During solvent removal operations, these polar substances form polymers, which lead to discoloration of the extract and poor flow characteristics. All this causes the extract to look different from the basic components in the raw material and hence it is more of a "pseudo" natural extract.	No such possibility exists since CO_2 is highly selective and no chance of polar substances forming polymers exists. In addition the operating temperature is only 40–80 °C.
Both polar and non-polar colours are extracted.	Only non-polar colours get extracted.
Solvent removal requires extra unit operations resulting in higher cost and lower recovery of useful material.	No extra unit operations needed and yield of useful material is very high.

Table 8.4: Summary of publications on applications of SFE in food industry

Sample	Analyte	Pressure (MPa)	Temperature (°C)
Extraction of lipid and cholesterol			
Egg	Cholesterol and lipid	31	45
Milk	Fat	10–35	50–70
		6.9–24.1	40–80
			30–50
			40
			30–50
Beef	Lipid and cholesterol	10.3–31	
		10.3–27.6	
Pork	Lipids	10–31	
Extraction of natural color			
Carrot	Carotenes	34.2–57	30–50
Alfalfa leaf	Carotene and lutien	10–70	40
Sweet potato	β-carotene	13.8–41.4	38–48
Bixa Orellana seeds	Bixin	20.7–48.3	40–55
Tomato paste waste	β-carotene & lycopene	20–30	35–65
Tomato skin	Lycopene	40.5	60–110
Red grape skin	Anthocyanins	7.0–15	80
Extraction of natural flavourings			
Oil from bitter orange peel	Terpens	7.7–12	40
Dried orange peel	Essential oils	10–28	40–50
Juice from citrus fruit	Limonin	20.7–41.4	30–60
Aromatic plants	Essential oils	20	40
Mushrooms	Oleoresins	11.5–14.9	–
Extraction of antioxidants			
Tamarind seed	Antioxidant	10–30	40–80
Eucalyptus leaves	Oil with high antioxidant activity	20	50
Decaffeination of coffee and tea			
Coffee	Caffeine	16–22	90
Tea	Caffeine	25–35	50–80
Mate	Caffeine, Theobromine Theophyline	25.5	70.2
Extraction of hops			
Hops	Humulone, lupulone and essential oils	20	40

As has been mentioned, the most important difference between pilot plants to process solid or liquid samples is in the use of an extraction cell or an extraction column. Solid processing is always done in batch in a discontinuous or semi-continuous process, while liquid processing is usually carried out under countercurrent conditions in a continuous mode. In liquid sample extractions, the supercritical fluid (usually CO_2) is moving upwards while the sample feed, introduced into the system from the top or centre of the column, is moving downwards by gravity [11].

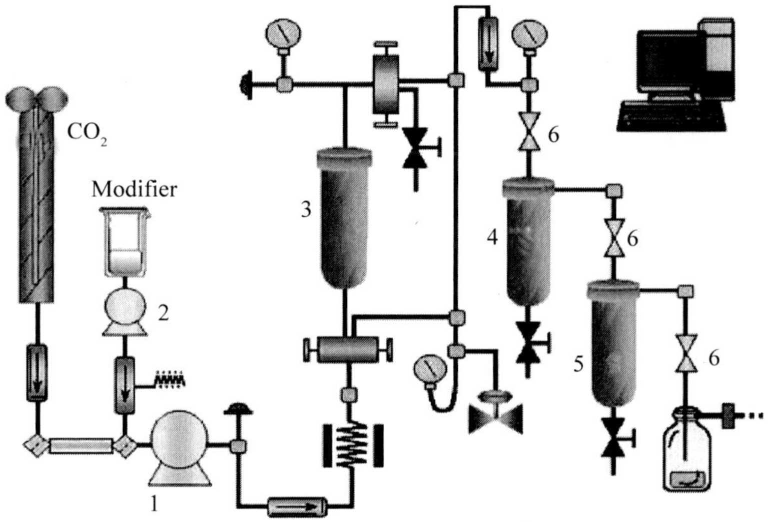

Fig. 8.2: Diagram of super critical fluid extraction pilot plant equipped with two fractionation cells. [(1) CO_2 pump, (2) modifier pump, (3) solid samples extraction cell, (4) fractionation cell 1, (5) fractionation cell 2, and (6) valve]

Extraction of Bioactive Ingredients from Plants

Supercritical fluid extraction of bioactive compounds from various natural product matrices with different temperature, pressure and co-solvent conditions has been done in the past. A summary of publications on applications of SFE in food industry is presented in Table 8.4. In all the past studies it was found that using an alcoholic co-solvent, such as ethanol or isopropanol, a greater yield of bioactive compounds was achieved. For example, only about 30–40 % of carotenes were recovered by extraction with SC-CO_2 at a temperature between 55 and 70 °C and pressure of 34.5 MPa, but when using 10 % ethanol as co-solvent, approximately 90 % of the carotenes could be extracted at the same temperature and pressure conditions [12]. It was established that pressure had a limited effect on the extraction of carotenes with supercritical CO_2 and 10% ethanol mixture. In a similar study, using response surface methodology models to quantify the carotene yield and using supercritical CO_2 and 10% ethanol mixtures at 50 °C and 34.2 MPa, it was found that there was a considerable decrease in the amount of solvent and time required to extract carotenes using the supercritical fluid mixture in comparison with organic solvents used in traditional extraction process [13]. However, in a study involving the extraction of carotenoids from leaf protein concentrates using SC-CO_2 in the absence of a co-solvent, it was found that no recovery of carotenoids occurred until around 30 MPa, above which approximately 90% of the carotene was recovered. For the extraction of lutein, a relatively polar carotenoid pigment, higher pressures (50–70 MPa) and SC-CO_2 volume were required for a maximum recovery of 70% [14]. In the processing of functional foods or nutraceuticals, application of subcritical water has been focused mainly on the extraction of polar compounds such as the antioxidants from natural product matrices. In order to study the effect of subcritical water as a solvent, however, it is also important to recognize its utility as a polar modifier for extraction of nutraceutical components using SC-CO_2. The effect of water and ethanol as co-solvents was evaluated in the treatment of tea leaves [15] using extraction conditions of 50–80 °C and 15–30 MPa. It was found that a greater amount of catechins was extracted in ethanol-modified SC-CO_2 in comparison with water-modified SC-CO_2 at these low co-solvent concentrations. In another study, water was employed as a co-solvent for the SFE of sunflower leaves to yield bioactive compounds of nutraceutical value [16]. In this study, both methanol and dimethyl sulfoxide were also used as modifiers. Water as a co-solvent yielded maximum extraction of bioactive compounds at 50°C and 50 MPa. Interestingly, lower extraction pressure and temperatures proved to be detrimental to the yield of bioactive compounds, especially when the sunflower leaf extract contained moisture. The maximum yield of bioactive compounds with water as a co-solvent at higher temperatures and pressures was rationalized in terms of higher density of water (when compared to SC-CO_2) under these extraction conditions, as well as the polar nature of the bioactive compounds being extracted.

Studies involving extraction of bioactive compounds from natural product matrices using subcritical water as a solvent are usually compared with traditional extraction techniques such as hydro-distillation and soxhlet extraction. Using subcritical water as solvent, bioactive compounds from kava roots [17] and anthocyanins from berry substrates [18, 19] have been extracted. All these studies reported a greater yield by a subcritical water extraction process in comparison with traditional extraction techniques. Continuous subcritical water extraction of nutraceutical components from laurel leaves has been reported [20]. This continuous extraction system allows faster solvent flow rates at higher temperature and pressure conditions resulting in greater solvent flux through the substrate matrix, thereby resulting in faster extractions.

Numerous vegetable matrices have been used as natural sources for compressed fluid extraction. Legumes, spices, aromatic plants, and even fruit beverages, such as natural orange juice [21] have been processed to obtain natural antioxidant compounds. Several studies have compared the antioxidant activities of plant extracts obtained using supercritical fluid extraction with those using traditional extraction methods. For example, eucalyptus [22] and *Lippia alba* [23] have been used to compare different extraction methods in terms of composition and activity. The research performed with eucalyptus demonstrated the differences that exist between the composition and functional properties of extracts obtained with SFE (using supercritical carbon dioxide at 200 bar and 50 °C) and hydrodistillation. Antioxidant activity was found to be greater in supercritical fluid extracts than in hydro-distillation extracts. In spite of the fact that the main compounds were the same in the studied extracts, their quantitative composition changes. For example, supercritical fluid extracts have a higher content of sesquiterpenes and oxygenated compounds. As for the antioxidant activity, this seems to be related to the concentration of both, p-cymen-7-ol and thymol in the extract. The concentrations of both compounds, mainly the amount of p-cymen-7-ol, in SFE extracts, was higher than that found in hydrodistillation extracts. Those values were in agreement with the higher antioxidant activity found in supercritical extracts compared to extracts from hydrodistillation [24].

Aromatic plants are among the most studied plants with antioxidant activity. The antioxidant activity of supercritical extracts from lemon balm (*Melissa officinalis* L.) was recently studied [25]. The plant material was subjected to pressures from 100 to 180 bar and temperatures from 35 to 40 °C. However, in this case, the solid residues were analyzed to determine their antioxidant activity (instead of on the supercritical carbon dioxide extracts). The results obtained showed that the best antioxidant activity was achieved by optimizing the extraction conditions at 100 bar and 35 °C for 4 h.

Ginger (*Zingiber officinale roscoe*) is another aromatic plant which has been widely studied due to its multiple functional activity. In one of the study the effect of temperature and pressure, as well as the addition of a co-solvent, on the kinetics of ginger extraction and extract antioxidant activity was studied [26]. By means of an experimental factorial design it was concluded that the addition of a co-solvent was not necessary in order to increase the mass transfer rate or yield. The factors selected to carry out the factorial design were extraction temperature (25–35 °C), extraction pressure (200– 250 bar) and solvents (CO_2, CO_2+ethanol, CO_2+isopropyl alcohol). The best results, in terms of antioxidant activity, were obtained when the extraction was carried out with a modifier at low temperatures, low pressures and relatively long extraction times; the antioxidant activity seemed to be related to the preferential extraction of gingerols and shogaols. A recent study has compared this aromatic plant with similar plants, such as rosemary (*Rosmarinus officinialis* L.) and turmeric

(*Curcuma longa L.*) [27]. Extracts of these three plants were obtained using supercritical carbon dioxide with/without modifier (ethanol and/or isopropyl alcohol). Experiments were carried out at pressures between 100 and 300 bar and temperatures between 30 and 40 °C. Once the antioxidant activity assays were performed, it was concluded that the extracts which lowered antioxidant activity were those from turmeric and ginger, whereas rosemary extracts provided higher values of antioxidant activity.

Rosemary is one of the plants with higher antioxidant activity and is, therefore, frequently studied [28, 29]. The antioxidant activity of supercritical extracts of rosemary is extremely high; even at low concentrations the resultant extracts are heat resistant and do not change the colour, taste or flavour of the food in which they are used. Rosemary extracts dissolve easily in different types of foods [30].

Extraction of vitamin E from natural sources has received increasing interest due to the high antioxidant activity associated with this family of compounds. Besides its well known antioxidant activity, recent studies have demonstrated that synthetic vitamin E, is less effective than natural vitamin E [31]. Several natural sources have been used to isolate vitamin E using supercritical carbon dioxide extraction. In past, the extraction of vitamin E rich oil from a plant (*Silybum marianum*) that naturally grows in the Mediterranean area was studied. It was pointed out that extractions at 60 ºC and 200 bar produced the most concentrated extracts in terms of α-tocopherol (0.08%), while the extraction yield was relatively high (19%). Another important source of vitamin E is wheat germ. The studies conducted with this raw material [32, 33], showed the capability of SFE coupled to supercritical fluid chromatography (SFC) for tocopherol enrichment. More recent studies have confirmed SFE (with CO_2) can achieve, extraction yields for tocopherols from wheat germ similar to traditional hexane extraction [34, 35]. Also, the production of wheat germ oil by SFE has been studied [35, 36], providing oils with a very high contents of tocopherols. Vitamin E has been extracted from wheat germ under the extracting conditions of 275 bar, 40 °C and CO_2 flow rate equal to 2 mL/min for 90 min [37, 38]. The amount of total vitamin E extracted under these conditions was higher than those obtained using traditional extraction methods (with n-hexane or chloroform/methanol mixtures). Nutraceuticals extracted and enriched using critical fluids and their therapeutic utility is provided in Table 8.5. A summary of studies on the supercritical carbon dioxide extraction of plant and botanical samples is presented in Table 8.6.

Table 8.5: Nutraceuticals extracted and enriched using critical fluids and their therapeutic utility

Nutraceutical	Therapeutic utility
Bixin	Hyperglycemia (diabetes)
Betulin	Anti-inflammatory, antimalarial
Lycopene	Anticarcinogenic (prostrate, gastrointestinal)
β-Carotene	Anticarcinogenic (lung, cardiovascular)
Limonene	Anticarcinogenic (mammary glands)
Terpenes	Skin
Flavonoids	Antioxidant
Lutein	Macular degeneration
Phospholipids	Cognitive
Isoflavones	Premenstrual syndrome, circulatory
Essential oils	Antioxidant, dermatological, circulatory
Phytosterols	Circulatory
Tocopherols	Antioxidant
Saponins	Anticarcinogenic, lower body cholesterol
Lipids	Anticarcinogenic, antioxidant
Carotenoid esters	Antioxidant
Hyperforin, adhyperforin (St John's wort)	Antidepression

Table 8.6: Summary of studies on the supercritical carbon dioxide extraction of plant and botanical samples

Substrate name	Bioactive compound (s)	SFE conditions (MPa/K/cosolvent)
Avocado	Quercetin, isoquercetin, rutin,vitexin	10/333/EtOH
Basil	Eugenol	10–30/313
Black pepper	Piperine, essential oils	20–32/318–338
Carrots	β-Carotene	34.5/328–343/EtOH
Chamomile	Bisabolol, chamazulene	10–20/303–313
Coriander	Volatile oil, phenolic compounds, β-Pinene, m-diethyl benzene, o-Diethyl-benzene, Curcumene, Zingiberene	20–30/298–331
Ginger	β-sesquiphellandrene	15–30/293–313
Jackfruit	Jackfruit Isoquercetin, vitexin	10/333/EtOH
Onion	Essential oils	24.5/310
Palm	Carotenoids, tocopherols, fatty acids,	15–30/318–328
Paprika	Lipids	35/323; 15/313
	Carotenoids and tocopherols	13.7–41.3/313
Pitanga	Isoquercetin, rutin	10–40/273–333
Rosemary	Camphor, carnosic acid	10–30/303–313
	Rosmarinic acids, phenolic diterpenes	10–16/310–320

Functional Ingredients from Food Industry By-products

Frequently, the processes that take place in the food industry generate products (the so called by-products) that are discarded, causing subsequent environmental problems. In recent years, companies have devoted effort to find value-added application for these food byproducts. Supercritical fluid extraction has been used in many of these enrichment studies. Thus, several studies have been developed to extract β-carotene and lycopene from by-products of tomato industry [39, 40].

These compounds are natural pigments, belonging to the carotenoid group, and their antioxidant properties are well known. In the past work the optimisation of β-carotene and lycopene extraction from tomato paste waste was done by employing a factorial design [21]. By previously determining the total amounts of these compounds in the tomato paste, it was possible to establish an optimum recovery of 54% of total lycopene using, as extraction conditions, CO_2 at 300 bar and 55 °C for 2 h.

The CO_2 flow rate was kept constant at 4 kg/h and ethanol was used as modifier (5%). The optimum value for β-carotene recovery was around 50%. Extraction conditions were similar to those previous selected for lycopene extraction, except for an increase in the extraction temperature up to 65 °C. Likewise, these authors suggested the use of higher extraction temperatures and pressures to increase extraction yields, although a 100% recovery would never be expected due to the degradation that these compounds can suffer during the extraction process.

The extraction of lycopene of tomato seeds and skins with supercritical CO_2 was recently studied [41]. In this work, more extreme extraction conditions were tested, with extraction temperatures ranging from 32 to 86 °C and extraction pressures from 138 to 483 bar. The amount of lycopene extracted increased at higher pressures and temperatures up to a maximum recovery of 61% at 86 °C and 345 bar. The CO_2 flow rate was 2.5 mL/min and 3g sample was used in this analytical system. The work demonstrated the possibility of lycopene extraction from tomato by-products using supercritical carbon dioxide without the addition of co-solvents.

Other interesting by-products are those from the wine industry. Their interest is related to the type and amount of phenolic compounds that are found in grape seeds and skins. Isolation of phenolic compounds from grape seeds has been attempted using supercritical carbon dioxide [42, 43]. Palma and Taylor (1999) observed that the recovery of catechin and other phenolic compounds from grape seeds was higher when using supercritical CO_2 with methanol as modifier, rather than when using traditional solid–liquid extraction. Supercritical fluid extraction is faster and allows fractionation of the phenolic compounds of the grape seeds by changing the pressure and adding co-solvents at different percentages. In a recent study, supercritical fluid extraction was employed to increase the added value of extracts from by-products of the wine industry obtained by extraction with ethyl acetate. These extracts had antioxidant activities similar to the synthetic antioxidant BHT [44].

The properties of the starting product were significantly improved by the subsequent supercritical CO_2 extraction that caused an increase of the antioxidant activity, allowing odourless and clearer extracts. These characteristics make the extracts more appropriate for use as natural antioxidants in the food industry. The selected parameters for performing the extraction were a pressure of 150 bar and a temperature of 45 °C. The addition of 0.5% of co-solvent did not significantly improve the results. The use of by-products from the olive oil industry can be used to extract tocopherols [45]. In the earlier study, the separation of tocopherols from the olive pomace (that is, the solid residue obtained using two phase olive oil production systems) was achieved by means of a supercritical carbon dioxide extraction with two step fractionation. In the second fractionation step, where complete depressurization took place, the CO_2 density was very low and enrichment of the extract with tocopherols was observed. The extraction was carried out at a pressure of 350 bar and at a temperature of 50 °C, while the fractionation conditions that gave the best results were 100 bar and 60 °C on the first separator and 10 bar and 25 °C on the second separator. Although the extract obtained using SFE was rich in tocopherols, these authors suggested a selective fractionation system of the different tocopherol isomers using supercritical fluid chromatography (SFC) with packed columns.

Another SFE work that suggested the utilization of food by-products to produce antioxidants, e.g. vitamin E, was performed [29]. It is based on soybean oil production and the by-product studied is obtained as a waste at the deodorization step during industrial production of soybean oil. Considering that soybean oil is the most widely consumed oil in the world, the use of this byproduct (also called deodorizer distillate or soybean sludge) is of economic importance. Different extraction temperatures (from 40 to 80 °C) and pressures (from 90 to 170 bar) were studied. The highest extraction efficiency was obtained under mild extraction conditions, fatty acids being extracted by the supercritical CO_2. This fact allowed tocopherol enrichment inside the reactor cell.

The extraction of natural pigments from food byproducts is of major importance, not only because of the increasing demand for natural ingredients by the food industry but also because some of these pigments may also have associated antioxidant activities that can even more, increase their added value. The possibility of extracting natural pigments with antioxidant properties using SFE has been already mentioned. However, sometimes pigments themselves are the target compounds. This is the case for the extraction of carotenoids from carrots and tomatoes, i.e., the sources most frequently used to obtain natural carotenoids. Among the different conditions studied in this work, the best results were obtained using an extraction pressure equal to 275 bar and an extraction temperature of 80 °C. The product obtained under these conditions had a composition of 65% lycopene and 35% carotene. However, taking into account that both compounds showed different solubility parameters on supercritical CO_2, it was possible to select a two step extraction to preferentially extract lycopene. For instance, on performing the extraction, in a first step, at 275 bar and 40 °C and, in a second step, at 275 bar and 80 °C, it was possible to obtain an end product with 87% lycopene and 13% beta carotene, because it is preferably extracted under the initial extraction conditions. The isolation of lycopene has been studied by using tomato skin [47]. After considering different extraction conditions (namely, different extraction temperatures, with/without co-solvents), they achieved an extremely high lycopene recovery when performing extractions with supercritical carbon dioxide at 1.5 mL/min flow rate, at 11 °C and 400 bar extraction pressure, for 50 min. Results were similar with and without acetone as modifier; leading to a lycopene recovery of 94% in just 15 min of extraction.

Extraction of Functional Ingredients and Other Compounds of Interest from Algae and Microalgae

In the search for feasible new sources of natural antioxidants that can be used in the food industry, algae and microalgae have been suggested as possible raw materials. Both organisms are widely known and consumed in certain countries, and numerous health benefits have been associated with their use. Therefore, algae and microalgae are potentially a great source of natural compounds that could be used as ingredients for preparing functional foods. Different compounds with antibacterial, antiviral and antifungical activity can be found in these type of organisms [48, 49], along with compounds with antioxidant activity. Starting from a complex matrix, such as an alga (*Dilophus ligulatus*), it was possible to obtain extracts with different compositions and yields by changing the extraction pressure [50]. Therefore, depending on the type of compounds of interest, optimum extraction conditions could be utilized for selective isolation of specific groups of compounds.

Many algae and microalgae are rich in polyunsaturated fatty acids. A consumption increase of these types of compounds has been associated with a decrease in the incidence of cardiovascular diseases [51]. The effect of the extraction conditions for obtaining fatty acids from *Hypnea charoides* algae, using supercritical CO_2, suggest the usefulness of this extraction technique for converting this alga species as a new source of ω-3 fatty acids [52]. Temperature ranges from 40 to 50 °C and pressure from 241 and 379 bar were studied. In general, the lipid recovery, as well as the ratio of unsaturated fatty acids, increased with extraction pressure and temperature. Extraction of omega-3 fatty acids was shown to depend on their chain length. Several microalgae species have been used to obtain natural compounds of interest for the food industry, using supercritical fluid extraction.

In the past several studies, this technique has been applied to several microalgae species to extract, for example, diolefines from *Botrycoccus brauniicells* [53-55]. This organism can store high amounts of long-chain hydrocarbons (i.e. 25–31 carbon atoms), that can be utilized as substitutes of paraffinic and natural waxes. The solubility of these types of compounds in CO_2 increased with pressure and 300 bar pressure provided the optimum with respect to yield and extraction speed. SFE has also been used to extract carotenoids from microalgae *Chlorella vulgaris*. High pressures allowed high extraction yield. When microalgae cells were crushed, the carotenoid extraction improved slightly. A temperature increase produced an opposite effect at low pressures in terms of solute recovery. The optimum conditions were 55 °C and 350 bar.

Dunaliella salinais, a microalga species, is capable of producing 14% of β-carotene relative to its dry weight. The optimum extraction yield for β-carotene extraction with supercritical CO_2 was found to be 300 bar and 40 °C [29]. Another microalga species studied has been Arthospira (Spirulina) maxima. This microalga is capable of producing high amounts of γ-linolenic acid (GLA). The recoveries of GLA with pure CO_2 and with CO_2 plus ethanol as co-solvent were compared with the results obtained using traditional organic solvents extraction. It was observed that, although both CO_2 and *n*-hexane provided similar extraction yields, the CO_2 allowed a higher recovery of GLA. The maximum extraction yield (0.44% GLA/dry biomass) was obtained using CO_2 with 10% of ethanol as modifier and performing the extraction at 350 bar and 60 °C. Cyanobacterium *Spirulina platensis* was studied by Qiuhui (1999) to determine the amount of lipids and GLA present in the microalgae [56]. The maximum extraction yield was obtained at 350 bar of pressure, in agreement with the results of Mendes et al. (2003). The temperature was set at 40°C and the CO_2 flow rate was fixed at 24 kg/h for 4 h for this SFE.

Production Scale Critical Fluid Based Nutraceutical Plants and Commercial Products

Irrespective of the high investment costs, there has been considerable application of super critical fluid technology for extraction as applied to functional foods. These critical fluid-treated functional food products have gained considerable consumer acceptance. The initial applications of critical fluids, especially SC-CO_2, in the food industry were centralized on the decaffeination of coffee and tea leaves. Further developments were made by extending this technology to extraction of aroma compounds from hops and essential oils from varied flower and fruit substrates. There are close to 110 critical fluid based plants concentrated mostly in Germany, United States, South Korea, France and Japan. Among plants conducting extractions and enrichments of nutraceutical based ingredients are those owned by US Nutraceuticals, GreenTek 21, Hitex, Flavex, Newly Wed Foods and Fuji Flavors, among others, which have invested in utilizing critical fluid technology for varied applications inclusive of decaffeination, hops processing and/or extraction of flavours or aroma compounds from spices. Prominent organizations working toward setting up super critical fluid based plants worldwide for varied applications, but not necessarily restricted to functional foods, are Separex, Natex, Uhde and Thar Technologies.

Of all the SCF processing plants, the Five King Cereals plant located in Taiwan is interesting because its purpose is not direct SFE. Established in 1999–2000 with an initial investment cost of US$10 million, this plant operates at 4 tons/hour through three 5800 liter supercritical CO_2 extractors. SC-CO_2 treatment here is principally used to facilitate a greater shelf life, reduced cooking time, reduction in metal and pesticide residue content, as well as reduced waxy and fatty acid content; the overall result being a shelf stable rice product exhibiting less rancidity and degradation over time. A similar plant was constructed by SNU-UMAX in South Korea, which produces commercial 'diet' rice. The principle and the commercial product produced by both these plants are technically the same, with slight differences in rice texture when the rice is exposed to SC-CO_2.

There has been considerable growth in the number of SC-CO_2 based food processing plants in South Korea and India over the last few years. South Korea perhaps leads the world in using critical fluid based technology partly due to an increase in health awareness in South Korea. Some of the leading organizations employing critical fluid processing in South Korea are Greentek, Hanhwa, UMAX and Ottogi Food Corporation. Greentek has developed a process for the removal of fat content from processed instant noodles. Using SC-CO_2, this company states a 20% reduction in fat and a decrease in the sodium content. UMAX, a commercial producer of sesame oil in South Korea, is another company producing a product, which provides numerous health benefits [57].

Recently, India has also been providing a growing market for critical fluid based processing plants. Initially restricted to pilot scale plants at educational institutions such as IIT Mumbai, other organizations have come forward to provide production scale units in India. There are two Natex critical fluid based spice processing plants in India. The most prominent feed stocks for these plants in India, apart from spices, are sandalwood and neem. Some of the functioning supercritical fluid based processing units in India are SAMI industries, Novo Agritech, Prerna Biotech, Sat Group, Vedic Supercriticals, South East Agro, RKS Agro, Indo Global Spices and Nisarga Biotech.

CFE has been applied for some time now in the extraction of specialty oils, such as evening primrose, borage, black currant and flax. Recently, there has been considerable research in employing critical fluids to obtain oil from fungi or marine sources, such as spirulina, which are devoid of cholesterol [24, 58]. Other nutraceutical ingredients that can be obtained similarly include extracts from chamomile, paprika, garlic and ginger. Most of the common spices and mint oils, including a commercially available extract of rosemary, can be obtained via SC-CO_2 extraction.

Summary

Irrespective of the high investment costs, there has been considerable application of critical fluid technology for extraction as applied to functional foods. These critical fluid treated functional food products have gained considerable consumer acceptance. The initial applications of critical fluids, especially SC-CO_2, in the food industry were centralized on the decaffeination of coffee and tea leaves. Further developments were made by extending this technology to extraction of aroma compounds from hops and essential oils from varied flower and fruit substrates and its application in food prodcuts.

References

1. Goldberg, I., *Functional foods: designer foods, pharmafoods, Nutraceuticals* 1994: Springer.
2. Bhat, V.B. and K. Madyastha, *C-Phycocyanin: A Potent Peroxyl Radical Scavenger in Vivo and in Vitro.* Biochemical and Biophysical Research Communications, 2000. **275**(1): 20-25.
3. Pinero Estrada, J., P. Bermejo Bescos, and A. Villar del Fresno, *Antioxidant activity of different fractions of Spirulina platensis protean extract.* Il farmaco, 2001. **56**(5): 497-500.
4. King, J., *Advances in critical fluid technology for food processing.* Food Science and Technology Today, 2000. **14**(4): 186-91.
5. Zosel, K., *Process for the decaffeination of coffee*, U.S. Patent No. 2157956.
6. Tena, M., *et al.*, *Supercritical fluid extraction of natural antioxidants from rosemary: comparison with liquid solvent sonication.* Analytical chemistry, 1997. **69**(3): 521-526.

7. Anklam, E., *et al.*, *Supercritical fluid extraction (SFE) in food analysis: a review.* Food Additives & Contaminants, 1998. **15**(6): 729-750.
8. Valle, J.D. and J. Aguilera, *Articles - High pressure CO_2 extraction. Fundamentals and applications in the food industry.* Food Science and Technology International-Frederick, 1999. **5**(1): 1-24.
9. Raventós, M., S. Duarte, and R. Alarcón, *Application and possibilities of supercritical CO_2 extraction in food processing industry: an overview.* Food Science and Technology International, 2002. **8**(5): 269-284.
10. Valcárcel, M. and M. Tena, *Applications of supercritical fluid extraction in food analysis.* Fresenius' Journal of Analytical Chemistry, 1997. **358**(5): 561-573.
11. Kislik, V.S., *Solvent extraction: classical and novel approaches* 2012: Elsevier.
12. Barth, M.M., *et al.*, *Determination of optimum conditions for supercritical fluid extraction of carotenoids from carrot (Daucus carota L.) tissue.* Journal of Agricultural and Food Chemistry, 1995. **43**(11): 2876-2878.
13. Vega, P.J., *et al.*, *Supercritical carbon dioxide extraction efficiency for carotenes from carrots by RSM.* Journal of Food Science, 1996. **61**(4): 757-759.
14. Favati, F., *et al.*, *Supercritical CO_2 extraction of carotene and lutein from leaf protein concentrates.* Journal of Food Science, 1988. **53**(5): 1532-1536.
15. Park, H.S., *et al.*, *Effects of cosolvents on the decaffeination of green tea by supercritical carbon dioxide.* Food Chemistry, 2007. **105**(3): 1011-1017.
16. Casas, L., *et al.*, *Supercritical fluid extraction of bioactive compounds from sunflower leaves with carbon dioxide and water on a pilot plant scale.* The Journal of Supercritical Fluids, 2008. **45**(1): 37-42.
17. Kubatova, A., D.J. Miller, and S.B. Hawthorne, *Comparison of subcritical water and organic solvents for extracting kava lactones from kava root.* Journal of Chromatography A, 2001. **923**(1): 187-194.
18. King, J., R. Grabiel, and J. Wightman. *Subcritical water extraction of anthocyanins from fruit berry substrates.* in *Proceedings of the 6th International Symposium on Supercritical Fluids*. 2003.
19. Ju, Z. and L.R. Howard, *Subcritical water and sulfured water extraction of anthocyanins and other phenolics from dried red grape skin.* Journal of Food Science, 2005. **70**(4): S270-S276.
20. Fernández-Pérez, V., M.M. Jiménez-Carmona, and M.D.L. de Castro, *An approach to the static–dynamic subcritical water extraction of laurel essential oil: comparison with conventional techniques.* Analyst, 2000. **125**(3): 481-485.
21. Señoráns, F.J., *et al.*, *Isolation of antioxidant compounds from orange juice by using countercurrent supercritical fluid extraction (CC-SFE).* Journal of Agricultural and Food Chemistry, 2001. **49**(12): 6039-6044.
22. Fadel, H., *et al.*, *Effect of extraction techniques on the chemical composition and antioxidant activity of Eucalyptus camaldulensis var. brevirostris leaf oils.* Zeitschrift für Lebensmitteluntersuchung und-Forschung A, 1999. **208**(3): 212-216.
23. Stashenko, E.E., B.E. Jaramillo, and J.R. Martýìnez, *Comparison of different extraction methods for the analysis of volatile secondary metabolites of Lippia alba (Mill.) NE Brown, grown in Colombia, and evaluation of its in vitro antioxidant activity.* Journal of Chromatography A, 2004. **1025**(1): 93-103.
24. Herrero, M., A. Cifuentes and E. Ibanez. *Sub-and supercritical fluid extraction of functional ingredients from different natural sources: Plants, food-by-products, algae and microalgae: A review.* Food Chemistry, 2006. **98**(1): 136-148.
25. Ribeiro, M., M. Bernardo-Gil, and M. Esquývel, *Melissa officinalis, L.: study of antioxidant activity in supercritical residues.* The Journal of Supercritical Fluids, 2001.

21(1): 51-60.

26. Zancan, K.C., *et al., Extraction of ginger (Zingiber officinale Roscoe) oleoresin with* CO_2 *and co-solvents: a study of the antioxidant action of the extracts.* The Journal of Supercritical Fluids, 2002. **24**(1): 57-76.
27. Leal, P.F., *et al., Functional properties of spice extracts obtained via supercritical fluid extraction.* Journal of Agricultural and Food Chemistry, 2003. **51**(9): 2520-2525.
28. López-Sebastián, S., *et al., Dearomatization of antioxidant rosemary extracts by treatment with supercritical carbon dioxide.* Journal of Agricultural and Food Chemistry, 1998. **46**(1): 13-19.
29. Mendes, M., F. Pessoa, and A. Uller, *An economic evaluation based on an experimental study of the vitamin E concentration present in deodorizer distillate of soybean oil using supercritical* CO_2. The Journal of Supercritical Fluids, 2002. **23**(3): 257-265.
30. Gerard, D., K. Quirin, and E. Schwarz, CO_2*-extracts from rosemary and sage.* Food Market Technology, 1995. **10**: 46-52.
31. Hadolin, M., *et al., High pressure extraction of vitamin E-rich oil from Silybum marianum.* Food Chemistry, 2001. **74**(3): 355-364.
32. Saito, M. and Y. Yamauchi, *Isolation of tocopherols from wheat germ oil by recycle semi-preparative supercritical fluid chromatography.* Journal of Chromatography A, 1990. **505**(1): 257-271.
33. Saito, M., *et al., Enrichment of tocopherols in wheat germ by directly coupled supercritical fluid extraction with semipreparative supercritical fluid chromatography.* Journal of Chromatographic Science, 1989. **27**(2): 79-85.
34. Ossa, E.M., *Quality of wheat germ oil extracted by liquid and supercritical carbon dioxide.* Journal of the American Oil Chemists' Society, 2000. **77**(9): 969-974.
35. Muñoz, B., A. Melero Gómez, and E. Martínez de la Ossa, *Optimisation of the extraction process of tocopherol from wheat germ by liquid and supercritical carbon dioxide.* Grasas y Aceites, 1999. **50**(4): 275-279.
36. Panfili, G., *et al., Extraction of wheat germ oil by supercritical* CO_2*: oil and defatted cake characterization.* Journal of the American Oil Chemists' Society, 2003. **80**(2): 157-161.
37. Ge, Y., *et al., Optimization of the supercritical fluid extraction of natural vitamin E from wheat germ using response surface methodology.* Journal of Food Science, 2002. **67**(1): 239-243.
38. Ge, Y., *et al., Extraction of natural vitamin E from wheat germ by supercritical carbon dioxide.* Journal of Agricultural and Food Chemistry, 2002. **50**(4): 685-689.
39. Baysal, T., S. Ersus, and D. Starmans, *Supercritical* CO_2 *extraction of β-carotene and lycopene from tomato paste waste.* Journal of Agricultural and Food Chemistry, 2000. **48**(11): 5507-5511.
40. Rozzi, N., *et al., Supercritical fluid extraction of lycopene from tomato processing byproducts.* Journal of Agricultural and Food Chemistry, 2002. **50**(9): 2638-2643.
41. Rozzi, N. and R. Singh, *Supercritical fluids and the food industry.* Comprehensive Reviews in Food Science and Food Safety, 2002. **1**(1): 33-44.
42. Murga, R., *et al., Extraction of natural complex phenols and tannins from grape seeds by using supercritical mixtures of carbon dioxide and alcohol.* Journal of Agricultural and Food Chemistry, 2000. **48**(8): 3408-3412.
43. Palma, M. and L. Taylor, *Extraction of polyphenolic compounds from grape seeds with near critical carbon dioxide.* Journal of Chromatography A, 1999. **849**(1): 117-124.
44. Louli, V., N. Ragoussis, and K. Magoulas, *Recovery of phenolic antioxidants from wine industry by-products.* Bioresource Technology, 2004. **92**(2): 201-208.

45. Ibanez, E., *et al.*, *Isolation and separation of tocopherols from olive by-products with supercritical fluids.* Journal of the American Oil Chemists' Society, 2000. **77**(2): 187-190.
46. Cadoni, E., *et al.*, *Supercritical* CO_2 *extraction of lycopene and β-carotene from ripe tomatoes.* Dyes and Pigments, 1999. **44**(1): 27-32.
47. Ollanketo, M., *et al.*, *Supercritical carbon dioxide extraction of lycopene in tomato skins.* European Food Research and Technology, 2001. **212**(5): 561-565.
48. Xue, C., *et al.*, *Molecular species composition of glycolipids from Sprirulina platensis.* Food Chemistry, 2002. **77**(1): 9-13.
49. Ötleº, S. and R. Pire, *Fatty acid composition of Chlorella and Spirulina microalgae species.* Journal of AOAC international, 2001. **84**(6): 1708-1714.
50. Subra, P. and P. Boissinot, *Supercritical fluid extraction from a brown alga by stagewise pressure increase.* Journal of Chromatography A, 1991. **543**: 413-424.
51. Cohen, Z. and A. Vonshak, *Fatty acid composition of Spirulina and spirulina like cyanobacteria in relation to their chemotaxonomy.* Phytochemistry, 1991. **30**(1): 205-206.
52. Cheung, P.C., *Temperature and pressure effects on supercritical carbon dioxide extraction of n-3 fatty acids from red seaweed.* Food Chemistry, 1999. **65**(3): 399-403.
53. Mendes, R.L., *et al.*, *Applications of supercritical* CO_2 *extraction to microalgae and plants.* Journal of chemical technology and biotechnology, 1995. **62**(1): 53-59.
54. Mendes, R.L., *et al.*, *Supercritical* CO_2 *extraction of carotenoids and other lipids from Chlorella vulgaris.* Food Chemistry, 1995. **53**(1): 99-103.
55. Mendes, R.L., *et al.*, *Supercritical carbon dioxide extraction of compounds with pharmaceutical importance from microalgae.* Inorganica Chimica Acta, 2003. **356**: 328-334.
56. Qiuhui, H., *Supercritical carbon dioxide extraction of Spirulina platensis component and removing the stench.* Journal of Agricultural and Food Chemistry, 1999. **47**(7): 2705-2706.
57. Smith, J. and E. Charter, *Functional food product development*2011: John Wiley & Sons.
58. Herrero, M., *et al.*, *Pressurized liquid extracts from Spirulina platensis microalga: Determination of their antioxidant activity and preliminary analysis by micellar electrokinetic chromatography.* Journal of Chromatography A, 2004. **1047**(2): 195-203.

9

Spray Drying Technology

Introduction

Spray drying process has been used for decades to encapsulate food ingredients such as flavors, lipids, and carotenoids. During this drying process, the evaporation of solvent, that is most often water, is rapid and the entrapment of the interest compound occurs quasi-instantaneously. This required property imposes a strict screening of the encapsulating materials to be used in addition to an optimization of the operating conditions. Likewise, if the encapsulated compound is of hydrophobic nature, the stability of the feed emulsion before drying should also be considered.

Thus, spray drying microencapsulation process must rather be considered as an art than a science because of the many factors to optimize and the complexity of the heat and mass transfer phenomena that take place during the microcapsule formation. Spray drying accomplishes drying while particles are suspended in the air and is one method in the family of suspended particle processing systems, along with fluid bed drying, flash drying, spray granulation, spray agglomeration, spray reaction, spray cooling, and spray absorption. This drying process is unique because it involves both particle formation and drying.

Spray Drying Process

Spray drying is a unit operation by which a liquid product is atomized in a hot gas current to instantaneously obtain a powder. The gas generally used is air or more rarely an inert gas as nitrogen. The initial liquid feeding the sprayer can be a solution, an emulsion or a suspension. Spray drying produces, depending on the starting feed material and operating conditions, a very fine powder (10–50 μm) or large size particles (2–3 mm). Water removal by spray drying solutions is a common engineering practice. By decreasing water content and water activity, spray drying is generally used in food industry to ensure a microbiological stability of products, avoid the risk of chemical and/or biological degradations, reduce the storage and transport costs, and finally obtain a product with specific properties like, for example, instantaneous solubility.

The spray drying process has been developed in connection with the manufacture of dried milk. However, when milk is spray dried, the process can be considered as a microencapsulation one; milk fat is being the core material that is protected against oxidation by a wall material consisting of a mix of lactose and milk proteins. In this mix, the carbohydrates provide structure through glass formation whereas the proteins provide emulsification and film forming properties. Schematic diagram of spray drying process is presented in Figure 9.1 [1].

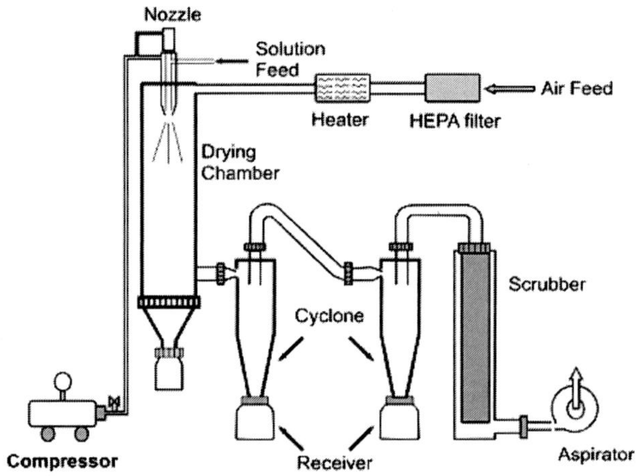

Fig. 9.1: Schematic of spray drying process

Some fundamental information concerning the process of spray drying are presented in this chapter. The knowledge of these considerations is very important to carry out the process of microencapsulation by spray drying.

Atomizing

Liquid atomization in small droplets can be carried out by pressure or centrifugal energy. Used atomizers include pneumatic atomizer, pressure nozzle, spinning disk configurations and recently introduced two fluid nozzle and sonic nozzle. The goal of this stage is to create a maximum heat transferring surface between the dry air and the liquid in order to optimize heat and mass transfers. The choice upon the atomizer configuration depends on the nature and viscosity of feed and the desired characteristics of dried product. The higher the provided energy is, the finer are the formed droplets. For the same energy amount, the size of formed particles increases with increasing feed rate. However, the size of particles increases when both viscosity and surface tension of the initial liquid are high. Different types of atomizers such as centrifugal, steam, and homogenizing are presented in Fig. 9.2.

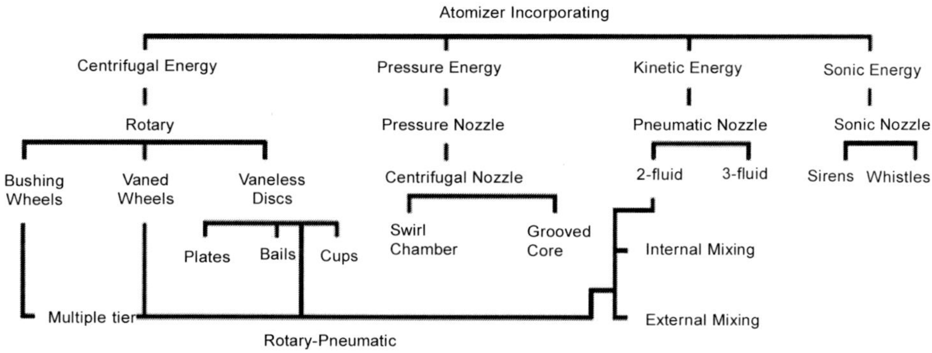

Fig. 9.2: Classification of atomizers

Droplet hot Air Contact

This contact takes place during atomization and initiates the drying stage. According to the atomizer emplacement compared to the hot air spreader, one can distinguish cocurrent drying and counter-current one. In co-current process the liquid is sprayed in the same direction as the flow of hot air through the apparatus, hot air inlet temperature is typically 150–152 °C, evaporation occurs instantaneously and for which dry powders will be exposed to moderate temperatures (typically 50–80 °C) which limits thermal degradations. Whereas, during counter-current drying, the liquid is sprayed in the opposite direction of the flow of hot air and for which the dry product is exposed to high temperatures which limits the applications of this process to thermo-sensitive products. However, the main advantage of the counter-current process is that it is considered as more economic in the terms of consumed energy.

There are three types of product-air flow pattern in spray dryers viz. co-current, counter-current, and mixed flow. In the co-current process, the droplets and air pass through the dryer in the same direction. The droplets meet the air at the highest temperature. This causes rapid surface evaporation, while it is still wet, providing safe conditions for heat-sensitive materials. The counter current configuration sprays the droplets in the opposite direction to the hot air flow, which exposes the dry product to high temperatures. This design can only be used for non-heat-sensitive materials and is less commonly used than the co-current configuration [2]. Mixed flow is a combination of co-current and counter-current flow patterns. A nozzle is positioned in the bottom of the chamber, forcing the spray to travel upward until overcome by gravity and the downward flow of the drying medium. Mixed flow is a good method for drying relatively coarse droplets in a small chamber at small production rates because the spray has a long path through the chamber [7]. Typical product-air flow patterns in spray dryers are shown in Fig 9. 3.

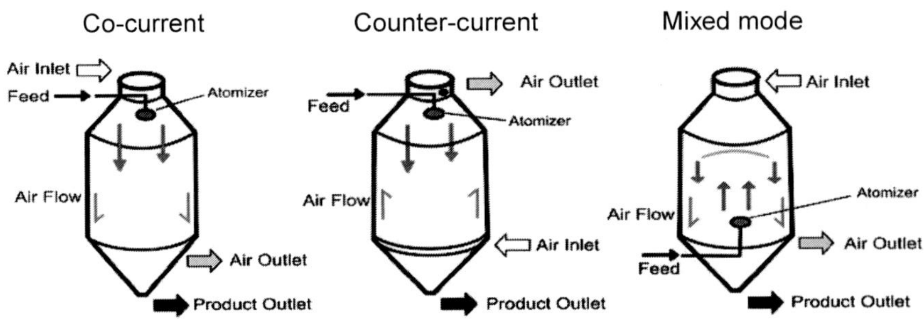

Fig. 9.3: Typical product air flow patterns in spray dryers

Evaporation of droplet water

At the time of droplets–hot air contact, balances of temperature and vapor partial pressure are established between liquid and gas phases. Thus, heat transfer is carried out from air towards the product as a result of temperature difference whereas water transfer is carried out in the opposite direction due to the vapor pressure difference. Based on the fundamental theory of drying, three successive steps can be distinguished. Just after the hot air–liquid contact, heat transfer principally causes the increase of the droplets temperature upto a constant value. This value is defined as the air drying humid thermometer temperature; the evaporation of droplets water is carried out at constant temperature and water vapor partial pressure. The rate of water diffusion from the droplet core to its surface is usually considered constant and equals to the surface evaporation rate. Finally, when the droplet water content reaches a critical value, a dry crust is formed at the droplet surface and the drying rate rapidly decreases with the drying front progression and becomes dependent on the water diffusion rate through this crust. Drying is theoretically finished when the particle temperature becomes equal to that of the air. These three steps have different durations depending on both product nature and air inlet temperature. In fact, if the air inlet temperature is high, the dry crust is rapidly formed because of the high water evaporation rate. Due to the large surface to volume ratio of the atomized droplets, drying of formed droplets in the hot atmosphere is a very rapid process. In the first stage, the hot gas causes an increase of the droplet temperature, which promotes liquid evaporation from the droplet surface and a corresponding shrinkage of the droplet. The rapid migration of the water to the droplet surface maintains a constant evaporation rate. At some point the suspended particles form a continuous network. Originally the use of spray drying was advocated as a rapid method for drying compounds which showed unacceptable levels of degradation when dried in the slower classic drying processes. Usually, drying times are of the order of 5–100 s.

However, in a well designed system 15-30 s is a fair time for the passage of the sprayed particle through the drying zone. The general spray drying process mechanism description proposed by past reseachers is still valid. In fact, at the moment of the mixing of the air with the atomized liquid, the drying takes place almost instantaneously and an intensive evaporation is taking place at the surface of each droplet. The evaporation is so rapid that the droplet remains cool until the dry state is reached.This is due to the absorption of heat in vaporizing the liquid. After the evaporation has ceased, the temperature of the particle rises to the general temperature of the drying chamber.

Dry Product-humid Air Separation

This separation is often done through a cyclone placed outside the dryer which reduces product losses in the atmosphere. Most dense particles are recovered at the base of the drying chamber while the finest ones pass through the cyclone to be separated from the humid air. In addition to cyclones, spray dryers are commonly equipped with (i) filters, called "bag houses" that are used to remove the finest powder, and (ii) chemical scrubbers to remove the remaining powder or any volatile pollutants (e.g. flavorings). The obtained powder is made of particles which originate from spherical drops after shrinking. Depending on the composition, the water and gas content of the drop, these particles can be compact or hollow.

The use of multi stage spray dryers makes it possible to increase the residence time and to decrease the drying temperature thus limiting thermal denaturation and improving thermal effectiveness (kJ needed per kg of dry product) [2]. Moreover, integration of a fluidized bed on the drier outlet side makes it possible to better control particle size and to manufacture powders with very low water contents [3].

Functional Ingredients Encapsulated by Spray Drying

Lactic acid starter cultures

The preservation of lactic acid cultures by spray drying has been widely studied as an alternate industrial process for the preservation of lactic acid starter cultures because of the high costs and energy consumption of freezing and freeze drying [4]. Spray drying is considered a good long term preservation method for lactic acid and probiotic cultures [1]. The spray drying of microorganisms dates back to 1914 to the study of Rogers on dried lactic acid cultures. The speed of drying and continuous production capability is very useful for drying large amounts of starter cultures. Since then, much research has been reported on the spray drying of bacteria without loss of cell activity in order to overcome the difficulties involved in handling and maintaining liquid stock cultures [5, 6].

The low production cost of spray drying makes it more energy efficient compared with freeze drying. Nevertheless, compared with other drying methods (e.g., freeze drying), spray drying of microbial cultures has been less developed commercially. The reasons for this are mainly low survival rates during drying of the cultures, low stability under storage and the difficulty in rehydrating the product [7, 8]. However, spray drying is an economical process for production of industrial scale quantities of viable microorganisms. Its application to generate preparations of lactic acid bacteria has recently received considerable interest. Therefore, in the present article we provide a survey of the state of knowledge of starter culture production at high levels of viability using spray drying. The spray drying parameters that affect the survival of lactic acid starter cultures are discussed.

Spray drying is regarded as a promising process for the preservation of lactic acid starter cultures. This process can replace the usual liquid bulk starter in the production of fermented food products. In general, spray dried cultures have relatively low viability and a long lag phase before onset of growth. Therefore, the preservation of lactic acid starter cultures during the drying process and improvement in the viability of dried cultures is important for fermented food industries and starter culture producers. Various parameters and conditions can influence the viability of the cultures from growing to storage. The viability of dried cultures depends on the intrinsic tolerance of the strains and the processing conditions. Thus, process conditions should be controlled and proper protectants should be added to obtain high viability during drying and storage. Conditions such as harvesting lactic acid cultures at the early stationary phase, slow rehydration (soaking) of dried cultures at 50 °C, using appropriate packaging materials and conditions will ensure high viability of lactic acid starter cultures. The bacterial cultures should be dried below 5% and stored at low temperatures. Dried cultures must be protected from heat, oxygen, light and moisture and so, proper packaging materials must be selected accordingly. In the case of anaerobic probiotics, vacuum sealed aluminium coated materials have been suggested.

Spray drying of lactic acid starter cultures presents a number of challenges. The challenges are to select appropriate protective substances and carrier media and also to optimize drying technology. As mentioned before, a major challenge is to improve the viability of lactic acid bacteria upon drying and subsequent storage. There are no commercial lactic acid cultures available that are stable at high temperatures. Therefore, discovering new strains of naturally thermo stable lactic acid bacteria or those have been genetically modified is important. Developing novel encapsulation systems may also improve thermal stability of lactic acid bacteria. Another important challenge is to maintain starter culture properties following spray drying and to evaluate the suitability of dried cultures in functional food developments. Experimental conditions recently optimized for the encapsulation of different food ingredients by spray drying are presented in Table 9.1.

Lipids and oleoresins

In general, lipids are difficult to disperse in food products, in addition and especially polyunsaturated fatty acids are susceptible to auto oxidation, which results in off-flavors and toxic compounds. Lipids can be used as solvents in which one can solubilize hydrophobic substances such as volatile aromatic compounds. Five advantages of lipid encapsulation are (i) retarding autooxidation, (ii) enhancing stability, (iii) controlling lipid soluble flavor release, (iv) masking bitter taste of lipid soluble substances, and (v) protecting dissolved substances against enzyme hydrolysis [9]. Spray drying is quite suitable in the encapsulation of oils and oleoresins. Microencapsulation of cardamom oleoresin by spray-drying using gum arabic, maltodextrin, and modified starch as wall materials has been reported [10]. This study showed that the stability of cardamom oleoresin decreased as the quantity of gum arabic decreased in its blends with maltodextrin and modified starch, and the blend (gum arabic: maltodextrin: modified starch) of (4/6:1/6:1/6) was considered as suitable. This ternary blend proved to be efficient for the encapsulation of cumin oleoresin [11]. Stability of black pepper oleoresin encapsulated in gum arabic and emulsifying modified starches by spray-drying has been reported [12]. Sodium caseinate/carbohydrate blends were used to encapsulate soya oil by spray drying and the obtained results showed that microencapsulation efficiency can be improved by increasing the dextrose equivalence of the carbohydrates [13].

Table 9.1: Experimental conditions recently optimized for the encapsulation of different food ingredients by spray drying

Encapsulated ingredient	Wall material	Feed temp (°C)	Air inlet (°C)	Air outlet (°C)
Anhydrous milk fat	Whey proteins/lactose	50	160	80
Ethyl butyrate ethyl caprylate	Whey proteins/lactose	5	160	80
Oregano, citronella and marjoram flavors	Whey proteins/milk proteins	NR	185-195	85-95
Soya oil	Sodium caseinate/carbohydrates	NR	180	95
Lycopene	Gelatin/sucrose	55	190	52
Fish oil	Starch derivatives/glucose syrup	NR	170	70
Cardamom essential oil	Mesquite gum	RT	195–205	105–115
Arachidonyl L-ascorbate	Maltodextrin/gum arabic/soybean polysaccharides	NR	200	100-110
Cardamom oleoresin	Gum arabic/modified starch/maltodextrin	NR	176–180	115–125
Bixin	Gum arabic/maltodextrin/sucrose	Ambient	180	130
D-Limonene	Gum arabic/maltodextrin/modified starch	NR	200	100–120
Black pepper oleoresin	Gum arabic/modified starch	NR	176–180	105–115
Cumin oleoresin	Gum arabic/maltodextrin/modified starch	NR	158–162	115–125
Fish oil	Sugar beet pectin/glucose syrup	NR	170	70
Short chain fatty acid	Maltodextrin/gum arabic	NR	180	80

NR: Not recommended

Study of spray dried powders with 50% butteroil encapsulated in sucrose and double encapsulated in a matrix of vegetable waxes shows that double encapsulation method could improve capsule resistance to moisture sorption, but also decreases powder flow properties [14, 15]. In a previous study, a research group suggested that encapsulated spray dried powders containing 40% butteroil had shown lower flow characteristics and were highly cohesive. These results had been explained by the fat content and the propensity of bridging, but flow properties were enhanced by adding a commercial anti-caking agent. The oxidative and thermal stabilities of crude squid oil have been effectively enhanced by microencapsulation using spray drying technology. Used wall materials were gelatin, sodium caseinate, and maltodextrin.

The addition of lecithin and carboxymethyl cellulose was reported to improve encapsulating effectiveness and oxidative and thermal stabilities. In general, microencapsulation of polyunsaturated acids or their acylglycerols limits or delays its oxidation. In fact, the esterification of polyunsaturated fatty acids with L-ascorbic acid and subsequent microencapsulation of the ester would be a useful technology for suppressing or retarding the oxidation of the polyunsaturated fatty acid. The 6-O-arachidonoyl-L-ascorbate was microencapsulated in various wall materials by spray drying and the oxidation of arachidonic acid under 37 °C and 12 % relative humidity was well retarded by its microencapsulation in gum arabic or a soluble soybean polysaccharide [16]. It was reported that saturated acyl-L-ascorbate can be used to stabilize (against oxidation) linoleic acid encapsulated using whey protein concentrate, maltodextrin or gum arabic by spray drying. Fat properties, such as crystal forms and habit, in spray dried microcapsules were also investigated as a function of storage time.

Vitamins

Vitamins are important micro-nutritional substances which are involved in many biochemical functions in the body. Vitamins cannot be synthesized by the body; they have to be supplied through diet. Insufficient supply of vitamins can lead to many deficiency diseases like scurvy, pellagra, ariboflavinosis, dermatitis, enteritis, etc. Use of multivitamin supplements has indicated a reduction in cases of certain diet related disorders. Vitamins are completely or partially denatured or damaged during the cooking and processing of foods. Vitamin C, folate and vitamin B_6 seem to be less stable during high temperature processing as compared to retinol, thiamine, riboflavin and niacin. Loss of vitamins also occurs during chilling, heating, reheating and storage. The losses of micronutrients are mainly due to excess use of heat, moisture changes, excessive trimming, cutting, chopping, slicing, washing, soaking, etc. Hence, it is important to take precautionary measures to preserve the vitamins during the subsequent processing steps in the preparation of quality foods.

Micro and nano encapsulation are best to preserve vitamins. There are many different methods available for the effective encapsulation of compounds such as spray drying, extrusion, pan coating, etc. However, since vitamins are thermolabile, they may get damaged during the drying and encapsulation processes. The efficacy of three different processes was studied and evaluated in terms of retention of β-carotene (pro-vitamin A). Spray drying, drum drying and freeze drying were considered for the experiment [17]. Pure β-carotene was encapsulated using maltodextrin (25 dextrose equivalent).

The inlet temperature of the spray dryer was set to 150 °C and outlet temperature was maintained at 100 °C. Encapsulated β-carotene stabilization was measured at different temperature and relative humidity (RH) for each drying process. During storage, RH was found to have no effect on encapsulated β-carotene. The least degradation of β-carotene was observed in freeze drying (8%), followed by spray drying (11%) and drum drying (14%).

Vitamin C not only acts as a dietary supplement but also possesses valuable antioxidant properties. It has been accepted and recognized by the US FDA as one of the four major dietary antioxidants along with vitamin E, vitamin A antecedent β-carotene and selenium. Lack of a sufficient supply of vitamin C leads to diseases like scurvy, while an adequate intake of vitamin C can be beneficial against many chronic diseases.

Ascorbic acid losses begin with harvesting and continue through storage. Vitamin C encapsulation was investigated in past with the assistance of methacrylate copolymers in drugs. The results showed that methacrylate is effective in encapsulating vitamin C [18]. The encapsulation efficiency was around 98% to 100% with good morphology and size distribution, but it showed poor controlled release of vitamin C. A similar research was conducted with different encapsulating materials for oral intake. They used chitosan as the encapsulating material [9] with the cross-linking agent tripolyphosphate for spray drying encapsulation of vitamin C. Inlet temperature was set to 175 °C, and flow rate was maintained at 3 mL/min. The final product was evaluated in terms of loading efficiency, surface morphology, particle size and distribution. The encapsulating efficiency was found to be around 45% to 55% with particle size of 6.1–9 μm. Ascorbic acid encapsulation process parameters were analysed with different methods like thermal phase separation, melt dispersion, solvent evaporation and spray drying [19]. The studied spray drying parameters were inlet and outlet temperatures kept at 200–300 °C and 70–95 °C, respectively. The loss of ascorbic acid during encapsulation was found to be minimum in spray drying (less than 2%) as compared to the other methods tested; however, the encapsulation percentage was less than 50%. Use of starch and β-cyclodextrinas encapsulating materials in spray drying delayed the degradation of ascorbic acid during storage and showed improved results over unencapsulated ascorbic acid. No colour change was observed in encapsulated ascorbic acid even after exposure to air for 1 month [19].

Minerals

Iron, iodine, potassium, sodium, phosphorus, magnesium and calcium are the main dietary minerals. The deficiency of minerals in diet can lead to many disorders and diseases like anaemia, hypokalemia, hypochloremia, hyponatremia, etc. The best method to overcome these mineral deficiencies is to fortify our food with a stable supply of these essential minerals. Mineral-fortified foods can help ensure adequate nutrition to consumers. The bioavailability of these mineral substances mainly depends upon the absorption mechanisms, reactivity with other substances and proper release of materials. Microencapsulation should ensure the reduction of mineral substance reaction with other ingredients and appropriate release and dietary uptake of necessary minerals [20].

Calcium micro particles were successfully encapsulated by spray drying with the help of derivatives of cellulose and neutral polymethacrylate. The spray dryer was set at inlet air temperature 120–128°C, outlet air temperature 91–95°C and a flow rate of 3 mL/min. The encapsulated final product was evaluated for its calcium content, size distribution, and morphology and release profiles. The results showed that the final product properties vary with the polymer concentration and its type. Microencapsulation efficiency of sodium caseinate with different dextrose equivalent carbohydrates was investigated by Hogan et al. (2001). The results showed that the encapsulation efficiency increased with an increase in dextrose equivalent but decreased with the higher core to encapsulating materials ratio [13].

Antioxidants' colors

Food drying processes usually lead to colour deterioration. This decolouration of foods is principally due to pigment degradation, Maillard reactions, oxidation, etc. These adverse reactions depend upon drying parameters such as temperature, air velocity, etc [21]. It was found that blanching of apples increased the retention of colour, total polyphenolic content and total flavonoid content when compared with fresh unblanched pomace. Temperatures 50 °C, 60 °C, 70 °C and 80 °C were used as drying temperatures in cabinet drying with air velocity of 3 m/s. However, drying process significantly reduced the polyphenol and flavonoid content with increasing process temperature, which leads to the reduction of total antioxidant activity. Lycopene is responsible for the red/orange colour familiar to tomatoes and has many established health benefits. Lycopene intake can help in the prevention of cardiovascular diseases and certain cancers.

Lycopene is a naturally occurring carotenoid, which has the highest physical quench rate constant with singlet oxygen, and it has a higher plasma level as compared to β-carotene. Lycopene was encapsulated using gelatin and sucrose,and the results showed that encapsulation yield and encapsulation efficiency were affected by the gelatin to sucrose ratio, homogenization pressure, spray drying inlet temperature, feed temperature, lycopene to encapsulating material ratio and lycopene purity. The optimum results were found at the feed temperature of 55 °C, air inlet temperature of 190 °C, homogenization pressure of 40 MPa, a gelatin to sucrose ratio of 3:7 and a lycopene to encapsulating material ratio of 1:4 with a lycopene purity of not less than 52% [22].

Anthocyanins are known as natural antioxidants and belong to the flavonoid family, and are widely dispersed in flowers, fruits and vegetables. Anthocyanin pigments from black carrot (*Daucus carota L.*) juice were microencapsulated by spray drying. Maltodextrins with different dextrose equivalents were used in the experiment as encapsulating materials. The tested inlet temperatures were 200 °C, 180 °C and 160 °C, and the outlet temperatures were 131 °C, 118 °C and 107 °C. Maltodextrin with dextrose equivalent of 20–21 retained higher anthocyanin content, while higher temperatures triggered the anthocyanin loss. During storage of encapsulated anthocyanin at 25 °C and 4 °C with maltodextrin (DE 20-21), the higher storage temperature (25 °C) led to higher anthocyanin loss (from 600 mg/100g to 400 mg/100g approximately) as compared to the lower temperature (4 °C) with minimal loss (from 600 mg/ 100 g to 500 mg/100 g approximately [23].

Watermelon juice with 3% and 5% maltodextrin was spray dried at different inlet temperatures of 145 °C, 155 °C, 165 °C and 175 °C in a study conducted in one of the earlier work [24]. The maltodextrin used in the experiment had 8–11 dextrose equivalents. Lycopene and β-carotene contents of the spray dried powder were determined and compared with raw fruits' lycopene and β-carotene contents. The results showed that the increase in inlet temperature decreased the lycopene and β-carotene contents. β-carotene was found to be more sensitive to the temperature than lycopene. The β-carotene loss was around 27% as compared to a 24% loss for lycopene with same inlet temperature [24].

The encapsulation of bixin, a pro-carotenoid found in annatto (*Bixa orellana* L.) was studied using spray drying with the help of different encapsulating materials such as gum arabic and maltodextrin [25]. The air inlet and outlet temperatures were maintained at 180 °C and 130 °C with an inlet pressure of 49.03 N/cm^2. The results indicated that bixin encapsulation with gum arabic can be three to four times more stable as compared to maltodextrins. Still today, very limited research has been done to use spray drying to produce natural food colours from food materials such as fruits and vegetables. Extensive research is needed to explore the preservation of natural colourants from agricultural materials using spray drying and combinations of encapsulating matrices.

Flavours

More than 90% of the flavours available in the market are produced or encapsulated by spray drying. Oleoresins which are good antioxidants are the substances responsible for most of the spice flavours. They are very reactive and unstable in light, temperature and oxygen. Their sensitivity can be overcome by effective encapsulation. Gum arabic, maltodextrin and modified starch have been used as encapsulating materials to encapsulate cardamom oleoresins [12]. The inlet and outlet temperatures of the spray dryer were maintained at 178 °C and 120 °C. Similar work was carried out on pepper oleoresins [12]. Gum arabic offered the best encapsulation results as compared to the other encapsulating material tested, namely modified starch. During the spray drying experiment, the inlet and outlet temperatures were maintained at 178±2 °C and 110±5 °C, and flow rate was maintained at 300 mL/h [26]. Encapsulation of menthol using modified starch and gum arabic was studied in the recent work [26]. The spray drying parameters set the inlet temperature at 180 °C, with corresponding outlet temperature of 100 ± 5 °C, feed rate of 45 mL/min and air flow rate of 100 kg/h. The results confirmed that the increase in encapsulating material concentration, i.e. modified starch or gum arabic, increased the retention of menthol. The highest I-menthol retention was found at the mass ratio of I-menthol/encapsulating materials of 1:9. At this ratio, gum arabic showed around 72% I-menthol retention, while modified starch showed a maximum retention of 87%.

Encapsulation of coffee extracts with cashew gum and gum arabic was studied [27]. The cashew gum (*Anacardium occidentale L.*) from Brazil was used in the experiment to study aroma protection, external morphology and size distribution. No significant difference was found between gum arabic and cashew gum, suggesting that cashew gum can be used as an alternative for gum arabic in encapsulation applications.

Sumac berries extract was spray dried at an inlet temperature of 200 °C, outlet temperature of 100 °C and feed rate of 125 mL/min The flavour retention in the spray dried powder was examined using different carriers, namely sodium chloride, sucrose, glucose and starch. Among the carriers studied, sodium chloride was found to be the best carrier for encapsulation of flavour from sumac berries as the other carriers caramelized, making them unsuitable. The carrier to core material, i.e. the sumac flavour extract, was fixed to 4:1 g/g for all samples. The solution was spray dried with different concentrations of sumac berries extract with soluble solids content starting from 5%, 10%, 15%, 20% and 25%. It was found that the solid concentration played an important role in flavour retention, and higher soluble solids content produced higher flavour retention [28].

D-Limonene was encapsulated by spray drying with gum arabic, maltodextrin and modified starch as encapsulating materials. A spray drying inlet temperature of 200 °C, outlet temperature of 110±10 °C and feed flow rate of 45 mL/min were maintained during the experiment. The d-limonene retention in the encapsulated powder was affected by emulsion size and powder size. Larger powder size showed higher stability and best retention of flavour [29]. Citrus peel oil is an important flavouring substance in many industries which include food, pharmaceutical, chemical and cosmetic manufacturing. This oil comprises highly volatile and nonvolatile components like terpene, oxygenated hydrocarbons, pigments and waxes. The oxygenated parts which include alcohols, aldehydes and ketones contribute more towards flavouring as compared to the terpenes.

The encapsulation of essential oil of oregano (*Origanum vulgare* L.), aroma extracts of citronella (*Cymbopogon nardus* G.) and sweet marjoram (*Majorana hortensis* L.) was studied with different encapsulating materials, namely, whey protein concentrate and skimmed milk powder [30]. The spray dryer was operated with an inlet air temperature of 190±5 °C and an outlet air temperature of 90±5 °C. Skimmed milk powder showed better results as compared to whey protein concentrate in terms of encapsulation efficiency and total oil recovery. The encapsulation efficiencies with skim milk powder of oregano essential oil, citronella aroma extract and marjoram aroma extract were 80.2%, 69.4% and 67.9% respectively as compared to whey protein concentrate with encapsulation efficiencies of 71.8%, 65.8% and 54.3%. The total oil content in the encapsulating material found in skim milk powder were 81.3%, 80.6% and 84.7%, while in whey protein concentrate, the oil contents were 73.2%, 81.0% and 76.4% [30].

Applications of spray drying in the production of flavours are well developed as they have wide applications in the food industry. Fewer applications have been studied for nutraceutical ingredients. However, with the increasing demand from consumers for healthier processed foods, novel spray drying applications are being studied and developed in this particular sector.

Essential fatty acids

Essential fatty acids like ω–6 fatty acids and ω–3 fatty acids are important for normal growth and development. The proper and adequate nutritional intake of fatty acids prevents many diseases such as coronary artery disease, hypertension, diabetes, arthritis, etc. To ensure availability of the right fatty acids, developments are taking place for their use as food ingredients for food fortification. Avocados are a rich source of many phytochemicals and fatty acids. They are rich in monosaturated fatty acids, vitamin E, lutein, glutathione, β-sitosterol, folate, potassium, magnesium and fibre. The consumption of these nutrients may help in the prevention of cardiovascular disease and cancer. Avocado oil was encapsulated with whey protein isolate and maltodextrins (dextrose equivalent of 5). The whey protein to maltodextrin ratio of 1:9 gave higher microencapsulation efficiency of around 66.18±3.82%. Since avocado oil is very stable in ambient and refrigerated conditions, no significant difference was found in terms of oil stability due to the encapsulation [31]. Fish oil is rich in many nutrients, especially ω–3 fatty acids. Consumption of fish oil has proven to help reduce several risk factors of cardiovascular disease and to lower blood pressure.

The use of alginate as encapsulating material increased the oil-holding capacity and decreased the oxidation of fish oil as compared to un-encapsulated fish oil [31]. Similar work was carried out on fish oil which had 33% ω–3 fatty acids with different encapsulating materials. Sugar beet pectin and glucose syrup with dextrose equivalent of 38 were used as encapsulating materials in this experiment. The sugar beet pectin used in the experiment was found to have acetylation of 18% and esterification of 58% with 5% protein content. The results indicated that 1–2% pectin content is sufficient to produce a stable emulsion during spray drying. Sugar beet pectin is recommended as a replacement for gum arabic in the encapsulation of lipophilic materials [32]. Short-chain fatty acids were successfully encapsulated using gum arabic and maltodextrin. The inlet and outlet temperatures used in the spray drying process were 180 °C and 90 °C. Five per cent gum arabic alone, 10% maltodextrin alone and combinations of 5% gum arabic and 5% maltodextrin were tested. The results showed that the size distribution and structure of the encapsulated substance were affected by the type and concentration of encapsulating materials used.

Summary

Spray drying is an important processing technology used in many functional food and nutraceutical applications. Spray drying is used to produce a large number of dried and encapsulated ingredients such as vitamins, minerals, flavours, etc. However, the contribution of spray drying to the production of antioxidant rich food products or incorporation of antioxidants in foods by encapsulation is still minimal as compared to other technologies, principally based on the important initial investment cost of the technology. Further limitation is explained by the high inlet temperature of spray drying which is not suitable for thermosensitive materials. However, the optimization of spray drying process, at lower temperature, for preserving antioxidant substances such as lycopene, anthocyanins and β-carotene has been possible. Further application development could strongly establish spray drying as an effective method to preserve functional and nutraceutical properties of particular food materials.

More research is required to explore the possibility to preserve or microencapsulate many different sources of antioxidants such as polyphenols, flavonoids, etc., playing with process parameters such as inlet and outlet temperatures, air flow and feed rates, nozzle design and complex encapsulating materials. The research should be done for effective preservation of functional components and exploration of efficacy of various encapsulating materials other than the normally used encapsulating materials. This approach would avoid the bottlenecks in availability of certain encapsulating materials such as gum acacia, and there would be more flexibility to choose different encapsulating materials for specific purposes.

References

1. Peighambardoust, S., A. Golshan Tafti, and J. Hesari, *Application of spray drying for preservation of lactic acid starter cultures: a review.* Trends in Food Science & Technology, 2011. **22**(5): 215-224.
2. Patel, R. and A. Suthar, *Spray drying technology: an overview.* Indian Journal of Science and Technology, 2009. **2**(10): 44-47.
3. Baker, C. and K. McKenzie, *Energy consumption of industrial spray dryers.* Drying Technology, 2005. **23**(1-2): 365-386.
4. Santivarangkna, C., Kulozik U. and P. Foerst, *Alternative drying processes for the industrial preservation of lactic acid starter cultures.* Biotechnology Progress, 2007. **23**(2): 302-315.
5. Kim, S.S. and S.R. Bhowmik, *Survival of lactic acid bacteria during spray drying of plain yogurt.* Journal of Food Science, 1990. **55**(4): 1008-1010.
6. Silva, J., *et al.*, *Bacteriocin production by spray dried lactic acid bacteria.* Letters in Applied Microbiology, 2002. **34**(2): 77-81.
7. Brian C. S. and M.R. Etzel, *Spray drying, freeze drying, or freezing of three different lactic acid bacteria species.* Journal of Food Science, 1997. **62**(3): 576-578.

8. Teixeira, P., H. Castro, and R. Kirby, *Spray drying as a method for preparing concentrated cultures of Lactobacillus bulgaricus.* Journal of Applied Bacteriology, 1995. **78**(4): 456-462.
9. Matsuno, R. and S. Adachi, *Lipid encapsulation technology-techniques and applications to food.* Trends in Food Science & Technology, 1993. **4**(8): 256-261.
10. Krishnan, S., R. Bhosale, and R.S. Singhal, *Microencapsulation of cardamom oleoresin: Evaluation of blends of gum arabic, maltodextrin and a modified starch as wall materials.* Carbohydrate Polymers, 2005. **61**(1): 95-102.
11. Kanakdande, D., R. Bhosale, and R.S. Singhal, *Stability of cumin oleoresin microencapsulated in different combination of gum arabic, maltodextrin and modified starch.* Carbohydrate Polymers, 2007. **67**(4): 536-541.
12. Shaikh, J., R. Bhosale, and R. Singhal, *Microencapsulation of black pepper oleoresin.* Food Chemistry, 2006. **94**(1): 105-110.
13. Hogan, S.A., McNamee B.F, Dolores E, Michael O.R, Sullivana O, *Emulsification and microencapsulation properties of sodium caseinate/carbohydrate blends.* International Dairy Journal, 2001. **11**(3): 137-144.
14. Gharsallaoui, A., Roudaut, G. Chambin O, Voilley A, Saurel R., *Applications of spray-drying in microencapsulation of food ingredients: An overview.* Food Research International, 2007. **40**(9): 1107-1121.
15. Onwulata, C., R. Konstance and V. Holsinger, *Properties of Single and Double Encapsulated Butteroil Powders.* Journal of Food Science, 1998. **63**(1): 100-103.
16. Watanabe, Y., *et al.*, *Oxidation of 6,arachidonoyl l-ascorbate microencapsulated with a polysaccharide by spray-drying.* LWT-Food Science and Technology, 2004. **37**(4): 395-400.
17. Desobry, S.A., F.M. Netto, and T.P. Labuza, *Comparison of Spray drying, Drum drying and Freeze drying for â Carotene Encapsulation and Preservation.* Journal of Food Science, 1997. **62**(6): 1158-1162.
18. Esposito, E., *et al.*, *Spray dried Eudragit microparticles as encapsulation devices for vitamin C.* International Journal of Pharmaceutics, 2002. **242**(1): 329-334.
19. Uddin, M., M. Hawlader, and H. Zhu, *Microencapsulation of ascorbic acid: effect of process variables on product characteristics.* Journal of Microencapsulation, 2001. **18**(2): 199-209.
20. Oneda, F. and M. Ré, *The effect of formulation variables on the dissolution and physical properties of spray-dried microspheres containing organic salts.* Powder Technology, 2003. **130**(1): 377-384.
21. Heras-Ramírez, M.E., *et al.*, *Effect of blanching and drying temperature on polyphenolic compound stability and antioxidant capacity of apple pomace.* Food and Bioprocess Technology, 2012. **5**(6): 2201-2210.
22. Shu, B., *et al.*, *Study on microencapsulation of lycopene by spray-drying.* Journal of Food Engineering, 2006. **76**(4): 664-669.
23. Ersus, S. and U. Yurdagel, *Microencapsulation of anthocyanin pigments of black carrot (Daucus carota L.) by spray drier.* Journal of Food Engineering, 2007. **80**(3): 805-812.
24. Quek, S.Y., N.K. Chok, and P. Swedlund, *The physicochemical properties of spray-dried watermelon powders.* Chemical Engineering and Processing: Process Intensification, 2007. **46**(5): 386-392.
25. Barbosa, M., C. Borsarelli, and A. Mercadante, *Light stability of spray-dried bixin encapsulated with different edible polysaccharide preparations.* Food Research International, 2005. **38**(8): 989-994.
26. Soottitantawat, A., *et al.*, *Microencapsulation of l-menthol by spray drying and its release characteristics.* Innovative Food Science & Emerging Technologies, 2005. **6**(2):163-170.

27. Rodrigues, R. and C. Grosso, *Cashew gum microencapsulation protects the aroma of coffee extracts.* Journal of Microencapsulation, 2008. **25**(1): 13-20.
28. Bayram, Ö.A., M. Bayram, and A.R. Tekin, *Spray drying of sumac flavour using sodium chloride, sucrose, glucose and starch as carriers.* Journal of Food Engineering, 2005. **69**(2): 253-260.
29. Soottitantawat, A., *et al.*, *Influence of emulsion and powder size on the stability of encapsulated D-limonene by spray drying.* Innovative Food Science & Emerging Technologies, 2005. **6**(1): 107-114.
30. Baranauskienė, R., *et al.*, *Properties of oregano (Origanum vulgare L.), citronella (Cymbopogon nardus G.) and marjoram (Majorana hortensis L.) flavors encapsulated into milk protein-based matrices.* Food Research International, 2006. **39**(4): 413-425.
31. Bae, E. and S. Lee, *Microencapsulation of avocado oil by spray drying using whey protein and maltodextrin.* Journal of Microencapsulation, 2008. **25**(8): 549-560.
32. Drusch, S., *et al.*, *Impact of physicochemical characteristics on the oxidative stability of fish oil microencapsulated by spray-drying.* Journal of Agricultural and Food Chemistry, 2007. **55**(26): 11044-11051.

10

Membrane Technology

Introduction

Since 1990, there has been a lot of interest in functional foods and nutraceuticals, particularly because of their impact effects on human health and the prevention of certain diseases. Apart from growing good quality natural products, there are challenges in extracting, refining, analyzing active ingredients, and evaluating safety before an effective, reliable product is made [1,2]. Special attention needs to be focused on separation and purification mainly because 40-80% of the cost of production is assigned to separation and purification unit operations in making functional foods and nutraceuticals. There are also several carotenoids and polyphenol products that have been identified from various plants and a single plant could contain highly complex profiles of these compounds. These are labile to heat, air, and light, and they may exist at low concentrations in the plants. This makes the separation and detection of nutraceuticals a challenging task. The membrane techniques used for the separation of nutraceuticals from the natural products, as well as their applications and influencing factors are discussed in this chapter [3, 4].

Process Description

In the past two decades, a number of separation processes have emerged in which membranes play a major role in the isolation and enrichment of functional compounds by concentrating, fractionating, and purifying solutions or colloidal food applications of membrane technology separations. Membrane separations have led to the development of innovative products with unique functional, nutritional, and health attributes. Membrane processing permits concentration and separation without the use of heat. Particles are separated based on molecular size, shape, and/or charge via pressure and specially designed semipermeable membranes. The most commonly used membrane processes in the food industry are pressure driven. Depending on the size of the particles to be separated, membrane filtration can be classified as microfiltration (MF), ultrafiltration (UF), nanofiltration (NF), and reverse osmosis (RO). Membrane separations operate at relatively low temperatures and pressures, resulting in minimal denaturation and degradation of temperature sensitive ingredients. Compared with thermal processes (e.g., evaporation), membrane filtration is an energy efficient method of clarifying, concentrating, and purifying food processing streams. Membrane filtration improves product quality by retaining flavors and vitamins and minimizing protein denaturation. Different memberane processes and their filtration spectrum are summarised in Figure 10.1. Driving force of different membrane processes their biological applications are presented in Table 10.1 and table 10.2.

Table 10.1: Classification of membrane technology by driving force

Driving force			
Pressure	Concentration	Temperature	Electrical potential
Gas and vapor separation Nanofiltration Ultrafiltration Microfiltration Pervaporation	Membrane extraction Dialysis	Membrane distillation	Electrodialysis

Membrane Selection

Membrane is the most important part of the separation process. The challenge for the process engineer is to develop a process giving the highest yield of the desired product with minimum investment and operational cost. Simple guidelines for implementing membrane filtration into new operational processes are outlined as follows.

- The characteristics of the particles and liquids (e.g., dissolved solid content, molecular weight of dissolved species, nature and loading of any suspended material) and the process conditions (e.g., temperature, pressure, and capacity requirements) should be known.
- Proper membrane pore size (MF, NF, UF, or RO) should be selected.
- Chemical compatibility between feed liquid, membrane, and potential membrane fouling must be ensured.
- Design criteria viz. capacity of the plant, operating costs, nature and desired composition of permeate and retentate should be identified.

- Pilot-scale tests using membrane material, pore size, configuration, and intended use on the production unit should be performed. Testing should establish the operating parameters (e.g., batch or continuous operation, permeation rate, pressure drop, and retention levels as concentration increases), process reproducibility, quality of end product, degree of membrane fouling, effectiveness of cleaning regimen, and estimating membrane life under repeated cycles of usage and cleaning.

Cellulosic membranes are most commonly used for bioprocessing applications because of their low fouling characteristics and low protein adsorption. Synthetic polymers such as polysulfone and polyvinylidene fluorides are widely used because of their greater chemical and mechanical stability.

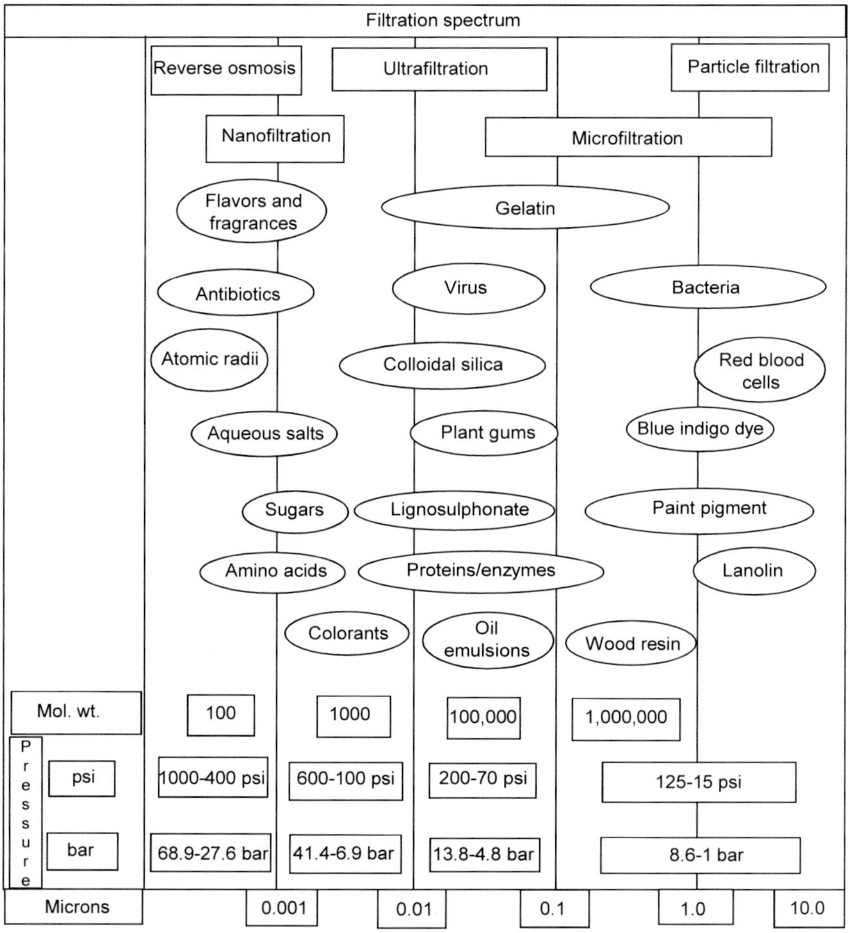

Fig. 10.1: Memberane processes and their filtration spectrum

Ultrafiltration

UF is used in a variety of industries mainly because of its low energy requirements, nonthermal nature, and simplicity. It covers the region between MF and RO to remove particles in the size range of 0.001–0.02 μm [5-7]. Solvents and salts of low molecular weight will pass through, whereas large molecules are retained. UF requires applied pressure of 1–7 bar to overcome viscous resistance of liquid permeation through the membrane. UF is ideal for the removal of antinutritional factors such as oligosaccharides, phytic acid, and some trypsin inhibitors from vegetable protein processing to produce purified protein isolates or concentrates with superior functional properties. Many commercial UF membranes are designed for cross-flow filtration. In cross-flow filtration, the liquid to be filtered flows at high speed along the membrane. The pressure on the membrane is the driving force for the passage of the liquid through the membrane. The high flow prevents the depositing of solids (fouling) and therefore the blockage of the membrane. The concentrated solid matter is called retentate, and the filtrate is called permeate.

Cross-flow membrane filtration technology is gaining acceptance as an important manufacturing step in many industries, from food and dairy to pharmaceutical and biotechnology sectors. Its ability to produce specific separations at low temperatures with no phase change makes the technology a cost-effective solution to conventional methods. In cross-flow filtration the feed flows parallel to the membrane surface, and with an appropriate driving force the permeating species passes through the membrane. This permeate is then collected as a "second" phase. The feed, gradually reduced in concentration of the permeating species, exits the unit as the retentate. UF technology was tested to be most promising for bioactive peptide separation. The common application of UF in the field of bioactive peptides is the peptide enrichment from protein hydrolysates. Adding UF membranes in the production process of bioactive peptides by the use of enzymatic membrane reactors allows the continuous production of specific peptide sequences with functional and nutritional properties.

Several research reports have described the continuous extraction of bioactive peptides, especially from milk proteins. Apart from their uses in the continuous production of bioactive peptides, UF membranes have also been used to fractionate proteins, peptides, and amino acids contained in hydrolysates. More recent studies reported the enrichment of antioxidant peptides by UF from an alfalfa leaf protein hydrolysate with a 3-kDa molecular weight cutoff (MWCO) polysulfone membrane and wheat gluten hydrolysates with UF membranes (10, 5, and 3 kDa). Other studies have also reported the feasibility of fractionation by UF to produce fractions exhibiting angiotensin converting enzyme inhibitory and antioxidant activities from potato protein hydrolysates, and UF also was used in the purification of peptides that had antimicrobial activity against several pathogens, such as *Candida albicans*, which is the most common cause of oral, esophageal, vaginal, and urinary candidiasis.

Application of ultrafiltration in functional food

Protein processing

One of the earliest applications of UF is the recovery and concentration of protein from cheese whey. There is considerable commercial interest in the preparation of whey proteins for nutraceutical applications. Recent developments in membrane filtration have provided exciting new opportunities for large scale protein fractionation. With UF, various types of amino acids can be processed more uniformly despite the strong characteristic differences of each kind of broth. UF in the field of bioactive peptides enriches peptides from protein hydrolysates. Furthermore, adding UF membranes in the production process of bioactive peptides using enzymatic membrane reactors allows production of specific peptide sequences with functional and nutritional properties. UF has become the primary means for protein concentration and buffer exchange in the production of therapeutic proteins and industrial enzymes (Table 10.2).

Table 10.2: Driving force of different membrane process along with biological activity

Driving force	Process	Protein hydrolysate	Biological activity
Pressure-driven membrane processes	Ultrafiltration	Alfalfa white protein	ACE inhibitor
		Alfalfa leaf protein	Antioxidant
		Wheat gluten	ACE inhibitor
		Soybean protein	ACE inhibitor
		Soybean β-conglycinin	ACE inhibitor
		Soybean glycinin	ACE inhibitor
		Sea cucumber gelatin	ACE inhibitor
		Potato	Antioxidant ACE inhibitor Antimicrobial

Diafiltration

Diafiltration is a specialized type of UF process where the retentate is diluted with water and reultrafiltered to improve the recovery of the material in the permeate or to improve the purity of the retained stream. Typical uses are recovery of antibiotics from the fermentation broth and enhancing the proteins. Recent experiments have demonstrated the production of protein products from partially defatted soybean meal and undefatted lupin meal using UF followed by diafiltration. In one recent experiment, peroxidase and Bowman-Birk inhibitor, valuable products that have potential medicinal uses, were produced from soybean hulls using UF, followed by discontinuous diafiltration, a membrane based method that removes low molecular weight solutes such as salts while continuously replacing the solvent lost with the permeate.

Microfiltration

MF is a low pressure (10-100 psig) process for separating larger size solutes from aqueous solutions using a semipermeable membrane[8]. This process is carried out by having a process solution flow along a membrane surface under pressure. MF removes suspended solids and colloids, bacteria, and viruses in the range of 0.05-10 microns and retains color and taste of compounds, alcohol, and sugars. The MF filter is usually made from a thin polymer film with a uniform pore size and a high pore density of approximately 80%. MF products are designed for cross-flow separation, where a feed stream is introduced into the membrane element under pressure and passed over the membrane surface in a controlled flow path [9]. A portion of the feed passes through the membrane and is called permeate. The rejected materials are flushed away in a stream called the concentrate. Cross-flow membrane filtration uses high cross-flow rates to enhance permeate passage and reduce membrane fouling. MF is used for fermentation, broth and biomass clarification, and recovery.

Application of microfiltration in functional food

From functional and nutritional points of view, MF can produce high purity whey protein isolates (WPIs). Microfiltered WPI tends to be low in fat content, which designates the "nutritional" or "nutraceutical" value. MF is used in the retention of fat and large molecular weight proteins from small whey proteins, lactose, and minerals of low molecular weight [10, 11].

Reverse Osmosis

RO is a moderate to high pressure driven process (80-1200 psig) for separating larger solutes from aqueous solutions by means of a semipermeable membrane [12]. Mechanisms are based on size, shape, ionic charge, and membrane interactions. The membrane pore size can vary from 0.1–1.0 nm [12, 13]. The process is carried out by flowing a process solution along a membrane surface under pressure. Retained solutes such as particulate matter and dissolved salts leave with the flowing process stream and do not accumulate on the membrane surface.This is achieved through cross-flow separation, where a feed stream is introduced into the membrane element under pressure and passed over the membrane surface in a controlled flow path [14]. A portion of the feed passes through the membrane and is called permeate. The rejected materials are flushed away in a stream called the concentrate. Cross-flow membrane filtration uses a high cross-flow rate to enhance permeate passage and reduce membrane fouling.

Nanofiltration

NF is selected when RO and UF are not efficient choices for separation. NF has an intermediate cutoff range of 200–1000 Da, between RO and UF [15, 16]. NF operates at low to moderately high pressure (50-450 psig) with higher flow rates of water. In NF, mechanisms underlying separation selectivity are more complex. The selectivity is regulated by both size exclusion and electrostatic repulsion mechanisms, which are dependent on membrane type and feed composition [17]. NF membrane allows diffusion of certain ionic species, predominantly monovalent ions. Larger ionic species, including divalent and multivalent and more complex molecules are highly retained.

The process is carried out by having a process solution flow along a membrane surface under pressure. Cross-flow membrane filtration uses a high cross-flow rate to enhance permeate passage and reduce membrane fouling. Retained solutes (such as dissolved salts) leave with the flowing process stream and do not accumulate on the membrane surface. It is difficult to predict the performance of NF membranes, especially if more than three solutes are present in the solution because membrane rejection is influenced by the size, structure, and charge of the components in solution. As a result, scaling up and pilot studies are highly recommended for NF applications.

Nanofiltration applications

NF is used in the partial desalination of whey reducing recovery costs of lactose and whey protein solids. It is also used in the desalination of concentrated antibiotics, increasing the value of the pharmaceutical product. The NF is used for color reduction or manipulation of food products and concentration of food, dairy, and beverage products. NF has also been used to fractionate complex protein hydrolysates that contain bioactive peptides. NF membranes has a range of MWCO between 1000 and 5000. These are used to separate peptides present in hydrolysates of whey proteins. More recently a research study used the combination of UF and NF to fractionate peptides contained in sweet whey (Table 10.3).

Table 10.3: Driving force along with biological activity

Driving force	Process	Protein hydrolysate source	Biological activity
Pressure-driven membrane processes	UF & NF	Bovine whey proteins Bovine whey β lactoglobulin	ACE inhibitor, opioid, antimicrobial ACE inhibitor, opioid, antimicrobial

Pressure Driven Processes Assisted by Electric Field

The ability to produce functional food ingredients from natural sources becomes increasingly attractive to the food industry. To improve the efficiency of conventional pressure driven membrane filtration, the application of an external electrical field, which acts as an additional driving force to the pressure gradient, has been carried out [18]. Electrically enhanced membrane filtration (EMF) seems to be a breakthrough technology for the isolation of charged nutraceutical ingredients from natural sources. EMF combines the separation mechanisms of membrane filtration and electrophoresis, and it is believed that EMF is more selective than membrane filtration and less costly than chromatography.

Besides many applications of EMF in functional foods and nutraceuticals, the production of bioactive peptides also finds interesting uses for EMF. Processes combining an electric field as the driving force to porous membranes have been developed for the separation of bioactive peptides.These pressure driven processes assisted by electric field have shown to be promising in the separation and purification of bioactive peptides. Most recently, electrodialysis (ED) with UF membranes has been developed to fractionate peptides using conventional ED cells.

Electrophoretic Membrane Processes

Electrophoresis

Electrophoresis is an electrically driven operation with no pressure applied and is a basic technique used in the separation of particles of colloidal size. Electrophoresis is generally carried out in supporting media like gels and is known for its high resolution [18]. However, when electrophoresis was used for production purpose, poor resolution was obtained compared with analytical scale, and finally it was found that the productivity could not be increased over a particular limit, typically about a few milligrams per hour.

To overcome these limitations, different technologies such as ED and fixed boundary electrophoresis were used. These electro membrane operations have two common characteristics. First, the selectivity comes from different solute flow rates through the membrane. Second, the productivity can be increased by increasing the number of separation chambers without significantly damaging the selectivity. Fixed boundary electrophoresis is another separation technology that has gained importance in the functional food industry, although this technique has never been tested for bioactive peptide separation. However, its operation conditions are limiting factors for large scale bioactive peptide production [18].

Electrodialysis

ED is a membrane separation process in which ionic (electrically charged) species are transported from one solution to another by crossing one or more selective permeable membranes under the influence of an electric potential. When the ED is running, the direct current field affects the flow of charged species in water solutions in such a way that cations moving toward cathode pass through the cation exchange membranes and cannot go through the anion exchange membranes, whereas the anions drawn to the anode pass through the anion exchange membranes but stop at cation exchange membranes.

Electrodialysis with filtration membranes

Recently the new technology ED with UF appears to be a more selective method for the separation of bioactive peptides from raw hydrolysates. ED with filtration membranes (EDFM) couples size exclusion properties of UF membranes with the charge selectivity of ED. EDFM acts by attracting the molecules, according to their charge, through filtration membranes with the proper MWCO range according to the fraction of interest.

The feed solution containing the peptides to be separated flows through a compartment, and the pH of the feed solution induces a charge dependent selective migration of the peptides under the influence of the electric field. At the same time, the filtration membrane performs the separation of the peptides by molecular weight exclusion or passage permission. EDFM has several advantages. It allows the selective separation at a large scale, and in one process we can separate one or more charged molecules of interest. However, the separation of one or more compounds in the same process is limited by the design of ED cells. Because EDFM stands as a reliable and cost-effective separation technique, it constitutes a major breakthrough in the purification and separation of bioactive peptides. The technology has already been tested for the separation of tobacco polyphenols, green tea catechins, proanthocyanidins, chitosan oligomers, and different bioactive.

Summary

It can be seen that the market of functional foods is growing at a rapid pace and membrane processing offers several advantages over conventional methods for separation, fractionation, and recovery of those bioactive components. The challenge for the process engineer is to develop a process giving the highest yield of the desired product with minimum investment and operational costs. The cost of production of functional food and nutraceuticals is mainly attributed to separation and purification stages and thus the process needs to be optimized for maximum productivity. In future, coupling of membrane separation and other technologies like supercritical fluid technologies can also offer combine unique advantages of each process resulting in a novel separation technology offering great potential for the nutraceutical and functional food industry. In summary, the prospect of membrane processing is very bright and it can offer a great advantage in development of functional food and nutraceutical.

References

1. Cuperus, F.P. and H.H. Nijhuis, *Applications of membrane technology to food processing.* Trends in Food Science & Technology, 1993. **4**(9): 277-282.
2. Baker, R.W., *Membrane technology* 2000: Wiley Online Library.
3. de Morais Coutinho, C., *et al., State of art of the application of membrane technology to vegetable oils: A review.* Food Research International, 2009. **42**(5): 536-550.
4. Pabby, A.K., S.S. Rizvi, and A.M.S. Requena, *Handbook of membrane separations: chemical, pharmaceutical, food, and biotechnological applications* 2008: CRC press.
5. Cheryan, M., *Ultrafiltration handbook*1986: Technomic Publishing Co. Inc.
6. Cui, Z. and H. Muralidhara, *Membrane technology: A practical guide to membrane Technology and Applications in Food and Bioprocessing* 2010: Elsevier.
7. Mohammad, A.W., *et al., Ultrafiltration in food processing industry: review on application, membrane fouling, and fouling control.* Food and Bioprocess Technology, 2012. **5**(4): 1143-1156.
8. Van der Horst, H. and J. Hanemaaijer, *Cross-flow microfiltration in the food industry. State of the art.* Desalination, 1990. **77**: 235-258.
9. Makardij, A., X. Chen, and M. Farid, *Microfiltration and ultrafiltration of milk: some aspects of fouling and cleaning.* Food and Bioproducts Processing, 1999. **77**(2): 107-113.
10. De Wit, J., *Nutritional and functional characteristics of whey proteins in food products.* Journal of Dairy Science, 1998. **81**(3): 597-608.
11. Singh, H., *The milk fat globule membrane—A biophysical system for food applications.* Current Opinion in Colloid & Interface Science, 2006. **11**(2): 154-163.
12. Sourirajan, S., *Reverse osmosis*1970: London, UK: Logos Press Ltd.
13. Sourirajan, S., T. Matsuura, and I. Chemist, *Reverse osmosis-ultrafiltration process principles*1985: National Research Council Canada Ottawa.
14. Cath, T.Y., A.E. Childress, and M. Elimelech, *Forward osmosis: principles, applications, and recent developments.* Journal of Membrane Science, 2006. **281**(1): 70-87.
15. Vatai, G., *Nanofiltration application in food technology and environmental protection*, in *Integration of Membrane Processes into Bioconversions* 2000, Springer.155-163.
16. Vincze, I. and G. Vatai, *Application of nanofiltration for coffee extract concentration.* Desalination, 2004. **162**: 287-294.

17. Van der Bruggen, B., M. Mänttäri, and M. Nyström, *Drawbacks of applying nanofiltration and how to avoid them: a review.* Separation and Purification Technology, 2008. **63**(2): 251-263.
18. Bazinet, L. and L. Firdaous, *Membrane processes and devices for separation of bioactive peptides.* Recent Patents on Biotechnology, 2009. **3**(1): 61-72.

Section IV: Functional Food Products: Packaging and Storage Stability

11

Choice of Suitable Packaging Material

Introduction

The emerging functional foods market presents a considerable challenge to all manufacturers to keep abreast of the use of adequate packaging materials for their products. The interest in functional foods is intrinsically linked to the potential health benefits of their bioactive components. Preserving these bioactive components is, therefore, a crucial factor to providing the anticipated health benefits of functional foods. However, recurring failure in maintaining the quality of functional foods has been reported. Functional foods include a wide variety of products such as prebiotics and probiotics, dried foods (fibers, tea, herbs, etc.), fermented products (yogurts, kefirs, vegetables), and fresh fruits and vegetables. Their bioactive components (minerals, vitamins, fibers, peptides, proteins, ω–3 polyunsaturated fats, antioxidants, enzymes, and symbiotics) range from very simple molecules to complex living organisms such as bacteria. The chemical nature of the bioactive components and the way functional foods are offered to consumers (in the form of dried extracts, fresh foods, or processed foods) dramatically influence the shelf life of these functional foods and their packaging requirements. Unfortunately, only a limited number of documents addressing this critical issue are available. Furthermore, the data are fragmented and dispersed within the literature on studies from other fields of research to which functional foods were incidental. A comprehensive review of the current literature would be useful to summarize the existing knowledge on what is required to preserve functional foods and to develop appropriate packaging guidelines.

Generally, packaging is discussed in terms of providing four basic functions [1-3]:

1. *Containment*: The containment function involves the ability of the packaging to maintain its integrity during the handling involved in filling, sealing, processing (in some cases, such as retorted, irradiated, and high-pressure processed foods), transportation, marketing, and dispensing of the food.

2 *Protection*: The need for protection depends on the food product but generally includes prevention of biological contamination (from microorganisms, insects, rodents), oxidation (of lipids, flavors, colors, vitamins, etc.), moisture change (which affects microbial growth, oxidation rates, and food texture), aroma loss or gain, and physical damage (abrasion, fracture, and/or crushing). Protection can also include providing tamper evident features on the package [4]. In providing protection, packaging maintains food safety and quality achieved by refrigeration, freezing, drying, heat processing, and other preservation of foods.

3 *Communication*: The information that a package provides involves meeting both legal requirements and marketing objectives. Food labels are required to provide information on the food processor, ingredients (including possible allergens in simple language), net content, nutrient contents, and country of origin. Package graphics are intended to communicate product quality and, thus, sell the product. Bar codes allow rapid check out and tracking of inventory. Other package codes allow determination of food production location and date. Various open dating systems inform the consumer about the shelf life of the food product. Plastic containers incorporate a recycling code for identification of the plastic material.

4 *Convenience:* Providing convenience (sometimes referred to as utility of use or functionality) to consumers has become a more important function of packaging. Range of sizes, easy handling, easy opening and dispensing, reclosability, and food preparation in the package are examples of packaging providing convenience to the consumer.

In addition to these four basic functions, three other functions have been defined that involve additional requirements of packaging.

i. *Production efficiency:* Another function often included is production efficiency (or machinability or economy), because of the requirement that packages perform well in rapid filling, closing handling, transportation, and storage operations [5-7].

ii. *Minimal environmental impact:* It is an additional function that reflects the need to consider, at the time of package design, the used package disposal [6, 7]. Reducing the amount of packaging (source reduction) to reduce packaging cost and waste has been a long time aspect of package development. To further reduce the use of natural resources, package development has increasingly considered additional approaches to minimizing the amount of packaging that ends up in land fills, such as package recyclability and energy recovery.

iii. *Package safety:* A function that involves the need to consider any possibility that the package might cause contamination of the food product [6, 7]. Increasing use of plastic packaging materials has brought increased attention to the possibility of migration of plastic monomers and additives into packaged foods. Thus, food packaging is highly regulated to ensure that packaging components do not migrate into food to produce a safety problem for consumers.

Different Classes of Packaging Material

Over $100 billion are spent annually on food packaging in the United States; ~40% on paper and paperboard packaging, ~18% on plastic film, ~12% on semi-rigid/rigid plastic packaging, ~14% for aluminum, ~6% for steel packaging, and ~10% for glass containers [8].

Glass

Glass is one of the oldest manufactured materials and one of the first manufacturing businesses in the New World. Nonetheless, glass still serves as an important packaging material for food. Glass is the most inert of the packaging materials and provides a total barrier to gases, water vapor and aromas, has good strength under compression, possesses good heat resistance to allow thermal processing of foods, allows viewing of the product, and is microwaveable. Other advantages of glass containers are customer perception that they add value to the food product and their recyclability. Because of their inertness, glass containers have the potential of being returned by the customer and refilled by the food manufacturer. This practice used to be common for milk, soft drink, juices, oil and beer, and is practiced to a limited extent [9, 10]. However, the weight of glass and the return distance to increasingly more centralized food manufacturers work against returning and refilling. The disadvantages of glass include its weight and vulnerability to fracture from thermal shock (rapid temperature change) and physical shock [11]. In recent years, advances in the science and technology of glass have resulted in lighter, stronger glass containers. For those food products vulnerable to light catalyzed reactions, glass's transparency to light is another disadvantage. Use of light absorbing colorants in the glass, as well as glass container labels and direct printing on the glass, will affect the transmission of light. Basic parts of glass container are shown in Fig. 11.1.

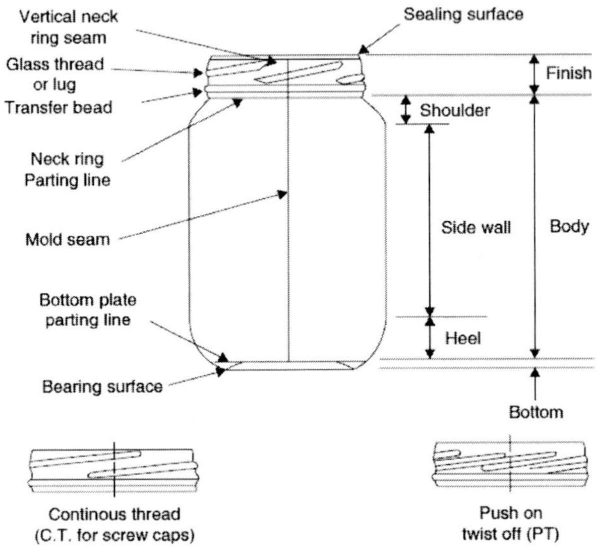

Fig. 11.1: Basic part of glass container

Metals

Tin plated iron cans have been used since the mid 18th century to preserve heat processed foods. Since that time, great advances in the production of steel and aluminum, coating of these metals, and forming them into containers have resulted in their being important packaging materials. Like glass, steel and aluminum are total barriers to gases, water vapor and aromas. Both also have good heat resistance and can withstand physical and thermal shock. Because of steel's greater strength, it is used more often in the thermal processing of foods. Neither steel nor aluminum is as inert as glass; thus both must be coated to avoid interactions with the foods they contain. Tin or chromium is used to coat steel, usually followed by a coating with a polymeric lacquer (enamel). Aluminum is coated directly with a lacquer. Other advantages of metal containers are exclusion of light from food products that are light-sensitive and their recyclability. The disadvantages of metal containers include their multi-step manufacture, weight (particularly steel), and (for some foods) lack of transparency [12]. In recent years, advances in the science and technology of these metals have resulted in lighter, stronger metal containers [3].

Steel is an alloy of iron and carbon, consisting of 99.5% iron with a small amount of carbon (0.02–0.3%). It is produced by reduction of iron ore (iron oxide, Fe_2O_3) with coke (carbon) to remove the oxygen as CO_2. Small amounts of contaminants such as manganese, silicon, sulfur and excess carbon are reacted with oxygen to form oxides that are removed as slag and CO_2. Small adjustments to the composition give the steel more ductility (D) or higher strength and stiffness (N), or make the steel appropriate for moderately-corrosive (MR) or strongly-corrosive (L) products. Slabs of steel are formed from molten steel and the slabs are hot rolled into sheets, which are then cold-rolled (CR) to approximately the thickness desired for containers. At this point, the sheets are approximately 0.01 inch thick. The CR also increases the steel sheet strength and stiffness. The sheets are then heat-annealed to increase ductility and then, possibly, cold-rolled again (2CR or DR) to increase strength and stiffness (i.e., increase temper). To protect the steel sheets from corrosion, they are electrolytically coated with either tin or chromium [13]. The coating thickness can be made different on the two sides of the sheet, with the thicker side commonly on the side facing the food product. When the steel sheets are coated with tin, the total tin thickness on the two sides of the sheet is approximately 1% of the sheet thickness. The resulting structure is often called tin plate. Depending on the food to be stored, tin plate steel is coated with an organic lacquer (enamel) to prevent interaction between the tin and food that could produce undesirable color or flavor change. Tin-free steel (TFS), also called electrolytic chromium coated steel (ECCS), has chromium coatings that are thinner than tin coatings which always have to be coated with a lacquer [14].

Normally stoved lacquers (i.e. organosol and epoxyphenolic) are used in food packaging for the inner surface of the can, while UV curing is used only for can outside because of adhesion problems. In fact UV lacquers show scarce dry and wet adhesion [15]. Different types of lacquers used for food packaging are as follows:

- *Epoxy-phenolic:* The most widely used at present; resistant to acids, with good flexibility, adhesion characteristics and high heat stability. Wide range epoxy-phenolic lacquers are available to cover a large variety of applications, such as fruits, fruit juices, vegetables, soup, meats, and marinated fish.

- *Acrylic:* Used when a good color retention and high resistance to heat are required. Formerly used only for exteriors, they are now providing internal protection as hard white coatings with an attractive ceramic and stain-free appearance.

- *Phenolic:* Outstanding chemical stability and low permeability, especially against sulfide ions, contains meat and fish products.
- *Vinyl:* Good adhesion and flexibility. It cannot resist sterilization at high temperatures, free from odor and flavor, functions as second lacquer layer for beer, wine and carbonated beverages and for cakes, sweets, dry foods and pharmaceutical products.
- *Oleoresinous:* It is usually modified with the addition of zinc oxide so as to prevent blackening of product due to reaction of thins with sulfur compounds. Subdivided to two groups viz. "R" or fruit enamels that protect the natural pigment of highly colored fruits such as dark berries, red beats and red cabbage, and "C" enamels which prevent the discoloration of foods containing proteins such as corn, peas, poultry and seafoods.

Aluminum composition and properties

Aluminum (Al) is made by electrolytic reduction of alumina (Al_2O_3) that is separated from bauxite. Small amounts of other elements are added to produce aluminum alloys which have different desirable formability and corrosion resistance properties. Aluminum is lighter and weaker than steel but is more easily formed into cans. Slabs made from molten aluminum are hot rolled into sheets for can making. The same lacquers lists above are used to coat the aluminum to prevent interaction with foods.

Major Types of Plastics Used in Food Packaging

Plastics are high molecular weight polymers that can be molded into desired shapes such as films, trays, bottles, and jars using heat and pressure. Two broad categories of plastics exist, those based on (i) thermoplastic polymers, and (ii) thermoset polymers. Thermoplastic polymers are linear or branched, but with no crosslinks between polymer chains. These polymers soften and become molten when heated, returning to their original condition upon cooling. Thus, they can be molded or extruded repeatedly. This property allows thermplastics to be recycled for many uses after their use in food packaging. Thermoplastic polymers constitute the most important category of plastics used for food packaging. These plastics exhibit a wide range of mechanical, optical, barrier, and thermal properties, depending on the specific polymer, polymer processing, and polymer additives. Additives can include plasticizers to improve plastic flexibility stabilizers to improve polymer resistance to degradation by heat, light and antistatic agents to prevent plastics from clinging to packaging equipment. Thermoset polymers crosslink into a set network when heated, often with addition of a crosslinking agent. Thus, after taking on their original cast shape, they can not be reheated for molding into new shapes. Thermoset polymers play an important role in food packaging, often used for making package closures [16].

The most commonly used thermoplastic polymers are inexpensive, and their conversion into food packaging is also relatively inexpensive. These plastics can be molded or extruded into a wide range of flexible, semi-rigid and rigid containers that are light weight, non-corrodible, shock-resistant, and heat-sealable. Most are transparent and some are microwaveable. Certain plastics have high enough heat resistance that they can be hot filled, retorted and/or used in a conventional oven. Finally, the most commonly used plastic semi-rigid and rigid containers are recyclable. Similar to glass and metal, plastic properties have improved over the years so that less material is necessary for making containers with acceptable integrity [17].

However, unlike glass and metal, plastics do not provide a total barrier to gases, water vapor, and aromas. The permeabilities of a given plastic material to water vapor, oxygen, carbon dioxide, and aromas depend on the particular polymer composition and structure [18]. This must be considered when selecting a plastic for a specific application and desired shelf life. Plastics are often combined in layers, to take advantage of the unique barrier properties of each polymer. Similar to glass, plastic container transparency to light can be detrimental to foods vulnerable to light catalyzed reactions. Pigmenting, labeling or direct printing of plastic containers can reduce this problem for sensitive food products. Plastic materials do not have the compressive strength of glass or metal, and only a few plastics have high enough heat resistance for heat processing or preparation of foods. Plastic additives and any residual monomers have potential for migrating into foods. Thus, much attention and testing are devoted to minimizing this possibility. On the other hand, food components such as aromas and flavors can absorb into plastic packaging, with resulting loss of food quality. Finally, most plastic materials used in food packaging are not recyclable. Fortunately, these are used in lower quantities than recyclable plastic containers [19].

Plastic materials and properties

The properties of a plastic polymer are influenced by its chemical composition, structure, additives, processing, and conditions of use [3, 20, 21]. The structure of thermoplastic polymers can include both organized crystalline regions, where polymer chains are parallel and closely packed, and disorganized amorphous regions, where the greater free volume results in lower polymer density than the crystalline regions. The relative amounts of crystalline and amorphous regions depend on the polymer structure and polymer processing. Thermoplastic polymers with amorphous regions have a characteristic glass transition temperature, T_g, at which the amorphous polymer regions transition from a stiff glassy state to a more flexible rubbery state. If the polymer has crystalline regions, they have a melting temperature T_m at ~1.5-2 T_g (°K). The T_g & T_m of a polymer have an important influence on the properties and uses of the polymer for food packaging [3, 22]. Some polymers are totally amorphous in nature.

Paper

More paper is used in food packaging than any other material. It can be found in all levels of packaging (primary, secondary, tertiary, and quaternary). Paper is a quite versatile material, utilized in flexible, semi-rigid, and rigid packaging. It is made into a wide variety of single and multi wall bags. It can also be made into a thicker/stronger structure (>0.012 in./0.03 cm) called paper board (PB) which is made into cartons and boxes that provide mechanical protection to many foods. The PB can be converted to an even stronger material called corrugated PB that is converted into boxes used for logistics (tertiary and quaternary packaging). Most types of paper provide a partial or complete barrier to light. It can also be manufactured into transparent and clear materials. The starting material for paper manufacture, wood, is a renewable resource; paper is recyclable and biodegradable. The main disadvantages of paper are that it provides negligible barrier against water vapor, and oxygen, and that it is not heat sealable. These disadvantages can be overcome by coating or laminating the paper or PB with wax, polyethylene, or other polymers (sometimes metalized) to improve the barrier properties and allow heat sealability. Coated or laminated paper and PB can not be recycled in most municipalities. But, the technology for separating the layers exists and is gradually being adopted for recycling.

Functional Food Shelf Life and Package Selection

Food shelf life is generally understood to be the elapsed time between times of manufacture and packaging of the food and time that the food becomes unacceptable to consumers. The shelf life of a food is influenced by three factors viz. (i) characteristics of the food (e.g., sensitivity to temperature, light, moisture and oxygen), (ii) environment (e.g., temperature, light, RH and O_2) to which the food will be exposed, and (iii) characteristics of the package. Accurate assessment of shelf life requires that an acceptance criterion be established by sensory evaluation and/or instrumental analysis. Several approaches are used to predict the packaged food shelf life [3, 23, 24]. These include:

- Assuming the same shelf life as that of a similar food product in the same packaging.
- Determining the shelf life by either exposure to real world conditions of warehousing, shipping, and retailing, or to long-term storage at warehouse conditions.
- Determining the shelf life by storage at controlled typical ambient conditions.
- Calculating the ambient shelf life based on results of accelerated shelf life testing (ASLT).
- Calculating the shelf life by use of a mathematical model that takes into account the properties of the food, package, and environment.

The first approach can result in under or over packaging of the product. The conditions of real world or warehouse testing in the second approach are difficult to control; the results are difficult to use for prediction.

Bioactive Packaging

Bioactive packaging is a novel set of technologies designed to give response to a number of issues related to the feasibility, stability and bioactivity of functional ingredients for the food industry. These technologies aim to integrate the bioactives within new packaging and coating material concepts and can greatly benefit from previous developments in the pharmaceutical and biomedical sectors and from the unique properties of synthetic and biomass derived biopolymers [25, 26].

The functional, or more precisely, bioactive packaging materials would thus be capable of withholding desired bioactive principles in optimum conditions until their eventual release into the food product either through controlled or fast release during storage, or just before consumption, taking into account the specific product/ functional substance characteristics or requirements. Curiously, when a relevant literature search was carried out under the term bioactive packaging only few calls were obtained, all of which, were totally unrelated to the food field or related to antimicrobial (active) packaging. It is worth stating at this point about the differences between active and bioactive packaging technologies. The main difference between the well known active packaging technologies and the concepts described here is that while active packaging primarily deals with maintaining or increasing quality and safety of packaged foods, i.e. shelflife of packaged food products, bioactive packaging has a direct impact on the health of the consumer by generating healthier packaged food.

Thus, the scope of the present work is to gather the basis for this novel technological approach, in which innovative functional packaging concepts are devised to give answer to the current barriers and limitations in the manufacturing of functional foods. The successful implementation of these developments requires multi-disciplinary efforts across various research fields. It is envisaged [26] that the development of these innovative concepts can be carried out by (i) integration and controlled release of bioactive components or nanocomponents from biodegradable and/or sustainable packaging systems, (ii) micro and nano encapsulation of these active substances either in the packaging and/or within foods, and (iii) packaging provided with enzymatic activity exerting a health promoting benefit through transformation of specific food borne components. The development of such novel functional hybrid food packaging systems (the functional concept including prebiotics, probiotics, phytochemicals, marine oils, lactose free foods, encapsulated vitamins, bioavailable flavonoids, etc.) will provide alternative, more efficient and, in some cases, unique industrial means to provide foods with improved impact on human health upon consumption.

Enzymatic Packaging

The immobilization of enzymes in materials was initially applied in food production lines [27, 28] as a result of the numerous technological advantages over the use of free enzymes such as reusability, improved stability to temperature, resistance to proteases and other denaturing compounds and improved activity. However, more recently, these technologies are also being considered for packaging applications [29]. The objective of these bioactive materials is to catalyse a reaction, which is considered beneficial from a nutritional point of view, i.e., decreasing the concentration of a non-desired food constituent, and/or producing a food substance beneficial to the health of the consumer. Many enzymes are currently being used in several food transformation processes. More recently, there is a number of trials reported in which these enzymes are also being immobilized in packaging materials. Soares and Hotchkiss [30] immobilized naringinase in a plastic package. The results indicated that the grapefruit juice reduced its bitterness by hydrolysis of naringine, a bitter principle of citrus juices.

Another enzymatically bioactive packaging concept already described elsewhere [31] consists the binding of β-galactosidase and cholesterol reductase in the package walls for the hydrolysis of lactose and cholesterol, respectively. The use of this type of packages would allow the production of a value added food without modifying the manufacturing procedure. For example, UHT milk produced by a conventional process, could be packaged in a β-galactosidase bioactive package and during storage, the product would transform into a low-lactose or free-lactose product. This processing plant inside the package appears to be a very promising technology. Within this specific area there are already some ventures to develop novel technologies and bring them into the food market [25, 26]. The techniques already used for the immobilization of enzymes or whole cells can be divided into five different categories, i.e., adsorption, ionic binding, covalent attachment, crosslinking, and entrapment/encapsulation.

The choice of the immobilization method and of the biomaterial support for the manufacturing of enzymatic packages will depend on the nature of the biocatalyst (e.g., whole cells or purified enzymes, from fungal or bacterial origin, either native or genetically modified), the envisaged storage conditions, the type of food to be packed and the specific application of the biocatalyst. Regarding the method of immobilization, covalent attachment and crosslinking often involve either the modification of the biopolymer surface or the use of toxic chemicals, such as glutaraldehyde, which is usually unsuited for applications in food. To overcome this handicap, mineral supports could be added to the biopolymer materials, which apart from improving their mechanical and barrier properties, have proved to be very efficient in enzyme binding. Within this area there are also some ventures to develop novel technologies based on the use of specific inorganic nano carriers for the production of bioactive enzymatic films. Several research studies have been carried out concerning the attachment of enzymes in diverse clay mineral supports. The mechanisms by which enzymes can be immobilized on clay minerals include cation exchange, physical adsorption and ionic binding [32]. What remains essential in all of these novel technologies is to maintain the enzyme functionality and here is where current research efforts are being focused.

Shelf Life from Storage Test at Controlled Typical Ambient Conditions

Determination of shelf life by storage at typical ambient conditions will take a long time, if the test is run to actual time of unacceptable product. However, if a criterion of food quality (e.g., a color, flavor, or vitamin that can be identified by instrumental analysis) has been established, the kinetics of quality change can be determined by measurement over a relatively short time. This involves determination of the order (usually zero or first order), and the reaction rate constant (*k*T) for the temperature studied. The reaction rate constant can then be used to calculate the time to reach the end of shelf life. Thus, for the conversion of attribute "*A*" to degradation product "*D*" (i.e., $A \rightarrow D$), the shelf life (*t*S) before *A* drops to an original level (*A*O) to an unacceptable level (*A*S) or *D* increases from an original level *(DO)* to an undesirable level (*D*S) can be determined from the following equations 1, 2 and 3, depending on the order of the reaction.

Zero order kinetics: $t\text{S} = (A\text{O} - A\text{S})/k\text{T}$ (1)

$t\text{S} = (D\text{S} \quad D\text{O})/k\text{T}$ (2)

First order kinetics: $t\text{S} = \ln(A\text{O}/A\text{S})/k\text{T}$ (3)

Assuming zero order kinetics in the determination of *k*T and then in the calculation of *t*S gives a conservative predicted shelf life, because with zero order kinetics the rate of quality change does not drop with time. Using data from storage at one temperature to calculate shelf life in the manner just described has limited use. It is possible to calculate shelf life at only the temperature of the testing. There is no way from this data to predict shelf time at another temperature or if the food product experiences temperature changes. Thus, use of ASLT or a mathematical model has distinct advantages [24].

Shelf Life from Accelerated Shelf Life Testing (ASLT)

ASLT involves determination of kinetics of packaged food quality at controlled conditions that accelerate deterioration of the relevant quality factor [3, 23, 33]. Using this approach requires determination of deterioration rate at three or more accelerating conditions to predict deterioration rate at a normal condition. Usually, kinetics of deterioration are measured at elevated temperatures and the Arrhenius equation or a linear model is used to predict the rate of deterioration at temperature(s) of normal handling and storage. For foods whose deterioration is determined by transport of moisture or oxygen, elevated levels of RH or pO_2 can be used to accelerate shelf life determination. A relationship between deterioration rate and RH or pO_2 must be established to predict deterioration at normal conditions. Use of the Arrhenius relationship is the most rigorous approach to relating deterioration rates at high temperatures to the lower temperatures of normal storage (Eq. 4):

$k\text{T} = k0 \exp(-E/RT)$ (4)

where *k*T is the rate constant at temperature T (K), *k*0 is the Arrhenius constant for deterioration of the selected quality factor, E is the activation energy for the degradation reaction, and R is the ideal gas constant. If the rate constants determined from testing at three elevated temperatures fall in a straight line on a plot of log k vs. ($1/T_o$ K), the Arrhenius relationship applies. One can then interpolate or extrapolate on this plot to predict the rate constants of the degradation reaction at any other temperature. Also, the values of *k*0 and E can be determined from the intercept and slope of the plot, respectively, and used in the Arrhenius equation to predict *k*T for any temperature. Shelf life can then be determined by use of above mentioned equation, depending on the order of the reaction.

Summary

Guaranteeing significant levels of beneficial bioactive components in functional foods is a great challenge to manufacturers. The shelf life of functional foods is highly dependent on the storage and packaging conditions that are applied throughout the entire distribution chain. These conditions must be precisely adapted to the specific needs of each and every functional food. Unfortunately, research on the effect of packaging conditions on functional foods is lacking. To optimize the preservation of these sensitive products, there is an urgent need for appropriate data. Further, the establishment of specific labeling regulations and standards to functional foods will, it is hoped, stimulate further research in this area. These studies should be undertaken to supply data supporting the determination of shelf life and expiration dates ensuring health benefit claims. Active packaging systems constitute a promising approach to improving the preservation of functional foods.

References

1. Marsh, K.S., *Looking at packaging in a new way to reduce food losses.* Food Technology 2001. **55**(2): 48-99.
2. Yam, K., J. Paik, and C. LAI III, *Packaging part III: material.* Encyclopedia of food science and technology. Chichester: John Wiley, 1992: 1978-1982.
3. Robertson, G.L., *Food packaging: principles and practice,* 2012: CRC press.London
4. Rosette, J.L., *et al., Packaging, Cosmetics and Pharmaceuticals.* Kirk-Othmer Encyclopedia of Chemical Technology.
5. Paine, F.A. and H.Y. Paine, *A Handbook of Food Packaging,* 1992: Springer.U.K
6. Steven, M.D. and J.H. Hotchkiss, *Package Functions.* Encyclopedia of Agricultural, Food, and Biological Engineering. New York: Marcel Dekker, 2003: p. 716-719.
7. Coles, R., D. McDowell, and M.J. Kirwan, *Food packaging technology.* Vol. 5. 2003: CRC Press.
8. Heldman, D.R., D.B. Lund, and C. Sabliov, *Handbook of food engineering,* 2006: CRC press.London, ISBN 9780824753313
9. Méndez, A.I. and E. Falqué, *Effect of storage time and container type on the quality of extra-virgin olive oil.* Food Control, 2007. **18**(5): 521-529.
10. Pérez Vicente, A., *et al., Influence of packaging material on pomegranate juice colour and bioactive compounds, during storage.* Journal of the Science of Food and Agriculture, 2004. **84**(7): 639-644.
11. Brody, A.L., Active *Packaging for food Application* 1997: CRC Press,U.S.A
12. Charbonneau, J.E., *Recent case histories of food product metal container interactions using scanning electron microscopy x-ray microanalysis.* Scanning, 1997. **19**(7): 512-518.
13. Catala, R., J. Cabanes, and J. Bastidas, *An impedance study on the corrosion properties of lacquered tinplate cans in contact with tuna and mussels in pickled sauce.* Corrosion Science, 1998. **40**(9): 1455-1467.

14. Zumelzu, E., F. Rull, and A. Boettcher, *Characterization and micro-and ultra-structural analysis of PET-based Co-rolled composite electrolytic chromium coated steel (ECCS).* Journal of Materials Processing Technology, 2006. **173**(1): 34-39.
15. Barilli, F., *et al.*, *Study on the adhesion of different types of lacquers used in food packaging.* Progress in Organic Coatings, 2003. **46**(2): 91-96.
16. Shaw, S., *Additives and modifiers for epoxy resins,* 1993: Springer.
17. Piringer, O.G. and A.L. Baner, *Plastic packaging: interactions with food and pharmaceuticals,* 2008: John Wiley & Sons.
18. Lange, J. and Y. Wyser, *Recent innovations in barrier technologies for plastic packaging— a review.* Packaging Technology and Science, 2003. **16**(4): 149-158.
19. Piringer, O.G. and A.L. Baner, *Plastic packaging materials for food: barrier function, mass transport, quality assurance, and legislation,* 2008: John Wiley & Sons.
20. Kirwan, M.J. and J.W. Strawbridge, *7 Plastics in food packaging.* Food packaging technology, 2003: 174.
21. Miltz, J., *Food packaging.* Handbook of Food Engineering, 1992: p. 667-718.
22. Armeniades, C. and E. Baer, *Transitions and relaxations in polymers.* John Wiley & Sons, Inc., Introduction to Polymer Science and Technology, 1977: p. 239-300.
23. Karel, M. and D.B. Lund, *Physical Principles of Food Preservation: Revised and Expanded.* Vol. 129. 2003: CRC Press.London
24. Steele, R., *Understanding and measuring the shelf-life of food* 2004: CRC Press.
25. Lopez-Rubio, A., R. Gavara, and J.M. Lagaron, *Bioactive packaging: turning foods into healthier foods through biomaterials.* Trends in Food Science & Technology, 2006. **17**(10): 567-575.
26. Lagaron, J. *Bioactive packaging: A novel route to generate healthier foods.* in *Second conference in food packaging interactions*, 2005.
27. Katchalski-Katzir, E., *Immobilized enzymes—learning from past successes and failures.* Trends in Biotechnology, 1993. **11**(11): 471-478.
28. Mosbach, K., *Immobilized enzymes.* Trends in Biochemical Sciences, 1980. **5**(1): 1-3.
29. Appendini, P. and J.H. Hotchkiss, *Immobilization of lysozyme on food contact polymers as potential antimicrobial films.* Packaging Technology and Science, 1997. **10**(5): 271-279.
30. Soares, N. and J. Hotchkiss, *Naringinase immobilization in packaging films for reducing naringin concentration in grapefruit juice.* Journal of Food Science, 1998. **63**(1): 61-65.
31. Brody, A. and J. Budny, *Enzymes as active packaging agents*, in *Active food packaging*1995, Springer. 174-192.
32. Sarkar, J.M., A. Leonowicz, and J.-M. Bollag, *Immobilization of enzymes on clays and soils.* Soil Biology and Biochemistry, 1989. **21**(2): 223-230.
33. Labuza, T., *An integrated approach to food chemistry: illustrative cases.* Food chemistry, 1985: p. 913-938.

12

Packaging Aspects of Different Functional Food Products

Introduction

The emerging functional foods market presents a considerable challenge to all manufacturers to keep abreast of the use of adequate packaging materials for their products. The interest in functional foods is intrinsically linked to the potential health benefits of their bioactive components. Preserving these bioactive components is, therefore, a crucial factor to providing the anticipated health benefits of functional foods. However, recurring failure in maintaining the quality of functional foods has been reported. Functional foods include a wide variety of products such as prebiotics and probiotics, dried foods (fibers, probiotics, tea, herbs, etc.), fermented products (yogurts, kefirs, vegetables), and fresh fruits and vegetables. Their bioactive components (minerals, vitamins, fibers, peptides, proteins, ω-3 polyunsaturated fats, antioxidants, enzymes, and symbiotics) range from very simple molecules to complex living organisms such as bacteria. The chemical nature of the bioactive components and the way functional foods are offered to consumers (in the form of dried extracts, fresh foods, or processed foods) dramatically influence the shelf life of these functional foods and their packaging requirements. Unfortunately, only a limited number of documents addressing this critical issue are available. Furthermore, the data are fragmented and dispersed within the literature on studies from other fields of research to which functional foods were incidental. A comprehensive review of the current literature would be useful to summarize the existing knowledge on what is required to preserve functional foods and to develop appropriate packaging guidelines.

Packaging for Different Functional Food Products

The effect of packaging on food composition during storage depends on the technology used and the storage conditions. Generally, functional foods packaging and preservation practices are based on the conditions applied to products of similar origin and format. This approach can give satisfactory results for several products but should not be taken as a universal rule regarding all functional foods. In some cases, the usual technologies and packaging materials could be detrimental to the shelf life of certain functional ingredients. In other cases, minor adaptations of the packaging technologies could likely improve the shelf life of functional foods greatly.

The effects of processing and preservation technologies on minerals and vitamins have been extensively documented for decades. In addition to vitamins and minerals, several other important functional phytochemicals such as flavanoids, prostaglandins, and sulfur compounds are present in fresh and processed plant products.

Bread and Bakery Products

The incorporation of various natural preservative and functional ingredeints has resulted in the requirement of specific packaing system for protection of active ingredient in bread and bakery prodcuts. Different types of packaging in combination with selected ingredients can give the baked bread a shelf-life of weeks and even months at ambient temperature [1]. This process makes it possible to produce through a direct method big batches without the need of throwing away the over production at the end of the day, because the breads can be sold the day after or for longer periods. Three aspects are important i.e. avoiding loss of water by the use of a good packaging, controlling process parameters and ingredients to have a slower aging of bread, and holding microbial parameters under control. As first parameter, moisture migration in and around the bread must be avoided. This moisture control starts during baking. A minimal but sufficient baking time gives a longer freshness to the bread. Cooling under conditioned parameters can diminish the water evaporation. Plastic packaging makes the barrier for water migration during long shelf-life. There is also water migration between the dry crust and the humid crumb. Decrusting the bread can eliminate this crumb-crust migration. If there is a filling with a lower water activity than the crumb, the latter will dry out. If the filling has a higher water activity than the crumb, moulds can start to grow. However, phenomena such as firming of the crumb, dry mouth feeling, and crumbling is not only a result of water migration. Some other molecular mechanisms of bread staling are completing the picture.

Different enzyme types are known to have an effect on softness/ freshness. Some amylases play an important role in staling by the very specific breakdown of starch [2, 3]. From the selective breakdown to destruction, the right balance of action on amylose and amylopectin has to be chosen to obtain the right level of softness, elasticity and stickiness. Maltogenic amylases leave the amylopectin backbone intact due to the partial debranching, which leads to less crystallisation and a softer granule/crumb. The minimal amylose breakdown of maltogenic amylases is resulting in a good elasticity of the crumb. Thermostable amylases result in an intensive breakdown and a sticky crumb. The thermostable amylases also have an intensive breakdown of the amylases resulting in a sticky non-elastic crumb. The long shelf-life also depends on the storage temperature. Frozen breads have a very long shelf life and are discussed in the next paragraph. Cold temperatures are speeding up the staling of the bread. Also, moulds prefer higher temperatures. To keep the breads longer fresh in bread automates, which allow the baker to sell bread in the street 24 h a day, these automates are heated during the winter period [4].

Fruits and Vegetables

Minimal processing is often a way to preserve the bioactive components of functional foods. Packaging, therefore, is a significant factor in extending the shelf life of minimally processed food products [2, 5]. However, the effect of minimal processing and of packaging technologies on the nutritional quality of functional foods is not well established. Modified atmosphere packaging (MAP) is a process well known for its capacity to enhance the shelf life of fresh fruits and vegetables. The levels of antioxidants (vitamins, phenolics, and carotenoids) in fresh fruits and vegetables depends on storage and handling conditions [6, 7]. Constant levels of flavanoids over a 7 days period for spinach maintained at 10 °C were observed; the total antioxidant capacity of the product decreased over the same storage period. The spinach subjected to MAP was more affected than the unpackaged spinach [8]. These results indicate that spinach packaged in the conventional manner but with a shorter shelf life could provide more functional benefits than MAP spinach. However, the lack of data in the literature prevents the extension of such a conclusion to the entire class of fresh fruits and vegetables. The improvement in shelf life obtained using MAP is usually associated with the preservation of normal appearance and firmness. The MAP process slows down or matches the metabolism of packaged fresh products and microflora to avoid fermentation, thus minimizing. Plant products contain phenolic compounds, which are closely associated with the nutritional quality of foods. Their properties are based on their antioxidant capability. As with other antioxidant compounds, these molecules easily oxidize in the presence of oxygen. Phenolic compounds are present in cereals, legumes, oilseeds, fruits, and beverages of plant origin (tea, cocoa, beer, wine, fruit juices, etc.). Processed plant products with medium to high water activity will be the most susceptible to browning and phenolic degradation. Conditioning and packaging these products require controlling the oxygen level to which they are exposed.

Probiotics Dairy Products

The incorporation and viability in foods throughout storage of probiotic bacteria which result in health benefits for the host is a constant challenge for the food industry and requires the understanding of all intrinsic and extrinsic factors associated with processing, including the selection of the type of packaging material. Packaging materials and the storage conditions are important factors for the quality of products containing probiotic microorganisms [9, 10]. In view of the ecology of the strains normally used in probiotic products, anaerobic and microaerophilic ones, the level of oxygen within the package during storage of the product should be as low as possible in order to avoid toxicity and death of the microorganism and the consequent loss of functionality of the product. Exposure to dissolved oxygen during processing and storage is highly detrimental to *B. bifidum* and *L. acidophilus*. Contrary to aerobic microorganisms, which completely reduce oxygen to water, the absorption system of this substance is minimal or even completely non-existing. The absence of an electron transport chain results in the incomplete reduction of oxygen to hydrogen peroxide. In addition, these probiotic bacteria do not produce catalase enzyme essential to breakdown hydrogen peroxide, a characteristic that consequently leads to the accumulation of derived toxic metabolites, such as superoxide anion (O^{2-}), the hydroxide radical (OH–), hydrogen peroxide (H_2O_2) in the cell, causing its death. This suggests that probiotic strains may be affected by H_2O_2 produced by other cultures present in the reaction medium. This has motivated several studies aimed at developing alternatives that minimize these negative effects, among which the most promising are those evaluating the addition of antioxidants, such as ascorbic acid and the elimination of peroxide producing strains [9].

Another important characteristic of the growth and stability of the probiotic species in fermented dairy products is the correlation between the synthesis of enzymes such as NAD-oxidase and NADH-peroxidase. High levels of these enzymes have been reported in aerotolerant species, which is highly significant in view of the fact that these enzymes are responsible for removing oxygen from the interacellular medium [11]. In general, being strictly anaerobic, *Bifidobacterium* sp. are more sensitive to oxygen than *L. acidophilus*, however this sensitivity is exclusively dependent on the specific strain used. This poses a serious problem, since almost all probiotic dairy products available on the market, mainly fermented milks and yogurts, are practically all contained in plastic packages, which by nature are materials with high oxygen permeability.

Studies evaluating the use of packagings for probiotic foods are generally restricted to "active" packages (oxygen absorbers) and plastic films with high oxygen barrier properties [12]. The viability of microorganisms may be affected by several factors, including packaging. The viability of both *Lactobacillus* and *Bifidobacterium* species may diminish drastically during storage, resulting in low numbers of living cells at the time of consumption of the food product. Oxygen is often reported to be a concern for the survival of these cell cultures. *Lactobacillus acidophilus* is microaerophilic, and surface growth on solid media is generally enhanced by anaerobiosis or reduced oxygen with 5–10% CO. The other popular probiotics, *Bifidobacteria*, are usually considered strictly anaerobic but some strains of *Bifidobacterium lactis* have demonstrated a relative tolerance to oxygen.

Yogurts, whether they contain probiotics or not, are usually packaged in regular high impact polystyrene (HIPS) cups [13]. This type of packaging is not ideal for probiotics because of their high oxygen transmission rate. Dave and Shah have demonstrated that for living cells contained in yogurts, oxygen concentration in the product and container significantly influences the survival of probiotics [14]. *Lactobacillus* and *Bifidobacterium* spp. are particularly sensitive to oxygen toxicity. Strains of these bacteria grown in defined oxygen concentration (0, 5, 10, 15, and 21% oxygen) showed a drastic decrease in lactate production with increased oxygen uptake. More recently, the Champagne research group study on probiotics in nonfermented milks indicated that the oxygen level in these types of products must also be controlled to ensure the survival of probiotics. Comparative packaging tests showed that the use of glass hermetic containers can significantly increase the survival rate of probiotics in yogurts. However, given the cost of this form of packaging, it is not a viable commercial alternative. The use of active packaging, conversely, could be an option worthy of consideration. Polystyrene based cups coated with a Nupak (gas barrier layer combined with an oxygen scavenger (Zero2™)) significantly improved the survival of *Lactobacillus* and *Bifidobacterium* spp. in yogurts [15].

Dried probiotic cultures are also affected by storage conditions such as temperature, moisture, light, and air. Other authors have also demonstrated that oxygen and moisture affect survival kinetics. Encapsulation could be used as a primary packaging system able to contribute to the preservation of dried probiotic cultures. As a general guideline, oxygen and water vapor should be carefully controlled to ensure the optimal packaging conditions of dried cultures.

Oils and Fats

Extracted natural oils and fats contain molecules with unsaturated molecules and natural antioxidants such as tocopherols. These molecules are easily oxidized when exposed to oxygen and light. Consequently, light and gas barrier packaging materials should be mandatory to preserve the functionality of these products. Conjugated linoleic acid (CLA) is a natural fatty acid recognized for its anticancer properties. When CLA is present in its original matrix, it remains quite well preserved within the usual shelf life of fresh products. The CLA can be found in dairy products and meat. Lavillonnière *et al.* showed that the CLA level in dairy products (e.g., yogurt, cheese, and ice-cream) stored and packaged in usual conditions does not vary significantly over a 6 week period [16].

Summary

The consumption of functional foods has greatly increased over the past years as a result of their benefits for human health. For that reason, it is necessary to optimize several technological and economical aspects of manufacturing process of these products. This includes the development of innovative packaging solutions that adequately protect and preserve the therapeutic activity of functional food ingredients in different food matrix over the defined shelf life of the product. The use of appropriate packaging materials and systems is of utmost importance to safeguard the improvements introduced in the manufacturing process as a whole and ensure that the product lives upto the expectations of the people that consume these products. Within this context, research efforts aimed at guaranteeing the preservation of the full therapeutic potential of the functional properties throughout the shelf life of the products should be encouraged.

References

1. Rodriguez-Mateos, A., *et al.*, *Bioavailability, bioactivity and impact on health of dietary flavonoids and related compounds: an update.* Archives of toxicology, 2014. **88**(10): 1803-1853.
2. Appendini, P. and J.H. Hotchkiss, *Review of antimicrobial food packaging.* Innovative Food Science & Emerging Technologies, 2002. **3**(2): 113-126.
3. Cooksey, K., *Effectiveness of antimicrobial food packaging materials.* Food Additives and Contaminants, 2005. **22**(10): 980-987.
4. Rice-Evans, C.A. and L. Packer, *Flavonoids in health and disease* 2003: CRC Press London.
5. Vermeiren, L., F. Devlieghere, and J. Debevere, *Effectiveness of some recent antimicrobial packaging concepts.* Food Additives & Contaminants, 2002. **19**(1): 163-171.
6. Han, J.H., *Antimicrobial food packaging.* Novel Food Packaging Techniques, 2003: 50-70.
7. Dainelli, D., *et al., Active and intelligent food packaging: legal aspects and safety concerns.* Trends in Food Science & Technology, 2008. **19**: S103-S112.

8. Rudkowska, I., *Functional foods for health: Focus on diabetes.* Maturitas, 2009. **62**(3): 263-269.
9. Vanschoonbeek, K., *et al.*, *Cinnamon supplementation does not improve glycemic control in postmenopausal type 2 diabetes patients.* The Journal of Nutrition, 2006. **136**(4): 977-980.
10. Stice, E., *et al.*, *Relation of reward from food intake and anticipated food intake to obesity: a functional magnetic resonance imaging study.* Journal of Abnormal Psychology, 2008. **117**(4): 924.
11. Seiders, K. and R.D. Petty, *Obesity and the role of food marketing: a policy analysis of issues and remedies.* Journal of Public Policy and Marketing, 2004. **23**(2): 153-169.
12. Talwalkar, A. and K. Kailasapathy, *A review of oxygen toxicity in probiotic yogurts: Influence on the survival of probiotic bacteria and protective techniques.* Comprehensive Reviews in Food Science and Food Safety, 2004. **3**(3): 117-124.
13. Kos, K. and J.P. Wilding, *SPARC: a key player in the pathologies associated with obesity and diabetes.* Nature Reviews Endocrinology, 2010. **6**(4): 225-235.
14. Menrad, K., *Market and marketing of functional food in Europe.* Journal of Food Engineering, 2003. **56**(2): 181-188.
15. Bigliardi, B. and F. Galati, *Innovation trends in the food industry: the case of functional foods.* Trends in Food Science & Technology, 2013. **31**(2): 118-129.
16. Saito, M., *Role of FOSHU (food for specified health uses) for healthier life.* Yakugaku zasshi: Journal of the Pharmaceutical Society of Japan, 2007. **127**(3): 407-416.

13

Innovative Packaging Technologies in The Production of Functional Foods

Introduction

Developments in innovative food packaging technologies have evolved in response to the need for protection of the food product containing bioactive component from both external and internal environments and in response to consumer expectations for health claims, convenience and product safety. Consumers around the world are willing to pay extra for guaranteed product freshness delivered by the type of packaging. Many new packaging developments have focused on extending the shelf life of the product and on delivering a higher quality product to the consumer. These developments would not be possible without significant advances in the materials used in packaging and the incorporation of various types of sensors into food packaging. This chapter outlines the innovative packaing solutions in the production of functional food.

Active Packaging

Active packaging, sometimes referred to as interactive or "smart" packaging, is intended to sense internal or external environmental change and to respond by changing its own properties or attributes and hence the internal package environment. Active packaging has been considered a component of the packaging discipline for several decades or since the first inclusion of desiccants in dry product packages [1, 2]. In their own moisture permeable sachets, desiccants absorb water vapor from the contained product and from the package headspace, and absorb any water vapor that enters by permeation or transmission through the package structure. As separate entities within packages, active packaging sachets, pouches, patches, coupons, labels, etc., are not often integral to the package [2]. The goal of active packaging, in conjunction with other food processing and packaging, is to enhance preservation of contained food and beverage products. For example, to optimize the effects of oxygen scavenging, oxygen should first be removed from the product during processing and packaging operations [3]. The oxygen must also be thoroughly removed from the package interior and the package materials, and the package structure, including materials and closure, must be barriers to further oxygen entry. In other words, oxygen scavenging complements good oxygen control practices.

In addition, oxygen is certainly not the only vector that can influence the quality of the contained food. For example, moisture gain or loss, light, non-oxidative reactions, microbiological growth, and enzymatic activity may all, individually or collectively, be involved in food products deterioration. Active packaging has reached the commercial market, some with notable success. High gas permeability films, including some that increase their oxygen permeability with increasing temperature, are used for packaging fresh cut produce. Use of these temperature sensitive package materials is expected to increase because the technology developer has acquired a fresh produce packager who, of course, uses the technology in its package materials (Table 13.1).

Carbon dioxide and ethylene scavengers for modified atmosphere (MA) or, more precisely, controlled atmosphere (CA) food preservation are common in large bulk shipments. Carbon dioxide emitters to suppress microbiological growth have experienced limited success in modified atmosphere packaging (MAP) [4]. Ethylene scavengers are among the more successful commercial active packaging technologies in the fresh fruit bulk shipment category. Antioxidants and oxygen interceptors incorporated into package materials, such as tocopherols (vitamin E), have emerged in recent years and are increasingly employed to combat odors generated in plastic processing. Tocopherols, which are non-volatile, have not replaced volatile butylated hydroxy anisole/ butylated hydroxy toluene (BHA/BHT) which migrate into foods in product antioxidant applications, but they appear to be new antioxidants of choice for mitigating the effects of oxygen. Entities such as oxygen scavengers/ interceptors react with oxygen to form new compounds. Oxygen absorbers may remove oxygen by any means, including physical. Antioxidants react with free radicals and peroxides to retard or block the actual oxidation reactions. Sequestering agents tie up inorganic catalysts that might otherwise accelerate adverse oxidative reactions. Members of the food technology and packaging communities have long regarded package materials as an ideal reservoir and delivery vehicle for antimicrobial compounds. For many years, sorbic acid has been incorporated sparingly on the interior of package structures as an antimycotic in a limited

Table 13.1: List of active packaging system with typical food application

Active	Packaging mechanisms	Food applications
Oxygen scavengers `	Iron based metal/acid nylon MXD6 metal (e.g. platinum) catalyst ascorbate/metallic salts, enzyme based	Bread, cakes, cooked rice, biscuits, pizza, pasta, cheese, cured meats and fish, coffee, snack foods, dried foods and beverages
Carbon dioxide scavengers/emitters	Iron oxide/calcium hydroxide' Ferrous carbonate/metal halide Ascorbate/sodium bicarbonate	Coffee, fresh meats and fish, nuts and other snack food products and sponge cakes
Ethylene scavengers	Potassium permanganate Activated carbon Activated clays/zeolites	Fruit, vegetables and other horticultural products
Preservative releasers	Organic acids Silver zeolite Spice and herb extracts BHA/BHT antioxidants	Cereals, meats, fish, bread, cheese, snack foods, fruit and vegetables
Ethanol emitters	Encapsulated ethanol	Pizza crusts, cakes, bread, biscuits, fish and bakery products
Moisture absorbers	PVA blanket Activated clays and minerals Silica gel	Fish, meats, poultry, snack foods, cereals, dried foods, sandwiches, fruit and vegetables
Flavour/odour absorbers	Cellulose triacetate Acetylated paper Citric acid Ferrous salt/ascorbate Activated carbon/clays/zeolites	Fruit juices, fried snack foods, fish, cereals, poultry, dairy products and fruit
Temperature control packaging	Non-woven plastics Double-walled containers Hydrofluorocarbon gas Quicklime/water Ammonium nitrate/water Calcium chloride/water	Ready meals, meats, fish, poultry and beverages

number of dry food packages. The obvious benefits of sorbic acid as a mold and yeast inhibitor have been one foundation by which numerous other antimicrobial agents have found their way into food package materials. Unfortunately, most antimicrobial agents also exhibit toxicity when they enter the food from the package and would be consumed as part of the food. Thus, actual commercialization has been proceeding slowly, except in Japan where several compounds have been reported to function effectively as antimicrobials in commercial packages (Table 13.2). As with oxygen scavengers, the major technological and commercial successes for antimicrobials have been achieved by Japanese organizations for packaging food products in Japan. Nevertheless, the concept of integrating microbistatic and microbicidal materials and plastic packaging has been very attractive. Numerous attempts have been and are being made to translate favourable laboratory results into safe and effective commercial food packaging. The growing list of successes in active packaging beyond oxygen scavenging has been noted by the food packaging community.

Table 13.2: Selected commercial antimicrobial packaging available for food applications

Antimicrobial compound	Trade name	Producer company	Packaging forms
Silver substituted zeolite	AgIon™	AgIon technologies	Bulk food storage containers, paper board cartons, plastic, or paper food wraps and milk containers
	Novaron™	Toagosei, Co. LTD	Many Japanese
Triclosan	Microban®		microban products Deliwrap, reheatable food containers
Allylisothio-cyanate	WasaOuro®	Lintec Corporation UK.	Pressure sensitive labels
Chlorine dioxide	Microsphere™	Bernard Technologies	Storage bags for produce paperboard coating, rigid containers
Carbon dioxide	Freshpax™	Multisorb Technologies	Sachets
	Verifrais™	SARL Codimer France	Sachets
Ethanol vapor	Ethicap®	Freund	
	Negamold®		
	Fretek®		
Glucose oxidase	Bioka®	Bioka Ltd. Finland	Sachets

Antimicrobial Packaging

Antimicrobial (AM) food packaging materials have to extend the lag phase and reduce the growth rate of microorganisms in order to extend shelf life and to maintain product quality and safety [4, 5]. Alternatives to direct additives for minimizing the microbial load are canning, aseptic processing and MAP. However, canned foods cannot be marketed as "fresh". Aseptic processing may be expensive and hydrogen peroxide, which is restricted in level by regulatory agencies, is often used as a sterilizing agent. In certain cases, MAP can promote the growth of pathogenic anaerobes and the germination of spores, or prevent the growth of spoilage organisms which indicate the presence of pathogens [6]. If packaging materials have self-sterilizing abilities due to their own AM effectiveness, the need for chemical sterilization of the packages may be obviated and the aseptic packaging process simplified [5].

Food packages can be made AM active by incorporation and immobilization of AM agents or by surface modification and surface coating. Present plans envisage the possible use of naturally derived AM agents in packaging systems for a variety of processed meats, cheeses, and other foods, especially those with relatively smooth product surfaces that come in contact with the inner surface of the package. This solution is becoming increasingly important, as it represents a perceived lower risk to the consumer [7]. Antimicrobial materials have been known for many years. However, antimicrobial packages have had relatively few commercial successes, except in Japan. Antimicrobial films can be classified in 2 types viz. those that (i) contain an AM agent that migrates to the surface of the food, and (ii) are effective against surface growth of microorganisms without migration.

In most, but not all, solid or semisolid foods, microbial growth occurs primarily at the surface. In prepared or mixed foods, microbiological growth can occur anywhere in the mass. Antimycotic agents are not uncommonly incorporated into waxes and other edible coatings used to package hard skin fresh produce items such as oranges and apples. More recently, the concept of incorporating antimicrobial agents directly into package films that contact the surface of the food has been developed, particularly in Japan, but also in other geographic regions.

Antimicrobial package materials may be classified into two types i.e. (i) those containing antimicrobial agents that migrate to the surface of the package material and thus can contact the food, and (ii) those that are effective against food surface microbiological growth without migration of the active agent(s) to the food. Several commercial antimicrobial films have been introduced, principally in Japan. One is a synthetic zeolite that has had a portion of its sodium ions replaced with silver ions, which can be antimicrobial under certain situations. Such zeolite may be incorporated directly into a food-contact film. The purpose of the zeolite apparently is to allow for the slow release of silver ions to the food. Only a few scientific descriptions of the effectiveness of this material have appeared, and the regulatory status of the intentional addition of silver ions to foods through the medium of package contact has not been clarified by U.S. regulatory authorities.

Several other synthetic and naturally occurring compounds have been proposed and/or tested for antimicrobial activity in packaging. For example, the commercial antimycotic (i.e., antifungal) agent "Imazalil" has been demonstrated to be effective when incorporated into low density polyethylene (LDPE) film for wrapping fruits and vegetables [8]. The same compound is effective at preventing mold growth on cheese surfaces when incorporated into LDPE films in contact. Although Imazalil is not approved for cheese packaging in the United States, the research results established that antimycotic films could be effective for control of surface mold on foods [9]. The antifungal agent benomyl, commonly used as a fungicide, may be chemically coupled to ionomer film to inhibit microbial growth. It is improbable that benomyl would be approved for food use because of the chemical's toxicology.

Reports have appeared that demonstrate the effectiveness of adding common food grade antimycotic agents to cellulose based edible films. For example, films were constructed of cellulose derivatives and fatty acids to control the release of sorbic acid and potassium sorbate. Such films would seem to have the greatest application as fruit and vegetable coatings. Cellulose films, e.g., cellophane, generally are not heat-sealable and are not good water vapor barriers in situations of high relative humidity.

Antimicrobial Packaging in Meat and Meat Products

Due to recent outbreaks of contaminations associated with meat products, as well as growing concerns regarding the safety of intermediate moisture foods, active packaging has been greatly developed in recent years. Although little developed in Europe, in the past decade active packaging has become one of the major areas of research in food packaging. Of these active packaging systems, the antimicrobial version is of great importance.

Edible Packaging

Edible packaging is rapidly advancing by utilizing edible compounds, such as proteins, polysaccharides, lipids and/or resins, and other edible components, derived from diverse renewable sources [10, 11]. Such edible packaging materials are intended to be integral parts of food products and to be eaten with the products, thus they are also inherently biodegradable in composting and other biological recycling [12]. Edible packaging generally consists of edible films, sheets, coatings, and pouches. Edible films (thickness < 254 μm) or sheets (thickness > 254 μm) are stand-alone structures that are preformed separately from the food and then placed on / or between food components or sealed into edible pouches. Whereas, edible coatings are thin layers of edible materials formed directly onto the surface of the food products [13]. Soft-gel capsules, hard-gel capsules, tablet coatings, and microcapsules made from edible materials could also be considered edible packaging. The main focus of this review is on edible films, coatings, and pouches.

Considerable interest and advanced research activity in edible packaging in the food and packaging industries have been driven by both increasing consumer demand for safe, high quality, convenient foods with long shelf lives and also ecological consciousness of limited natural resources and the environmental impact of packaging waste. Edible packaging materials have been considered as attractive alternatives for some applications because of their unique properties, including the ability to protect foods with their barrier and mechanical properties, enhance sensory characteristics, control-release active ingredients, and control mass transfer between components of heterogeneous foods.

However, edible packaging materials are not normally meant to entirely replace conventional packaging. Rather, the efficiency of food preservation can be improved by using primary edible packaging together with non-edible packaging as secondary packaging to add additional protection from the atmosphere and prevent contamination from microorganisms or foreign particles. The utilization of edible packaging can reduce the complexity of overall packaging requirements by allowing conversion from multilayer or multilevel packaging to a single component package, resulting in source reduction and improved recyclability of the simplified packaging system without compromising protective functions.

Edible packaging materials can also be used for non-edible packaging as an oxygen or grease barrier layer to improve protective functions and biodegradability of multilayer packaging. The unique advantages and versatility of edible packaging materials are envisioned to create green, innovative packaging for reducing waste and/or for improving product stability, quality, safety, variety, and convenience for consumers. This chapter presents a review of recent developments in biopolymer and lipid based food packaging materials. It includes discussion of edible packaging functions, properties, promising applications, techniques used for developing edible packaging, and detailed descriptions of their film-forming ability and their associated applications.

Carrier of Antimicrobials, Antioxidants, and other Food Additives

The ability of edible packaging to carry and control-release active compounds is a promising active food packaging function attracting current research interest and much attention from both food and pharmaceutical industries [14]. Formulations can be designed to carry desired food additives (including antioxidants, antimicrobials, pigments, flavors, spices, salts, nutrient, light absorbers) as well as pharmaceutical or nutraceutical ingredients in the form of hard capsules, softgel capsules, microcapsules, soluble films/strips, flexible pouches, or coatings [13]. Depending on individual application, specific controlled-release rates of active solutes are required to fully perform their assigned functions. For example, limiting the migration rate of a preservative can reduce surface microbial growth. By slowing diffusion from the food surface into the food bulk, effective surface concentration of preservative can be maintained while decreasing the total concentration required. In certain cases, modification of an edible film matrix such as cross-linking could be used to reduce bioactive compound mobility. A controlled rate of release of active solute could be designed based on the chemical affinities between active ingredient, film forming materials, and food product and the conditions (e.g., temperature, pH, a_w, time) to which the edible packaging containing the active ingredient would be subjected. Antimicrobial edible packaging is a promising application that can be used alone or synergistically in combination with other preservation such as refrigeration, modified atmosphere packaging or irradiation to improve microbial stability. Extensive effort has been given to incorporating natural and synthetic antimicrobial agents and antioxidants into various edible film forming materials.

Testing the Effectiveness of Antimicrobial Packaging

There are a variety of official test methods to determine the resistance of plastic materials to microbial degradation. There is, however, no agreement upon standard methods to determine the effectiveness of antimicrobial polymers [5]. In Japan, a method referred to as 'Film Contact Method' is used as a standard to assess the ability of products containing antimicrobials to impart antimicrobial properties to products. The method was developed for inorganic antimicrobials such as silver substituted zeolites. It is suitable for films and sheets and consists of inoculating bacteria on the test specimen and incubating and counting the bacteria under specified conditions. The intent is to determine the resistance of the plastic to microbial growth, but it may also serve to determine if polymers are 'self-sterilizing'. To assess if antimicrobial packaging have an effect on microorganisms present in the foods, agar plate methods, minimally inhibitory concentrations (MIC) and dynamic shake flask tests have been used following similar methods to those used to evaluate antimicrobials alone. MIC can indicate the antimicrobial strength of the polymer and allows the comparison of the polymer's antimicrobial activity to that of the antimicrobial alone [15]. The method consists in seeding a series of tubes containing growth medium with the target microorganism and with polymers containing different concentrations of antimicrobial. The tubes are incubated for a predetermined period of time and visually inspected for microbial growth. MIC is the lowest concentration of an antimicrobial in a polymer resulting in the complete inhibition of growth of a test microorganism. Results should include polymer dimensions, composition and other relevant characteristics that vary from specimen to specimen [7].

In the agar plate test, antimicrobial film is placed on a solid agar medium containing the test microorganism. The agar plates are incubated until growth is visible. A clear zone surrounding the film indicates antimicrobial diffusion from the film and subsequent growth inhibition (Figure 13.1). Lack of growth under a film may indicate inhibition, but appropriate controls must be included; this may be due to simple restriction to oxygen [5]. The agar plate tests method simulates wrapping of foods and may suggest what can happen when films contact contaminated surfaces and the antimicrobial agent migrates from the film to the food. The method can be quantitative if the diameter of the clear zones around the films is measured. Shake flasks tests provide more detailed information on antimicrobial kinetics. Liquid media (buffer, growth media or foods) are seeded with the target microorganisms and the antimicrobial polymer. The flasks are incubated with mild agitation. Samples are taken over time and enumerated unlike the MIC test, this method measure reduction in growth rate even if substantial grow occurs. Tests in buffer provide information on the microcidal properties of the polymers while tests in broth provide information on microbial growth kinetics and the antimicrobial mode of action of the polymers. Tests in buffer may be misleading since cells susceptible in nutrient poor media may recover if nutrients are present.

When testing antimicrobial films by the shake flask test, the ratio of film surface area to volume is considered. Previous examples show that increasing the surface area volume ratio increases the activity of bioactive molecules incorporated into polymer films. From an antimicrobial standpoint, high surface volume ratios may seem adequate. But in real packaging applications, surface area-volume ratios of one are considered optimal, and values higher than that may be impractical [15]. By accounting the area and volume ratio, the feasibility of such films for practical applications may be assessed. As its name implies, the shake flask test includes agitation, which enhances the contact between the antimicrobial polymer and the cells. The test may not be indicative of the degree of agitation that packaged foods receive and, therefore, studies should simulate agitation during storage and transportation.

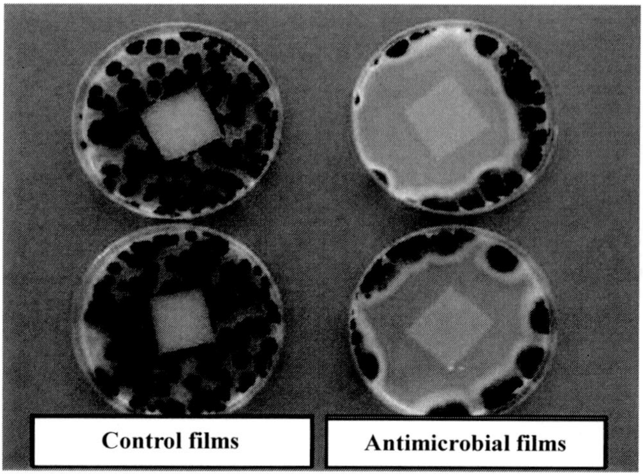

Fig. 13.1: Effect of antimicrobial plastic films on *Aspergillus niger* agar diffusion method

Regulatory Aspects

Materials used in formation of edible packaging must be generally recognized as safe (GRAS) for intended use or sanctioned by the United States Food and Drug Administration (FDA) Code of Federal Regulations or the U.S. Pharmacopoeia/National Formulary [16]. The edible packaging materials and additives must be used in accordance with good manufacturing practice (GMP) (i.e., food grade, prepared and handled as a food ingredient) and within any limitations specified by the FDA. The list of materials to be used and their permissible limits can be found at FDA's "Everything Added to Food in the United States (EAFUS)" data base. The ingredients of edible packaging must be declared on the label under regulation of the Federal Food, Drug and Cosmetic Act (21 USC 343). Edible packaging made from common allergens (i.e., milk, eggs, peanuts, tree nuts, fish, shellfish, soy, wheat) must be clearly labeled to provide information to consumers with allergies or intolerances to particular food components according to Food Allergen Labeling and Consumer Protection Act of 2004.

Modified Atmosphere Packaging

Modified atmosphere packaging (MAP) is defined as 'the packaging of a perishable product in an atmosphere which has been modified so that its composition is other than that of air' [17-19]. Whereas controlled atmosphere storage (CAS) involves maintaining a fixed concentration of gases surrounding the product by careful monitoring and addition of gases.The gaseous composition of fresh MAP foods is constantly changing due to chemical reactions and microbial activity. Gas exchange between the pack headspace and the external environment may also occur as a result of permeation across the package material [17].

Packing foods in a modified atmosphere can offer extended shelf life and improved product presentation in a convenient container, making the product more attractive to the retail customer [20]. However, MAP cannot improve the quality of an inferior food product. It is therefore essential that the food is of the highest quality prior to packing in order to optimise the benefits of modifying the pack atmosphere. Good hygiene practices and temperature control throughout the chill chain for perishable products are required to maintain the quality benefits and extended shelf life of MAP foods.

Packaging material for MAP

Selection of the most appropriate packaging materials is essential to maintain the quality and safety of MAP foods. Flexible and semi-rigid plastics and plastic laminates are the most common materials used for MAP foods. Plastic materials account for approximately one-third of the total materials demand for food packaging applications, and their use is forecast to grow. Relative ease of forming, light weight, good clarity, heat sealing and strength are some of the properties of plastics that make them suitable as food packaging materials. Advances in polymer processing have enabled the development of plastics that are better suited to particular food packaging applications. However, no single plastic possesses the properties that make it suited to all food packaging applications.

Plastic packaging materials may consist of a monolayer formed from a single plastic, but most, if not all, MAP films are multilayer structures formed from several layers of different plastics. Using coextrusion, lamination or coating technologies, it is possible to combine different types of plastic to form films, sheets or rigid packs. By carefully selecting each component plastic, it is possible to design a material which possesses the key properties of packaging importance to best match the requirements of the product/package system. Plastics packaging for MAP applications is most commonly found in the form of flexible films for bags, pouches, pillow packs and top webs or as rigid and semi-rigid structures for base trays, dishes, cups and tubs. Commonly used plastic flexible laminates are produced from polyethylene (PE), polypropylene (PP), polyamide (nylons), polyethylene terephthalate (PET), polyvinyl chloride (PVC), polyvinylidene chloride (PVdC) and ethylene vinyl alcohol (EVOH). Rigid and semi-rigid structures are commonly produced from PP, PET, unplasticised PVC and expanded polystyrene.

Emerging Packaging Technologies: Case Studies

Bioactive packaging for meat products

Among some commercial systems based on bioactive compounds directly incorporated into the packaging, an alternative to gaseous generator sachets is the incorporation of the scavenger into the packaging structure itself (e.g. Cryovac OS2000 systems, Sealed Air Corporation, USA; ZERO2, CSIRO-FSA, Australia). The Cryovac OS system was based on the incorporation of an oxygen smart packaging scavenging organic compound into a polymer for use as a layer in a laminated packaging film. The scavenger is designed to be activated by ultraviolet light just before filling and sealing on the packaging line. In MAP systems, with less than 2% initial oxygen levels, Cryovac OS films can reduce oxygen to less than 0.1% typically in 3 days. The ZERO TM 2 from CSIRO was also based on a UV light activated oxygen scavenging polymer. It should be noted that the speed and capacity of O_2-scavenging films are considerably low compared with iron based O_2 scavenger sachets. The O_2-scavenging film of Ageless-OMAC (Mitsubishi Gas Chemical Co., Japan) is activated by heat.

Today, it is used in particular for the preservation of pasteurized cooked rice and could potentially be used for meat preservation. Another good example of commercial systems, is the AgION antimicrobials which are based on the use of silver, a powerful antimicrobial metal ion, as the active ingredient. AgION compounds are compatible with a wide range of polymers and they will not volatilize during processing. AgION antimicrobial products can be compounded into bulk polymers, incorporated into fibers during spinning, and mixed into a variety of coatings. Another example is Triclosan, the antimicrobial substance used by Microban (UK). For more than 20 years this chemical compound has been used effectively in personal hygiene products such as toothpaste, mouthwash, deodorant and soap, as well as an antibacterial agent in the hospital environment. This protection is achieved by combining Triclosan with any of the major polymers (e.g., PE, PP, PVC). The Triclosan fits into the empty spaces of the polymer and migrates to the surface to start its work against any developing bacteria. During washing, the molecules closest to the surface are cleaned away but are immediately replaced by other protective molecules.

Such systems are mainly elaborated by adding the antimicrobial compound to the extruder when the film or coextrusion is produced. A disadvantage is that the high temperatures and shearing associated with the extrusion process can deteriorate the antimicrobial additives. If this category of bioactive films has not made it into the supermarket in larger numbers, it is not through any lack of research. Around the world, a range of agents have been selected with the aim of prolonging shelf-life and ensuring food safety. But this is one area where those in R&D may be left feeling that the commercial and regulatory world still has some catching up to do. Different antimicrobial molecules, such as bacteriocins, enzymes, chelatants and organic acids (i.e. propionic and sorbic acids), included in different films, have been tested to suppress pathogenic or saprophyte bacteria growth in meat product preservation. For the systems described here and with a non-volatile biocide, a contact between the food and the package is obviously necessary and a migration process from the packaging materials is expected.

Smart technologies for bakery products

The use of antimicrobial food packaging films can offer advantages compared to the direct addition of preservatives to the food product, as the active agents are applied to the packaging material in a way such that only low levels of preservative come into contact with the food. An example of this technology is a cellulose acetate film containing 4 % sodium propionate that effectively reduced mould growth in bread. Other volatile antimicrobial substances from spices and herbs have proved efficient in the control of fungal spoilage by common bread spoiling fungi. Through entrapment of these substances in cyclodextrins, as explained above, microbial control through packaging using natural substances could be obtained, but there is a lot of research needed prior to commercialization of these technologies. Oxidation of fats and oils from some high-fat content bakery products such as snacks, crackers, biscuits and cereal products, can lead to the formation of malodorous aldehyde substances such as hexanal and heptanal, which upon the opening of the package will cause rejection of the product.Aldehyde scavengers incorporated into polymers are another kind of smart packaging technology for this specific problem.

Smart technologies for fresh produce

In conjunction with a_w, the relative humidity (RH) of the storage environment is important in determining the growth of microorganisms in foods. The excess water development inside a food package usually occurs due to the respiration of fresh produce. Humidity absorbers, another kind of smart polymer technology that will be further described below, could be used to avoid the presence of water inside the package of fresh produce. Several antimicrobial films have also been developed specifically for fruit and vegetables. In the case of fruits mainly, the antimicrobial compound can be incorporated into an edible film or coating applied by dipping or spraying onto the fruit. A commercial antifungal coating produced from chitosan is sold as a shelf-life extender for fresh fruit.

Reducing lactose and cholesterol content: enzymatically active packages

Some immobilized enzymes that were initially applied in food production lines are currently being considered for food packaging applications. The objective of these systems is to catalyse a reaction that is considered beneficial from a nutritional point of view (i.e., decreasing the concentration of a non-desired food constituent and/or producing a food substance attractive for the consumer). Lactase, for example, can be immobilized by covalent attachment to functionalized surfaces of, for instance, low density polyethylene (LDPE), retaining 10 % of the free enzyme activity. The lactase-active package is meant to reduce the lactose content of milk during storage by splitting this disaccharide sugar into glucose and galactose. Similarly, the enzyme cholesterol reductase could be immobilized into polymers to convert cholesterol into coprostanol and coprosterol. These converted compounds are very poorly absorbed by the digestive system and pass through intact. The use of this type of package allows the production of a value-added product without modifying the manufacturing procedure. UHT milk produced by a conventional process can be packaged in a lactase-active or cholesterol-active package, and through storage, the product arrives at the market as a low/free-lactose or low-cholesterol product, respectively. This processing plant inside the package appears to be a very promising technology.

Summary

Implementation of some of the existing smart polymer packaging systems can make these technologies a major partner in protection and shelf-life extension of foods. Given the variety of smart concepts and the versatility of polymers, packaging structures can, in principle, be designed to enhance quality and safety while prolonging the shelf-life of potentially any food product in the market. The general trends seem to point towards the use of natural substances (which, in case of migration, do not pose any health risk), incorporated into biodegradable biomass derived matrixes. The latter ones, apart from being environmentally friendly and better vehicles for the release of substances, will be a good economical option when the shortage of oil will cause an increase in the prices of its derived products. Several raw materials originating from agricultural and marine sources can be used to fabricate biodegradable films such as a number of polysaccharides (starch, cellulose and derivatives, alginates, carrageenan, chitosan, etc.), proteins (corn zein, soy protein isolates, wheat gluten, milk proteins, gelatine, etc.) and lipid-based matrixes (waxes or glycerides, seedoils).

Among the active polymer packaging technologies, the area of antimicrobial packaging is probably the one receiving most of the attention and a number of natural substances are being tested against the most common food-borne pathogens. However, looking a step further in what the role of packaging can be in the commercialization of food products, apart from the increasing number of natural substances that can be added to the package walls to exert an active role in preservation, polymeric sustainable and/or biodegradable biomaterials could also be vehicles of functional or bioactive substances, like vitamins, pre- and probiotics, overcoming some of the existing drawbacks in the fabrication of functional foods. This is a new concept in packaging, termed bioactive packaging that seeks to transform foods into functional food upon packaging. The future will also see the application of various nanotechnologies, such as the use of nanoclays and electrospun or electrosprayed active or bioactive nanostructured materials. These technologies will offer invisibility within the package (i.e. transparency and required mechanical properties) and will also become excellent carrying systems where the active principle can be better dispersed and, therefore, be more effective. These novel nanotechnologies will also enhance the plastic or bioplastic properties and will act as functional barriers against unintended migration or as scavengers of potential toxic byproducts, thus increasing the quality and safety of fast moving consumer goods.

References

1. Brody, A.L., E. Strupinsky, and L.R. Kline, *Active packaging for food applications* 2001: CRC Press.
2. Vermeiren, L., *et al.*, *Developments in the active packaging of foods.* Trends in Food Science & Technology, 1999. **10**(3): 77-86.
3. Ahvenainen, R., *Novel food packaging techniques,* 2003: Elsevier.
4. Suppakul, P., *et al.*, *Active packaging technologies with an emphasis on antimicrobial packaging and its applications.* Journal of food science, 2003. **68**(2): 408-420.
5. Appendini, P. and J.H. Hotchkiss, *Review of antimicrobial food packaging.* Innovative Food Science & Emerging Technologies, 2002. **3**(2): p. 113-126.
6. Cha, D.S. and M.S. Chinnan, *Biopolymer-based antimicrobial packaging: a review.* Critical Reviews in Food Science and Nutrition, 2004. **44**(4): 223-237.
7. Vermeiren, L., F. Devlieghere, and J. Debevere, *Effectiveness of some recent antimicrobial packaging concepts.* Food Additives & Contaminants, 2002. **19**(S1): 163-171.
8. Weng, Y.-M. and J. Hotchkiss, *Inhibition of surface molds on cheese by polyethylene film containing the antimycotic imazalil.* Journal of Food Protection (USA), 1992.
9. Ito, Y., *et al.*, *Application of ion-exchange cartridge clean-up in food analysis: I. Simultaneous determination of thiabendazole and imazalil in citrus fruit and banana using high-performance liquid chromatography with ultraviolet detection.* Journal of Chromatography A, 1998. **810**(1): 81-87.
10. Coma, V., *et al.*, *Antimicrobial edible packaging based on cellulosic ethers, fatty acids, and nisin incorporation to inhibit Listeria innocua and Staphylococcus aureus.* Journal of Food Protection®, 2001. **64**(4): 470-475.
11. Janjarasskul, T. and J.M. Krochta, *Edible packaging materials.* Annual Review of Food Science and Technology, 2010. **1**: 415-448.

12. Gontard, N. and S. Guilbert, *Bio-packaging: technology and properties of edible and/or biodegradable material of agricultural origin*, in *Food packaging and preservation,* 1994, Springer. p. 159-181.
13. Debeaufort, F., J.A. Quezada-Gallo, and A. Voilley, *Edible films and coatings: tomorrow's packagings: a review.* Critical Reviews in Food Science, 1998. **38**(4): 299-313.
14. Quintavalla, S. and L. Vicini, *Antimicrobial food packaging in meat industry.* Meat Science, 2002. **62**(3): 373-380.
15. Cooksey, K., *Effectiveness of antimicrobial food packaging materials.* Food Additives and Contaminants, 2005. **22**(10): 980-987.
16. Dainelli, D., *et al.*, *Active and intelligent food packaging: legal aspects and safety concerns.* Trends in Food Science & Technology, 2008. **19**: S103-S112.
17. Sivertsvik, M., J. Rosnes, and H. Bergslien, *Modified atmosphere packaging.* Minimal processing technologies in the food industry, 2002: p. 61-86.
18. Mir, N. and R.M. Beaudry, *Modified atmosphere packaging.* Agriculture Handbook, 2004. **66**.
19. Inns, R., *Modified atmosphere packaging*, in *Modern processing, packaging and distribution systems for food,* 1987, Springer. p. 36-51.
20. Church, N., *Developments in modified-atmosphere packaging and related technologies.* Trends in Food Science & Technology, 1994. **5**(11): 345-352.

Section V: Functional Food Products: Prevention, Disease Control and Bioavailability

14

Functional Food in Reducing The Risk of Cardiovascular Disease

Introduction

Cardiovascular disease (CVD) is still a major cause of death in Western populations and is becoming an important cause of morbidity and mortality worldwide. Thanks to advanced medical knowledge and treatments, many patients survive an initial event. Because of that, prevention of secondary CVD is a growing task for nutritionists and other health professionals. Cardiovascular risk can be reduced by lifestyle changes, one of which is diet. There is now substantial evidence from epidemiological and clinical studies that a diet rich in fruits, vegetables, unrefined grains, fish and low-fat dairy products, and low in saturated fats and sodium, can reduce the risk of coronary heart disease and hypertension.

Many functional foods have been found to be potentially beneficial in the prevention and treatment of cardiovascular disease, the leading cause of mortality in the United States. These foods include soybeans, oats, psyllium, flaxseed, garlic, tea, fish, grapes, nuts, and stanol and sterol ester enhanced margarine. When eaten in adequate amounts on a consistent basis, these foods may aid in decreasing the risk of cardiovascular disease by several potential mechanisms viz. lowering blood lipid levels, improving arterial compliance, reducing low density lipoprotein oxidation, decreasing plaque formation, scavenging free radicals, and inhibiting platelet aggregation.

Diet Habits and its Role in Prevention of Cardiovascular Disease

Active prevention of coronary heart disease (CHD) is usually started immediately after its first clinical manifestation. Secondary prevention focuses on risk reduction in people with established CHD who are at high risk of recurrent cardiac events and death [1]. It is important to remember that the two main causes of death in these patients are sudden cardiac death (SCD) and heart failure (HF), often resulting from myocardial ischaemia and subsequent necrosis [2].

Whatever the specific clinical aims of the programme, nutritional evaluation and counselling of each individual with CHD must be a key point of the preventive intervention. Nutrition is, however, only one component of such a programme. Exercise training, behavioural interventions (particularly to help the patient abstain from smoking) and drug therapy have equally important roles. The control of risk factors has been seen traditionally in the perspective of prevention. The dietary prevention programme is commonly initiated during hospitalisation for a first CHD event. With the shortening of stay in the coronary care unit (CCU), dietary intervention is initiated during the following days at hospital, then continued in secondary prevention centres and included in cardiac rehabilitation programmes. An individual dietary prevention programme should be developed under the guidance of a specialised dietician and in close collaboration with the patient's cardiologist and primary care physician, so that there is no discontinuity or discrepancy in dietary counselling between the hospitalisation and post-hospitalisation phases of the rehabilitation programme. Numerous epidemiologic studies indicate that an increase in the consumption of fruits and vegetables is associated with a decrease in the incidence of cardiovascular disease (CVD), CHD, and stroke [3].

Criteria for Demonstrating Functional Effects

Evidence of the effects of functional foods has to be based on objective criteria developed and accepted by the scientific community. Functional foods have to beneficially modulate target functions of the body, which are relevant to improved state of health or well-being, or for reduction of disease risk. As a result, based on current knowledge, the target function to be modulated has to be clearly defined. A set of markers can be used to define the function, demonstrate its modulation and demonstrate the effects of the modulation (Fig. 14.1)[4]. Different types of markers may be chosen viz. biochemical, physiological or behavioural. Markers may directly represent either an event of interest or correlated events.

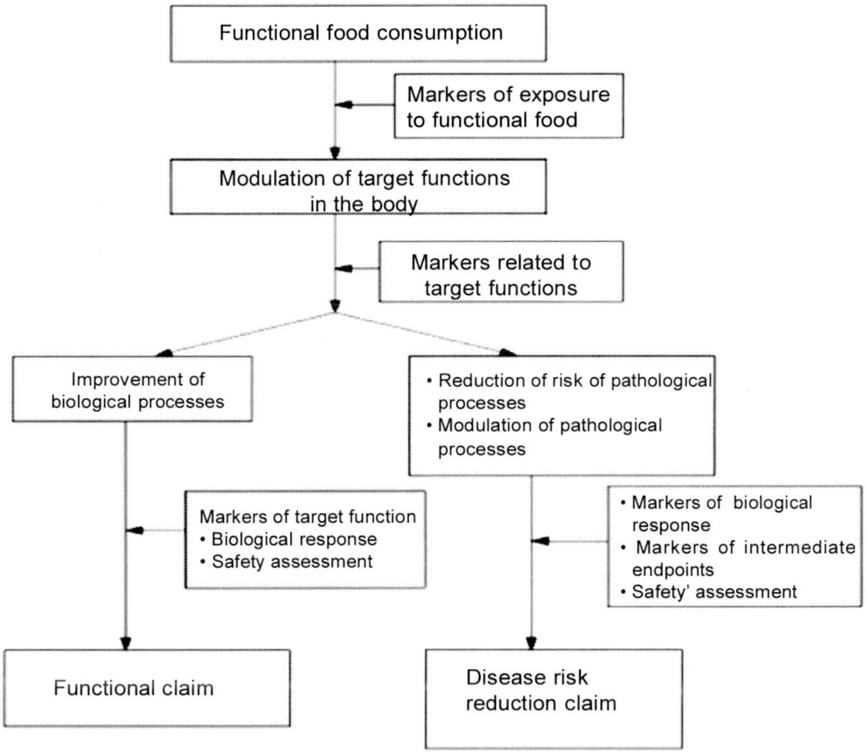

Fig. 14.1: Basis for functional claims and disease risk reduction claims for functional food

Markers must be specific, sensitive, reproducible, validated and biologically significant. They should be related to the target functions, function improvement or biological response improvement, or to appropriate intermediate end-points of improved health or reduced disease risk [4]. Markers should register short and long term impact of the foods. It should also be possible to use these markers in safety assessment of functional foods. Markers of exposure to the foods (food components) are also needed. All markers must fulfil standard quality control criteria [5] .

Using a defined set of markers the modulation of relevant target functions and the improvement of the target biological process must be documented. The functional effects have to be demonstrated in human nutrition studies for all members of a population or for particular group clearly defined by specific markers for example by age, sex or genetic markers. Based on data on dose-dependent effects, the functional effects of realistic amounts of foods consumed as part of the daily diet has to demonstrate. The assessment of the safety of the amount of foods needed for functional effects has also to be performed, documented and presented according to accepted standards.

Bioactive Compounds in Reducing Risk of Cardiovascular Disease

"Bioactive compounds" are extra nutritional constituents that typically occur in small quantities in foods. They are being intensively studied to evaluate their effects on health. The impetus sparking this scientific inquiry was the result of many epidemiologic studies that have shown protective effects of plant based diets on CVD and cancer (Table 14.1). Many bioactive compounds have been discovered [6]. These compounds vary widely in chemical structure and function and are grouped accordingly. Phenolic compounds, including their subcategory, flavonoids, are present in all plants and have been studied extensively in cereals, legumes, nuts, olive oil, vegetables, fruits, tea, and red wine [7]. Many phenolic compounds have antioxidant properties, and some studies have demonstrated favorable effects on thrombosis and tumorogenesis and promotion.

Table 14.1: Various bioactive compounds helpful in reducing the risk of coronary heart disease

Bioactive compound	Examples	Sources
Flavonoids		
Flavonols	Quercetin, kaempferol, catechin	Onion, apple, tea, berries, olives, broccoli, lettuce, red wine, cocoa/chocolate
Flavonols	Epicatechin Epigallocatechin Epicatechin-3-gallete Epigallocatechin-3-gallete	Green/black tea, cocoa/ chocolate
Phytoestrogens		
Lignans, coumestran	Enterolactone, enterodial, coumestrol	Flaxseed oil, lucerne, clover
Isoflavones	Genistein, daidzein	Soybeans, legumes
Resveratrol		Grapes, red wine, peanuts
Organosulfur compounds	Allicin, diallyl sulfide, diallyl disulfide, allyl mercaptan	Garlic, onion, leek
Soluble dietary fibers	Glucan, pectin, psyllium	Oats, barley, yeast, fruit,vegetables, psyllium seed, fortified cereals and grains
Isothiocyanates (ITC)	Phenethyl, benzyl, sulforaphanes	Cruciferous vegetables (watercress, broccoli)
Monoterpenes	d-Limonene, perillic acid	Essential oils of citrus fruit, cherries, mint, herbs
Plant sterols	Sitostanol, stigmasterol, campesterol	Tall oil, soybean oil, rice bran oil
Olive oil	Tyrosol, hydroxytyrosol, oleoeuropeine, caffeic acid, cumaric acid	Extra virgin olive oil

Although some epidemiologic studies have reported protective associations between flavonoids or other phenolics and CVD and cancer, other studies have not found these associations. Numerous bioactive compounds appear to have beneficial health effects. Much scientific research needs to be conducted before we can begin to make science based dietary recommendations. Despite this, there is sufficient evidence to recommend consuming food sources rich in bioactive compounds. From a practical perspective, this translates to recommending a diet rich in a variety of fruits, vegetables, whole grains, legumes, oils, and nuts. Collectively, the epidemiologic studies evaluating associations between intakes of a variety of plant-based foods indicate a protective effect, both on CVD and certain cancers. Possible constituents in food that account for these protective effects are discussed in subsequent sections of this chapter.

Phenolic compounds

Phenolic compounds, commonly referred to as polyphenols, are present in all plants and, thus, are in the diet. There are about 8,000 phenolic structures that have been identified that vary structurally from being simple molecules (e.g., phenolic acids with a C_6 ring structure) to being highly polymerized compounds (i.e., tannins). More than 10 classes of polyphenols have been defined on the basis of chemical structure. The flavonoids are the most common polyphenolic compounds present in plant food.

Several population studies have reported an inverse association between flavonoid intake and risk of coronary disease and cancer [8, 9]. In the Zutphen, a high intake of flavonoids (approximately 30 mg/day) was associated with approximately a 50% reduction in CHD mortality rate compared with individuals who had a low flavonoid intake (19 mg/day) [10].

Red wine is a rich and concentrated source of polyphenolic substances and around 200 individual phenolic compounds have been identified to date. Studies have shown that red wine inhibits oxidation of LDL *in-vitro* and increases antioxidant capacity of plasma [11]. The antioxidants identified in red wine include phenolic acids, flavonols, monomeric catechins, and polymeric anthocyanidins. Collectively, there is evidence emerging that phenolic compounds have antithrombotic effects that appear to be the result of reduced susceptibility of platelet aggregation, reduced synthesis of prothrombotic and proinflammatory mediators, decreased expression of adhesion molecules, and tissue factor activity. In addition, there is some evidence that wine polyphenols can modulate the production of nitric oxide by the vascular endothelium, resulting in vasorelaxation [12].

Olive oil phenolics

Olive oil has been shown to reduce LDL oxidizability in the post-prandial state rather than the fasting state [13]. In part, this reflects its fatty acid profile (high in monounsaturated fatty acids) that is less susceptible to lipid peroxidation than polyunsaturated fatty acids. Tocopherol also may contribute to the antioxidant effects of olive and other vegetable oils. Antioxidant effects of olive oil also appear to be the result of phenolic compounds [14]. Olive oil, in particular, the first pressed or better known "extra virgin" type, has a high content of phenolic components, which have been shown to be powerful antioxidants. Olive oil phenols are a complex mixture of compounds that include 3,4-dihydroxyphenylethanol (hydroxytyrosol), 4-hydroxyphenylethanol (tyrosol), 4-hydroxyphenylacetic acid, protocatechuic acid, syringic acid, vanillic acid, caffeic acid, and p-coumaric acid. The concentration of the phenolic fraction in olive oil varies depending on the cultivar, climate, and degree of ripeness of the fruit; the average concentration is 500 mg/L in extra virgin olive oil [15]. Of the various phenolic constituents of olive oil, hydroxytyrosol seems to be among the most important. It is present in free form and also as a constituent of complex molecules (i.e., oleuropein). Its *in-vitro* antioxidant potency is greater than butylated hydroxytoluene. In addition, olive oil phenols have greater antioxidant potency than other vegetable oils [16]. High phenolic olive oil compared with high oleic sunflower oil reduced LDL peroxidation in hypercholesterolemic postmenopausal women [15].

Phytoestrogens

Phytoestrogens, or estrogenic compounds in plants, are divided into 3 main classes viz. isoflavonones, coumestans, and lignans. Structurally, all are diphenolic compounds that are similar to estrogen and, as might be expected, bind to the estrogen receptor. Surprisingly, however, they act both as partial estrogen agonists and antagonists, thereby having similar and opposing actions compared with estrogen [17]. The isoflavones, genistein and daidzein, are found predominantly in legumes, namely soybeans. Coumestrol is found in lucerne, alfalfa, and clovers and at low concentrations in beans and peas. Lignans are the most ubiquitous phytoestrogens, because they exist as minor constituents of many plants, where they are involved in plant cell wall formation. The primary dietary source of lignans is flaxseed oil, but it can be found in varying concentrations in soybeans, seaweed, whole grains, fruits, and vegetables.

Isoflavonoids are the most extensively studied phytoestrogens with respect to CVD. Because soy foods are the most significant dietary source of isoflavones, many studies with humans and nonhuman primates have been conducted evaluating the effects of soy foods and constituents of soy foods on numerous

CVD risk factors [18, 19]. Soy foods have been shown to have favorable effects on plasma lipids and lipoproteins. A meta analysis of 38 clinical studies reported that total cholesterol was decreased by 9%, LDL cholesterol by 13%, and triglycerides by 11% when an average of 47 g of soybean protein was consumed, 100 with a greater response observed in subjects having a higher baseline cholesterol level [20]. High-density lipoprotein (HDL) cholesterol was increased modestly (*i.e.*, 2.4%; not significant). The active components of soy protein that are thought to account for these effects are the isoflavones genistein and daidzein. Research to date shows multiple beneficial effects of soy that appear to be the result of isoflavones. However, further studies are needed to resolve the biological effects that individual isoflavones have on CVD risk and the mechanisms that account for these effects. Furthermore, it will be important to reconcile these beneficial effects of soy protein with recent evidence indicating potentially adverse effects of soy protein with isoflavones on lipoprotein and brachial artery flow mediated dilation. The adverse effects reported in the latter study occurred despite favorable effects on LDL cholesterol and the LDL cholesterol:HDL cholesterol ratio [20].

Resveratrol

Resveratrol is a polyphenol (3,5,4'-trihydroxystilbene), thought to be a phytoalexin, one of a group of compounds produced during times of environmental stress or pathogenic attack [21]. It is found principally in the skin of grapes and is produced in other plants, including peanuts. Red wine is a rich source of resveratrol and is thought to confer the cardioprotective effects associated with moderate consumption of wine. There is evidence to suggest that resveratrol inhibits both LDL oxidative susceptibility *in-vitro* and platelet aggregation as well aseicosanoid synthesis. However, resveratrol attracted little interest until 1992, when it was suggested to explain some of the cardioprotective effects of red wine [22]. Resveratrol, a key compound implicated in the cardiovascular benefits associated with red wine consumption, exerts cardiovascular protection by diverse mechanisms, resulting in the identification of multiple direct targets of resveratrol. Since that time, many studies have shown that resveratrol can prevent or slow the progression of a variety of conditions, including cancers, cardiovascular diseases, and ischemic injuries [21, 23, 24], as well as enhance stress resistance and extend lifespan. However, it should be kept in mind that at present, the literature on resveratrol is contradictory and confusing. The wide ranges of concentrations and doses used to achieve the various effects reported for resveratrol in both in-vitro cell culture and animal studies raise many questions about the concentrations that are achievable *in-vivo*. Furthermore, resveratrol has a short initial half-life and is metabolized extensively in the body.

Lycopene

Lycopene is a hydrocarbon carotenoid that has recently received attention for its potential role in preventing prostate cancer and cardiovascular disease in humans. Although similar in structure to the more studied β-carotene, lycopene does not have provitamin A activity. The many conjugated double bonds of carotenoids make them potentially powerful antioxidants, and lycopene is no exception. Lycopene is an acyclic carotenoid found primarily in tomatoes and tomato products (about 80% of dietary lycopene in the United States). Other minor food sources include apricots, grapefruit, guava, watermelon, and papaya. Tomato lycopene levels vary widely among different varieties and stages of ripeness. Bioavailability is enhanced by cooking food sources of lycopene, particularly in the presence of oil or fats. Considerable evidence suggests that lycopene, a carotenoid without provitamin A activity found in high concentrations in a small set of plant foods, has significant antioxidant potential *in-vitro* and may play a role in preventing prostate cancer and cardiovascular disease in humans [25-27].

Organosulfur compounds

Epidemiological studies have shown an inverse relationship between consumption of fruits and vegetables and the risk of cardiovascular disease. Phytochemicals are non-nutritional chemical compounds found in small quantities in fruits and vegetables with known health benefits. Among them, organosulfides are present mainly in garlic and onion characterized by their antioxidant and anti-inflammatory properties, and isothiocyanates in cruciferous vegetables have anticarcinogenic effects in experimental models.

Dietary factors play a key role in the development or prevention of various human diseases, including cardiovascular disease. Epidemiologic studies have shown an inverse relationship between diets rich in fruits, vegetables and spices, and risk of all-cause cancer and cardiovascular disease death [28]. These foods contain phytochemicals that have beneficial properties in cardiovascular disease. One source of such phytochemicals is organosulfur compounds whose role in cardiovascular disease protection has been extensively studied. Organosulfur compounds have been shown to exert diverse biological effects such as antioxidant effects, anti-inflammatory properties, inhibition of platelet aggregation, reduction of systolic blood pressure, and reduction of cholesterol levels [29]. The organosulfur compounds best studied in cardiovascular protection are those found in alliums mainly garlic and onion. On the other hand, the isothiocyanates, particularly sulforaphane (best known for its anticancer properties) have recently been attributed antioxidant and anti-inflammatory properties in cardiovascular disease.

Most of the CVD research conducted with food sources of organosulfur compounds has evaluated the effects of garlic oil and garlic on various risk factors. A number of studies have shown, in general, that garlic favorably affects important risk factors for CVD [30]. Garlic oil and garlic consumption have been shown to decrease total and LDL cholesterol and triglyceride levels. Consumption of 0.5 to 1 clove of garlic per day lowers cholesterol levels approximately 10%. Mechanisms that explain the observed effects of garlic include a decrease in cholesterol and fatty acid synthesis and cholesterol absorption. However, there is some evidence that garlic powder does not lower cholesterol levels, which may reflect either a loss of active compound(s) during processing or an inhibition in the release of active components in garlic. The formation of these active compounds is influenced by crushing garlic, duration of the drying process, the temperature at which garlic is dried, and humidity. Garlic extracts also have been shown to elicit antithrombotic effects and modestly decrease blood pressure (i.e., 5.5% decrease in systolic blood pressure and a slight decrease in diastolic blood pressure). There also is evidence that aged garlic extracts, including water and lipid soluble compounds, have antioxidant effects. Aged garlic extract results from the prolonged extraction of fresh garlic at room temperature and contains allyl amino acid derivatives, stable lipid soluble allyl sulfides, flavonoids, and saponins.

There is much evidence from *in-vitro* and *in-vivo* studies supporting the beneficial effects of organosulfide consumption in the prevention of cardiovascular disease. Existing evidence to date suggests that organosulfides have the ability to prevent ROS production, increase antioxidant status, increase the bioavailability of nitric oxide, prevent vascular inflammation, reduce lipid content and inhibit platelet aggregation. The results obtained in human are inconsistent, probably due to differences in garlic preparations, unknown active constituents and their bioavailability, differences in the population selected and duration of trials. Also, organosulfur compounds from alliums and cruciferous vegetables are usually cooked before being consumed, and it is known that cooking inactivates the enzyme and thus the production of its bioactive constituents [31]. Therefore, it is also important to apply the best cooking conditions in order to prevent the loss of the bioactive compounds.

Plant sterols

Phytosterols are naturally occurring plant sterols that are present in the non-saponifiable fraction of plant oils. Structurally, plant sterols are similar to cholesterol except that there always are some substitutions on the sterol side chain at the C-24 position. They are not synthesized in humans, are poorly absorbed, and are excreted faster from the liver than cholesterol, which explains their low abundance in human tissues. The primary plant sterols in the diet are sitosterol, stigmasterol, and campesterol. Typical consumption of plant sterols is approximately 200 to 400 mg/day. The most abundant plant sterol in western diets is β-sitosterol. Studies with sitosterol or mixtures of plant sterols (approximately 1 g/day) have shown that they reduce serum cholesterol levels in humans by approximately 10 %. This discovery has resulted in subsequent research to evaluate the effects of sitosterol derivatives on cholesterol absorption and serum cholesterol levels. Sitostanol was shown to be more potent in reducing cholesterol absorption and serum cholesterol levels than sitosterol. These findings provided the basis for the current era of research evaluating the effects of sitostanol and sitostanol esters from different plant oil sources. Special margarines are the primary food source of plant sterols/stanols. The plant sterol mixtures are derived from different oil sources, including pine tree wood pulp (tall oil), soybean oil, rice bran oil, and shea nut oil. The European Atherosclerosis Society (EAS) recommendations also provide information on the use of plant sterols. Currently, foods with added plant sterols/stanols may be considered in individuals with high cholesterol levels but equally in those with intermediate or low global CV risk who do not qualify for pharmacotherapy. On the basis of critical appraisal of the evidence base above, this EAS Consensus Panel therefore considers that there is a place for these products, in conjunction with other lifestyle interventions, in patients receiving lipid-lowering therapy with statins or other agents who do not achieve LDL-C targets, or in those with statin intolerance. Finally, given the increasing importance of early preventive strategies in hypercholesterolaemia, the potential for inclusion of plant sterol or plant stanol enriched foods in the diet of adults and children (>6 years) with familial hypercholesterolaemia (FH), as an adjunct to lifestyle and pharmacotherapy, may be considered.

Epidemiological studies in the UK (n = 22,256), Sweden (n = 77,652) and China (n = 3940) observed that naturally occurring dietary plant sterol intake is inversely related to plasma total- and LDL-C levels [32, 33]. However, in a well controlled study in healthy subjects, low (126 mg plant sterols/2000 kilocalories) or high intake of plant sterols (449 mg plant sterols/2000 kilocalories) did not affect plasma LDL-C concentrations in spite of modulating cholesterol metabolism. In contrast to the minimal effects of variation in consumption of naturally-occurring plant sterols/stanols in the diet, early studies such as that by Farquhar et al. (1956) demonstrated that betasitosterol supplementation lowered both total serum cholesterol and LDL-C (as beta-lipoprotein lipid) in young men with atherosclerotic heart disease [34]. Subsequently, Miettinen et al. (1995) demonstrated for the first time that foods (such as margarine) enriched with sitostanol ester lowered both total serum cholesterol and LDL-C in mildly hypercholesterolaemic subjects [35].

Subsequent data show consistent support for the LDL-C lowering effects of foods with added plant sterols/stanols [36, 37]. In the most recent meta-analysis, consumption of foods with added plant sterols/stanols (2 g/day) lowered LDL-C to a similar extent (8.2% and 9.3%, respectively) [36]. Regrettably, there is a paucity of data relating to the potential of higher plant sterol/stanol doses to further lower LDL-C levels, and thus CV risk [37]. It has been suggested that maximal LDL-C lowering may be greater with plant stanols (up to 16%) and plant stanol esters (17%), but this conclusion relies on limited studies using doses of 4-9 g/day, none of which involved direct head-to-head comparisons with plant sterols. Whether there is a position for plant sterol and/or plant stanol intakes to be raised higher than those currently recommended (2-3 g/day) in prevention strategies for the general population remains open. The fact that there is consistent, robust evidence indicating that lowering of LDL-C by different mechanisms (statins, diet, partial ileal bypass, bile acid sequestrants) results in reduction in CV risk, underlies the rationale for inclusion of plant sterols/stanols in international clinical guidelines for the management of dyslipidaemia. In this context, it must be recognised that the use of functional foods enriched with plant sterols/stanols is currently not advised for children under 6 years.

Dietary fibers

Despite remarkable improvements in its prevention, diagnosis and therapy, cardiovascular disease (CVD) remains the leading cause of morbidity and mortality in the United States and most Western countries [38]. One well-established way to reduce the risk of developing CVD is to lower serum LDL cholesterol levels by making dietary changes. In addition to reducing saturated fat and cholesterol intake, and increasing cisun saturated fat intake, the importance of other dietary approaches, such as increasing the intake of water-soluble dietary fibers has become increasingly recognized. Dietary fiber is a collective term for a variety of plant substances that are resistant to digestion by human gastrointestinal enzymes. Dietary fiber can be classified as either water soluble or water insoluble. The structural or nonviscous fibers (lignins, cellulose, and some hemicelluloses) are water insoluble. Vegetables and cereal grains are especially rich in water insoluble fiber, with the highest amounts in wheat and corn. Water insoluble fiber is responsible for increased stool bulk and help to regulate bowel movements. The natural gel forming or viscous fibers (pectins, gums, mucilages, algal polysaccharides, some storage polysaccharides, and some hemicelluloses) are water soluble. Foods rich in water soluble fiber are dried beans, oats, barley, and some fruits and vegetables [38]. The mean total daily fiber intake amongst adults in most industrialized countries is well below 25g, the minimal amount recommended by various health organizations. Of total dietary fiber intake, approximately 20% is water soluble and 80% is water insoluble [39]. Defining the role of dietary fiber in coronary heart disease risk is still vitally important for three reasons. First, an understanding of the role of fiber in coronary heart disease risk can help clarify and reinforce dietary recommendations and goals. Second, the mechanisms of how dietary fiber may reduce cardiovascular disease risk are not yet fully understood. It appears that the protective effect of dietary fiber is independent of its effect on serum cholesterol levels. The cholesterol lowering effect of dietary fiber has been shown to be modest at best, and the reduced risk of coronary heart disease associated with whole grain intake may be only partly due to its contribution to total fiber intake [40]. Insights into the mechanisms of fiber's protective effect may lead to better dietary recommendations and new prevention strategies [41]. Third, increasing the fiber content in the diet is relatively low-cost, safe and associated with few side effects, making it a viable option for a large proportion of the population. If its effect is truly protective, even if only modestly so, it has the potential to have a large impact on disease rates at the population level because the prevalence of low-fiber diets in the U.S. is substantial.

Epidemiologic studies have suggested that dietary fiber protects against CHD. In the Scottish Heart Health study, increased fiber intake was associated with both a reduced risk of CHD and decreased mortality. In the Nurses' Health Study, a 10-g/day increase in total fiber intake was associated with a 20% reduction in CHD events. In the latter study, only cereal fiber was strongly associated with decreased risk of CHD. In contrast, based on a meta-analysis, Anderson *et al*., 1990 reported that cereal fiber derived from refined cereal products was not associated with a cardioprotective effect [38].

β-glucan, psyllium, and pectin are soluble dietary fibers that have been shown to lower total and LDL cholesterol levels. β-glucan is found in oats, barley, and yeast; pectin is present in fruits and vegetables; and psyllium is derived from the husks of blond psyllium seed and currently is being added to some foods, including cereals and other grain products. Fiber from 2 servings of oats (approximately 2.6 g of soluble fiber) elicits a 2% to 3% cholesterol lowering effect beyond what is achieved by a serum cholesterol lowering diet. A meta analysis of 8 studies has shown that approximately 10 g/day of psyllium lowers total cholesterol 4% and LDL cholesterol 7%. Another soluble fiber, flaxseed gum, has been shown to lower LDL cholesterol by 7% to 8% when 5 to 6 g/day is consumed. Another meta analysis has shown that the cholesterol lowering effects of β-glucan, psyllium, and pectin are comparable. Because different foods have varying amounts of these soluble fibers, the cholesterol lowering response expected will depend on the quantity of soluble fiber consumed. Thus, soluble fiber has a modest cholesterol lowering effect that goes beyond what can be achieved by lowering saturated fatty acid and cholesterol.

Summary

Impressive progress is being made in defining the role of bioactive compounds in reducing the risk of major chronic diseases and the underlying biological mechanisms that account for these effects. An ever expanding list of bioactive compounds is being scientifically evaluated. As discussed herein, numerous bioactive compounds appear to have beneficial health effects. On the basis of a large population database, there is sufficient evidence to recommend a diet high in food sources rich in bioactive compounds. From a practical perspective, this means recommending a diet rich in a variety of fruits, vegetables, whole grains, legumes, and nuts that are prepared in different ways.

As CVD continues to be the leading cause of mortality in the United States, researchers, practitioners, and consumers are searching for alternatives and additions to traditional therapies for prevention and treatment. Consistent consumption of effective levels of functional foods, including soybeans, oats, psyllium, flaxseed, garlic, tea, fish, grapes, nuts, and stanol and sterol ester enhanced margarine, may provide an opportunity for many consumers and patients to decrease the risk of cardiovascular disease. These foods may potentially provide benefit alone, in combination, or in addition to cholesterol lowering medications and other therapies. Although additional studies are needed, the research in the area of functional foods and CVD reduction has been growing significantly and will continue to uncover practical, food-based recommendations for heart health

References

1. Pearson, T.A., *et al.*, *AHA guidelines for primary prevention of cardiovascular disease and stroke: 2002 update consensus panel guide to comprehensive risk reduction for adult patients without coronary or other atherosclerotic vascular diseases.* Circulation, 2002. **106**(3): 388-391.
2. Hasler, C.M., S. Kundrat, and D. Wool, *Functional foods and cardiovascular disease.* Current Atherosclerosis Reports, 2000. **2**(6): 467-475.
3. Roberfroid, M.B., *Concepts and strategy of functional food science: the European perspective.* The American Journal of Clinical Nutrition, 2000. **71**(6): 1660s-1664s.
4. Pearson, T.A., *et al.*, *Markers of inflammation and cardiovascular disease application to clinical and public health practice: a statement for healthcare professionals from the centers for disease control and prevention and the American Heart Association.* Circulation, 2003. **107**(3): 499-511.
5. Bigal, M., *et al.*, *Weight change and clinical markers of cardiovascular disease risk during preventive treatment of migraine.* Cephalalgia, 2009. **29**(11): 1188-1196.

6. Kris-Etherton, P.M., *et al., Bioactive compounds in foods: their role in the prevention of cardiovascular disease and cancer.* The American Journal of Medicine, 2002. **113**(9): 71-88.
7. Erdmann, K., B.W. Cheung, and H. Schröder, *The possible roles of food-derived bioactive peptides in reducing the risk of cardiovascular disease.* The Journal of Nutritional Biochemistry, 2008. **19**(10): 643-654.
8. Morton, L.W., *et al., Chemistry and biological effects of dietary phenolic compounds: relevance to cardiovascular disease.* Clinical and Experimental Pharmacology and Physiology, 2000. **27**(3): 152-159.
9. Voutilainen, S., *et al., Carotenoids and cardiovascular health.* The American Journal of Clinical Nutrition, 2006. **83**(6): 1265-1271.
10. Hertog, M.G., *et al., Dietary antioxidant flavonoids and risk of coronary heart disease: the Zutphen Elderly Study.* The Lancet, 1993. **342**(8878): 1007-1011.
11. Wollin, S.D. and P.J. Jones, *Alcohol, red wine and cardiovascular disease.* The Journal of Nutrition, 2001. **131**(5): 1401-1404.
12. Lippi, G., *et al. Moderate red wine consumption and cardiovascular disease risk: beyond the "French paradox".* in *Seminars in thrombosis and hemostasis.* 2010. Thieme Medical Publishers.
13. Visioli, F. and C. Galli, *The effect of minor constituents of olive oil on cardiovascular disease: new findings.* Nutrition Reviews, 1998. **56**(5): 142-147.
14. Covas, M.-I., *et al., The Effect of Polyphenols in Olive Oil on Heart Disease Risk FactorsA Randomized Trial.* Annals of Internal Medicine, 2006. **145**(5): 333-341.
15. Visioli, F., A. Poli, and C. Gall, *Antioxidant and other biological activities of phenols from olives and olive oil.* Medicinal Research Reviews, 2002. **22**(1): 65-75.
16. Tuck, K.L. and P.J. Hayball, *Major phenolic compounds in olive oil: metabolism and health effects.* The Journal of Nutritional Biochemistry, 2002. **13**(11): 636-644.
17. Cos, P., *et al., Phytoestrogens: recent developments.* Planta Medica, 2003. **69**(7): 589-599.
18. Cassidy, A. and L. Hooper, *Phytoestrogens and cardiovascular disease.* British Menopause Society Journal, 2006. **12**(2): 49-56.
19. Anthony, M.S., *Phytoestrogens and Cardiovascular Disease Where's the Meat?* Arteriosclerosis, Thrombosis, and Vascular Biology, 2002. **22**(8): 1245-1247.
20. Clarkson, T.B., *Soy, soy phytoestrogens and cardiovascular disease.* The Journal of Nutrition, 2002. **132**(3): 566-569.
21. Juturu, V., *Resveratrol and Cardiovascular Disease.* Current Nutrition & Food Science, 2008. **4**(3): 227-230.
22. Siemann, E. and L. Creasy, *Concentration of the phytoalexin resveratrol in wine.* American Journal of Enology and Viticulture, 1992. **43**(1): 49-52.
23. Nakata, R. and H. Inoue, *Resveratrol and Cardiovascular Disease.* Current Nutrition Reports, 2014: 1-7.
24. Wang, H., *et al., Resveratrol in cardiovascular disease: what is known from current research?* Heart Failure Reviews, 2012. **17**(3): 437-448.
25. Arab, L. and S. Steck, *Lycopene and cardiovascular disease.* The American Journal of Clinical Nutrition, 2000. **71**(6): 1691-1695.
26. Petr, L. and J. Erdman, *Lycopene and risk of cardiovascular disease.* Carotenoids and retinoids: Biological actions and human health, 2005:204-217.
27. Sesso, H.D., *et al., Dietary lycopene, tomato-based food products and cardiovascular disease in women.* The Journal of Nutrition, 2003. **133**(7): 2336-2341.
28. Genkinger, J.M., *et al., Fruit, vegetable, and antioxidant intake and all-cause, cancer, and cardiovascular disease mortality in a community-dwelling population in Washington County, Maryland.* American Journal of Epidemiology, 2004. **160**(12): 1223-1233.

29. Rose, P., *et al.*, *Bioactive S-alk (en) yl cysteine sulfoxide metabolites in the genus Allium: the chemistry of potential therapeutic agents.* Natural Product Reports, 2005. **22**(3): 351-368.
30. Tapiero, H., D.M. Townsend, and K.D. Tew, *Organosulfur compounds from alliaceae in the prevention of human pathologies.* Biomedicine & Pharmacotherapy, 2004. **58**(3): 183-193.
31. Cavagnaro, P.F., *et al.*, *Effect of cooking on garlic (Allium sativum L.) antiplatelet activity and thiosulfinates content.* Journal of Agricultural and Food Chemistry, 2007. **55**(4): 1280-1288.
32. Andersson, S.W., *et al.*, *Intake of dietary plant sterols is inversely related to serum cholesterol concentration in men and women in the EPIC Norfolk population: a cross-sectional study.* European Journal of Clinical Nutrition, 2004. **58**(10): 1378-1385.
33. Klingberg, S., *et al.*, *Inverse relation between dietary intake of naturally occurring plant sterols and serum cholesterol in northern Sweden.* The American Journal of Clinical Nutrition, 2008. **87**(4): 993-1001.
34. Farquhar, J.W., R.E. Smith, and M.E. Dempsey, *The effect of beta sitosterol on the serum lipids of young men with arteriosclerotic heart disease.* Circulation, 1956. **14**(1): 77-82.
35. Miettinen, T.A., *et al.*, *Reduction of serum cholesterol with sitostanol-ester margarine in a mildly hypercholesterolemic population.* New England Journal of Medicine, 1995. **333**(20): 1308-1312.
36. Katan, M.B., *et al. Efficacy and safety of plant stanols and sterols in the management of blood cholesterol levels.* in *Mayo Clinic Proceedings*. 2003. Elsevier.
37. Demonty, I., *et al.*, *Continuous dose-response relationship of the LDL-cholesterol–lowering effect of phytosterol intake.* The Journal of Nutrition, 2008: p. jn. 108.095125.
38. Anderson, J.W., *et al.*, *Dietary fiber and coronary heart disease.* Critical Reviews in Food Science & Nutrition, 1990. **29**(2): 95-147.
39. Bazzano, L.A., *et al.*, *Dietary fiber intake and reduced risk of coronary heart disease in US men and women: the National Health and Nutrition Examination Survey I Epidemiologic Follow-up Study.* Archives of Internal Medicine, 2003. **163**(16):1897-1904.
40. Liu, S., *et al.*, *Whole-grain consumption and risk of coronary heart disease: results from the Nurses' Health Study.* The American Journal of Clinical Nutrition, 1999. **70**(3): 412-419.
41. Marlett, J., S. Cho, and M. Dreher, *Dietary fiber and cardiovascular disease.* Handbook of Dietary Fiber, 2001: 17-30.

15

Functional Food in Reducing The Risk of Diabetes

Introduction

Diabetes shows a wide range of variation in prevalence around the world and it is expected to affect 300 million by the year 2025. In a prevention framework where banning policies and educational strategies lead the interventions, functional foods (FFs) with their specific health effects could, in the future, indicate a new mode of thinking about the relationships between food and health in everyday life. Functional ingredients, such as stevioside, cinnamon, bitter melon, garlic and onion, ginseng, *Gymnema sylvestre* and fenugreek, have been addressed for their specific actions towards different reactions involved in diabetes development. New strategies involving the use of FF should be validated through large scale population trials, considering validated surrogate end points to evaluate the effect of FF in prevention of chronic diseases such as type 2 diabetes mellitus.

Diet and its Role in Prevention of Diabetes

Diabetes mellitus is one of the most common endocrine disorders affecting almost 6% of the world's population. The number of diabetic patients will reach 300 million in 2025 [1]. More than 97% of these patients will have type 2 diabetes. Research shows that the prevalence of type 2 diabetes is more common in females than in males [2]. In addition, the risk for cardiovascular disease (CVD), the most common complication attributable to diabetes, is more serious among women then men. Notably, women with diabetes lose their premenopausal protection from CVD and have risk for this condition as great as, or greater than that of diabetic or non-diabetic men. Therefore, a unique challenge exists between menopause and the years leading up to it and diabetes because of the additional need for glycemic control. Studies have shown that modest weight loss of 5-10% body weight is associated with improvements in cholesterol, blood pressure and insulin sensitivity, known risk factors for CVD and type 2 diabetes [2]. In addition, current guidelines for the prevention of type 2 diabetes include the increase in total dietary fiber intake [2], since prospective studies have observed reduced diabetes risk with high cereal fiber and whole grain consumption. It has been suggested that the benefits of increased fiber intake result principally from the greater consumption of soluble forms due to effects on gastric emptying, macronutrient absorption, and reduced postprandial glucose response. Overall, current nutritional recommendations for diabetes frequently include modest weight loss and increase in fiber intake.

Clearly, weight control is an effective diabetes management technique and functional foods promoting weight loss could be aimed towards those with type 2 diabetes. However, it may also be possible to incorporate functional foods in the diet that impact insulin action independently of weight loss. In general, a functional food is similar in appearance to, or may be, a conventional food that is consumed as part of a usual diet, and is demonstrated to have physiological benefits and/or reduce the risk of chronic disease beyond basic nutritional function.

Functional Ingredients in Reducing the Risk of Diabetes

There are three key defects in the onset of hyperglycaemia in diabetes mellitus (DM), namely increased hepatic glucose production, diminished insulin secretion and impaired insulin action [3]. Functional food act comprehensively increasing insulin secretion, enhancing glucose uptake by adipose and skeletal muscle tissues and inhibiting intestinal glucose absorption and hepatic glucose production. Hereafter, the potential of functional ingredients will be considered in animal and human models, presenting the different mechanisms of action of the single elements. In the following section, different functional ingredient will be addressed for reduction of the risk of diabetes.

Stevioside

A natural sweetener has widely caught attention, due not only to its properties of sucrose's substitute, but also for its potential functional properties. Stevioside is a sweet glycoside extracted from stevia rebaudiana Bertoni, a small shrub originally grown in South America, particularly in Brazil and Paraguay where it is known as stevia or honey leaf [4]. Extracts are being used commercially in many countries for sweetening a variety of products including pickled vegetables, sea foods, soft drinks, and soy sauce and confectionary products.

Stevioside is an intense sweetener and the extract of its source finds extensive use in countries such as Japan, China, Russia, Korea, Paraguay, Argentina, Indonesia, Malaysia, Australia, New Zealand, South America and others, to sweeten local teas, medicines, food and beverages [5]. Oral stevioside is taken up by the human body at extremely low levels and none of the digestive enzymes from the gastrointestinal tract of different animals and human body are able to degrade stevioside into steviol.

The use of stevioside as a functional ingredient has been addressed due to its biological effects and, therefore, to its potential applications. *In-vitro* studies with incubated mouse pancreatic islets have indicated that hypoglycaemic effects of stevioside and steviol result from the stimulation of insulin secretion via the direct action of these compounds on β-cells and the β-cell line INS-1 [6]. In a recent study, results from Gregersen et al. (2004) showed a positive association between reduced postprandial blood glucose (PPG) response compared to placebo, in Type 2 diabetic (T2DM) patients, questioning whether significant results can be obtained only in high glycaemic level subjects [4]. Thomas (2010) has widely discussed the use of stevioside in different conditions of body weight (BW), T2DM and MS, stressing the deep need of experimental studies on humans in order to confirm the results seen in animals, primarily in rats, but highlighting the various utilities and potential benefits seen with stevioside [7].

Several studies on animals and humans have been performed in order to assess the potential toxicology of this compound. Taken together, most studies are in agreement demonstrating that oral stevioside, at an acceptable daily intake (5 mg/kg BW), is safe and not carcinogenic or teratogenic [8]. An ADI of 0–2 mg/kg of BW for steviol, equivalent to 0–5 mg/kg BW of stevioside, as suggested by Joint FAO/WHO Expert Committee on Food Additives (JECFA 2006), was estimated to yield a stevioside concentration in colon at about 0.05– 0.2 mM, a concentration at which no toxicity to intestinal cells is expected [9]. A recent study showed that after maximum oral administration of single dose of 4.2 mg of stevioside/kg BW in humans, the highest mean concentration in plasma of steviol glucuronide and free steviol in plasma was 1.89 mg/mL (3.7 mM) and 0.19 mg/mL (0.38 mM), respectively [10], stressing how these levels are not toxic on human cell and how steviol does not undergo any conversion in the process [10].

Nuts

Nuts, including peanuts, are documented as having the potential to improve the blood lipid profile in humans [11]. Several epidemiological studies have revealed that people who consume nuts regularly are less likely to suffer from CVD [12, 13]. More recently, interest has developed in the potential value of including nuts in the diets of individuals with diabetes. Data from the Nurses Health Study indicate that there are benefits to consuming higher nut and peanut butter to lower risk of type 2 diabetes in women [14] . In this study, the relative risk of developing diabetes was reduced by 27% in women who ate nuts five or more times per week compared with those who rarely or never ate nuts.

In general, nuts contain 70–80% fat, and most FAs in nuts are unsaturated—being either polyunsaturated and monounsaturated, which may be beneficial for glucose and insulin homeostasis. Several studies have shown that a high amount of monounsaturated and polyunsaturated fat vs. saturated fat improves glucose homeostasis [15, 16]. The mechanisms by which specific types of dietary fat affect insulin sensitivity are not well understood. It has been shown that the fatty acid composition of phospholipids in skeletal muscle cell membranes is directly related to insulin sensitivity in humans. Therefore, changes in dietary FAs might result in compositional changes of membrane phospholipids, perhaps in turn modulating insulin action and sensitivity in skeletal muscle. In addition, other components of nuts such as fiber and magnesium may decrease insulin demand and resistance. Nuts are a rich source of vitamins, minerals, antioxidants and plant protein, which could be also beneficial. In this respect, nuts contain many bioactive components, which may have potential advantages for glycemic control [17].

However, acute postprandial studies suggest that nuts should be combined with the carbohydrate portion of the meal to reduce postprandial glycemia. A lower glycemic response is often thought to equate to a lower insulin demand, better long-term blood glucose control and a reduction in blood lipids. Josse et al. demonstrated that almonds may reduce the glycemic impact of carbohydrate foods with which they are eaten [18]. Additionally, another study concluded that reductions in 24 h insulin secretion appear to be a further metabolic advantage of nuts, which the investigators suggest may in the longer term help to explain the association of nut consumption with reduced CVD risk. Overall, acute feeding studies have established the ability of nuts, when eaten with carbohydrate, to depress postprandial glycemia, but long term benefits of consuming nuts have not yet been established.

In conclusion, more long term clinical trials should evaluate the effect of nuts on glycemic control. However, nuts may have favourable effects on the postprandial response in both people with normal blood glucose and those with type 2 diabetes. Thus, nuts can be incorporated into a healthy diet to lower the glycemic index of a meal or snack for individuals with diabetes or those at risk for diabetes, on condition that overall caloric intake is regulated to maintain a healthy body weight.

Fenugreek

The anti-diabetic efficacy of *Trigonella foenum graecum* (Fenugreek) has been tested in several preliminary human trials, wherein whole seeds, seeds extract and leaves were screened for activity. Although leaves did not show any beneficial hypoglycaemic effect, it is believed that dietary fibres in the seeds may act in delaying gastric emptying and suppressing release of gastric inhibitory peptides and insulinotropic hormones [19]. Fenugreek was also found to act on the metabolic clearance rate of glucose and erythrocyte insulin receptors, as reported by a study carried out on a Type 2 diabetic (T2DM) patient who was given 25g fenugreek seeds for 15 days [20]. In a trial aimed at elucidating the therapeutic role of fenugreek on alloxan-induced diabetic rats, oral administration of whole seed powder (5% in the diet for 21 days) improved glucose homeostasis as a result from altered glycolytic, gluconeogenic and lipogenic enzymes activity, whose levels were restored to control values [21]. Even though further clinical studies are needed to confirm the therapeutic power of this herb, preliminary evidence and substantial lack of toxic effects at proposed use levels make Fenugreek a promising candidate as a FF for diabetes management [22].

Omega-3-fatty acids (FAs)

Dietary or supplemental ω-3 polyunsaturated FAs, mainly eicosapentaenoic acid (EPA, 20:5 n-3) and docosahexaenoic acid (DHA, 22:6 n-3), have well documented potent triglyceride (TG) lowering effects in healthy and diabetic subjects. An epidemiological study estimated that replacing 2% of energy from trans-isomer FAs isoenergetically with polyunsaturated fat would lead to a 40% lower risk for type 2 diabetes [23]. Another prospective cohort study of 35,988 older women supported this inverse relationship between incident type 2 diabetes and vegetable fat, and the benefit of substituting polyunsaturated FAs for saturated FAs and cholesterol. Overall, the usefulness of ω-3 FAs in diabetes management includes reducing levels of TG, increasing high density lipoprotein cholesterol (HDL-C), increasing glucose induced insulin secretion, and possibly lowering insulin resistance. The effects of ω-3 FAs on insulin sensitivity are through to be mediated by different possible mechanisms viz. competition for the same enzymes and/or transcription factors, production of eicosanoids with different anti-inflammatory potency and change in membrane fluidity [24].

In conclusion, ω-3 FAs including fish oil may be able to modulate insulin action more than revert an established insulin resistance; the beneficial effects may be more evident in healthy people and not in diabetic patients. However, the effects of fish oil supplements on glycemic control or insulin sensitivity are insufficiently clear in type 2 diabetes. Future, long-term intervention studies are recommended. In addition, it is possible that the consumption of ω-3 FAs in fish matrix may modulate the glycemic control differently than a capsule of ω-3 FAs. Overall, modest amounts of ω-3 FAs in the range of 1–2 g/d appear to pose no significant risk to glucose control in persons with type 2 diabetes. The ADA recommends two or more servings of fish per week (except commercially fried fish filets) to provide the ω-3 FAs.

Cinnamon

Cinnamon extracts containing polyphenol type-A polymers, have demonstrated insulin mimetic properties (Fig.15.1). *In-vitro* and *in-vivo* animal studies have reported strong insulin like or insulin potentiating effects after cinnamon administration. The clinical trial showed that relatively low amounts of cinnamon powder supplementation (1 g/d) reduced fasting blood glucose (FBG) concentration and improved the blood lipid profile in post menopausal patients with type 2 diabetes [25]. Likewise, in a study with men and women with the metabolic syndrome, the results support the efficacy of cinnamon supplementation in reducing FBG and improving body composition. Also a high dose (6 g/d) of cinnamon with rice pudding reduces postprandial blood glucose and delays gastric emptying without affecting satiety. Yet, some studies failed to replicate these findings. Past report suggests that cinnamon supplementation (1.5 g/d) did not improve whole body insulin sensitivity, oral glucose tolerance or blood lipid profile in post menopausal patients with type 2 diabetes [26]. Secondly, Tang *et al.* also demonstrated no changes in FBG and blood lipids after a 4-week supplementation with cinnamon in healthy young patients. Finally, a recent meta-analysis examining five prospective randomized controlled trials indicates that the use of cinnamon does not confirm to improve FBG or lipid parameters in patients with type 1 or type 2 diabetes. More research is needed before cinnamon can be recommended for treatment of type 2 diabetes. Therefore, studies should be conducted to determine how specific variables, such as diet, population type, body mass index (BMI), glucose levels, cinnamon type/dose, and concurrent medication affect cinnamon.

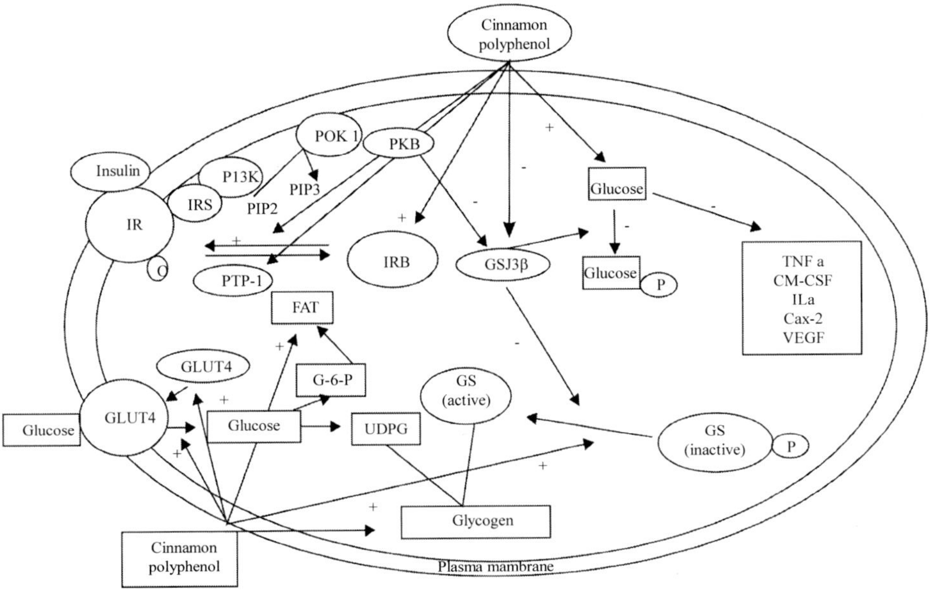

Fig. 15.1: Mechanism of cinnamon polyphenols in reducting the risk of diabetes

Citrus fruits – A Case Study

Citrus fruits are among the most consumed fruits after mangoes, tomatoes and bananas. Citrus fruits not only provide ample supply of vitamins, minerals, dietary fibers and pectins but they contain active phytochemicals including phytophenolics (e.g. flavanones, flavones, flavonols, phenolic acids, etc.) that are widely suggested to protect health due to their biological properties, which include anti-atherogenic, anti-inflammatory and anti-tumor activities, inhibition of blood clots, antimicrobial activity and strong antioxidant activity. Oxidative stress (the imbalance of the physiologic antioxidant protective mechanism and the extent of oxidative burden) induced cell damage has been implicated in a variety of chronic diseases.

The mechanisms of action of naturally occurring antioxidants can be diverse *in-vivo*, a demonstration of which requires assessment of a multiplicity of biomarkers. Chemoprevention by dietary means continues to attract major attention for the management of chronic degenerative diseases primarily due to the dramatic rise of cancer and type 2 diabetes mellitus. The level of antioxidant enzymes along with potential antioxidant vitamins are decreased in diabetic experimental animals and in humans, thus weakening the endogenous antioxidant defense homeostasis.

For example, there is demonstrated evidence for the inactivation of glutathione synthesis in erythrocytes from type 2 diabetic patients and also a total reduction in antioxidant activity in both type 1 and type 2 diabetic patients. A free radical is any chemical species (capable of independent existence) possessing one or more unpaired electrons, an unpaired electron being one that is alone in an atomic or molecular orbital. Free radicals are formed from molecules via the breakage of a chemical bond such that each fragment keeps one electron free radicals may also be formed by collision of the non-radical species by a reaction between a radical and a molecule which must then result in a radical since the total number of electrons is odd, by cleavage of a radical to give another radical and, finally via redox reactions. Radicals are generally less stable than non-radical species, although their reactivity varies [27].

The mechanism of action of citrus phytophenolics in models of diabetes suggest that naringenin is able to reduce glucose uptake and inhibit intestinal and renal Na+-glucose co-transporter (Figure 15.2) and that both naringin and hesperidin significantly increased the glucokinase mRNA level, while naringin reduced the mRNA expression of phosphoenolpyruvate carboxy kinase and glucose-6-phosphatase in the liver [21]. The extracts from Dangyuja (a Korean citrus fruit), containing high levels of flavanone glycosides, could be used to control the blood glucose level of diabetic patients by inhibiting α-amylase and β-glucosidase in the intestinal tract. The neuroprotective action of naringenin has been reported and this could benefit the management of diabetic neuropathies.

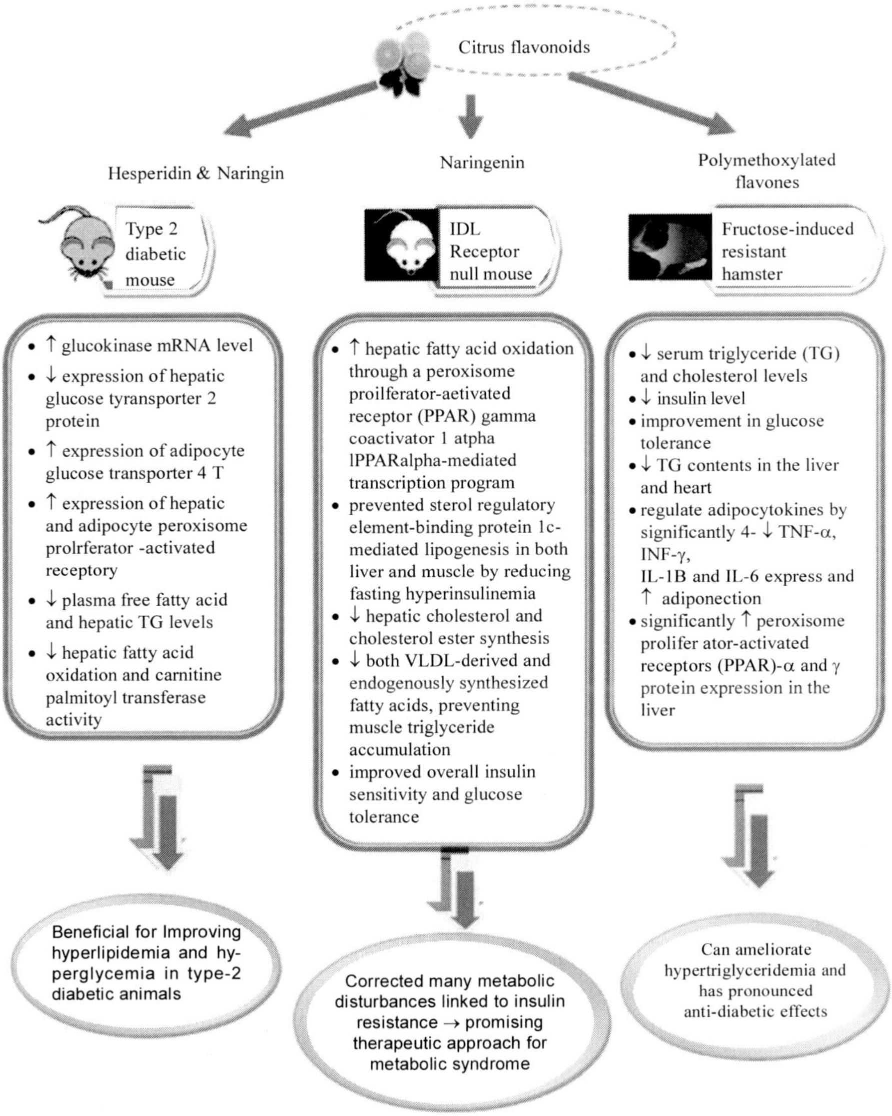

Fig. 15.2: Mechanisms of action of selected citrus flavonoids in animal models of diabetes

Summary

In over weight and obese insulin-resistant individuals, modest weight loss has been shown to improve insulin resistance. Thus, weight loss is recommended for all such individuals who have, or are at risk for diabetes. In addition, people with diabetes are encouraged to consume a variety of fiber-containing foods.

Studies also show that in individuals consuming a high-glycemic index diet, a low glycemic diet can produce a modest benefit in controlling postprandial hyperglycemia. To lower the glycemic index, nuts and peanuts can be potentially included in a healthy diet. However, more long term studies are needed to demonstrate the effects of nuts and peanuts on glycemia. Also, given the potential benefits of ω-3 FAs on CVD risk, regular fish consumption is recommended to assure the intake of ω-3 FAs. Finally, studies of patients with type 2 diabetes also show that cinnamon may have the potential to lower glucose levels. Yet, more research on the proposed health benefits of cinnamon supplementation is warranted before unambiguous recommendations can be made. In conclusion, a growing number of individuals are adding functional foods and natural health products to their diet to help control blood glucose; a large amount of research still needs to be done before benefits can be confirmed, and before these foods can be recommended routinely for glycemic control.

References

1. Ballali, S. and F. Lanciai, *Functional food and diabetes: a natural way in diabetes prevention?* International Journal of Food Sciences and Nutrition, 2012. **63**(S1): 51-61.
2. Chandalia, M., *et al.*, *Beneficial effects of high dietary fiber intake in patients with type 2 diabetes mellitus.* New England Journal of Medicine, 2000. **342**(19): 1392-1398.
3. Chiasson, J.-L. and R. Rabasa-Lhoret, *Prevention of type 2 diabetes insulin resistance and beta-cell function.* Diabetes, 2004. **53**(suppl 3): S34-S38.
4. Gregersen, S., *et al.*, *Antihyperglycemic effects of stevioside in type 2 diabetic subjects.* Metabolism, 2004. **53**(1): 73-76.
5. Geuns, J., *Stevioside.* Phytochemistry, 2003. **64**(5): 913-921.
6. Jeppesen, P.B., *et al.*, *Stevioside acts directly on pancreatic â cells to secrete insulin: Actions independent of cyclic adenosine monophosphate and adenosine triphosphate—sensitivie K-channel activity.* Metabolism, 2000. **49**(2): 208-214.
7. Thomas, J.E. and M.J. Glade, *Stevia: it's not just about calories.* benefits, 2010. **35**: 36.
8. Chatsudthipong, V. and C. Muanprasat, *Stevioside and related compounds: therapeutic benefits beyond sweetness.* Pharmacology & Therapeutics, 2009. **121**(1): 41-54.
9. Boonkaewwan, C., *et al.*, *Specific immunomodulatory and secretory activities of stevioside and steviol in intestinal cells.* Journal of Agricultural and Food Chemistry, 2008. **56**(10): 3777-3784.
10. Wheeler, A., *et al.*, *Pharmacokinetics of rebaudioside A and stevioside after single oral doses in healthy men.* Food and Chemical Toxicology, 2008. **46**(7): S54-S60.
11. Marsh, K., *Nuts and Diabetes.* 2011.
12. Kris-Etherton, P.M., *et al.*, *The role of tree nuts and peanuts in the prevention of coronary heart disease: multiple potential mechanisms.* The Journal of Nutrition, 2008. **138**(9): 1746S-1751S.
13. Kris-Etherton, P.M., *et al.*, *Nuts and their bioactive constituents: effects on serum lipids and other factors that affect disease risk.* The American Journal of Clinical Nutrition, 1999. **70**(3): 504s-511s.
14. Jiang, R., *et al.*, *Nut and peanut butter consumption and risk of type 2 diabetes in women.* JAMA, 2002. **288**(20): 2554-2560.

15. Li, T.Y., *et al.*, *Regular consumption of nuts is associated with a lower risk of cardiovascular disease in women with type 2 diabetes.* The Journal of Nutrition, 2009. **139**(7): 1333-1338.
16. Lovejoy, J.C., *The impact of nuts on diabetes and diabetes risk.* Current Diabetes Reports, 2005. **5**(5): 379-384.
17. Jenkins, D.J., *et al.*, *Possible benefit of nuts in type 2 diabetes.* The Journal of Nutrition, 2008. **138**(9): 1752S-1756S.
18. Kendall, C.W., *et al.*, *Health benefits of nuts in prevention and management of diabetes.* Asia Pacific Journal of Clinical Nutrition, 2010. **19**(1): 110.
19. Grover, J., S. Yadav, and V. Vats, *Medicinal plants of India with anti-diabetic potential.* Journal of Ethnopharmacology, 2002. **81**(1): 81-100.
20. Sharma, R., T. Raghuram, and V.D. Rao, *Hypolipidaemic effect of fenugreek seeds. A clinical study.* Phytotherapy Research, 1991. **5**(3): 145-147.
21. Raju, J., *et al.*, *Trigonella foenum graecum (fenugreek) seed powder improves glucose homeostasis in alloxan diabetic rat tissues by reversing the altered glycolytic, gluconeogenic and lipogenic enzymes.* Molecular and Cellular Biochemistry, 2001. **224** (1-2): 45-51.
22. Yeh, G.Y., *et al.*, *Systematic review of herbs and dietary supplements for glycemic control in diabetes.* Diabetes Care, 2003. **26**(4): 1277-1294.
23. Popp-Snijders, C., *et al.*, *Dietary supplementation of omega-3 polyunsaturated fatty acids improves insulin sensitivity in non-insulin-dependent diabetes.* Diabetes Research (Edinburgh, Scotland), 1987. **4**(3): 141-147.
24. Simopoulos, A.P., *Omega-3 fatty acids in health and disease and in growth and development.* The American Journal of Clinical Nutrition, 1991. **54**(3): 438-463.
25. Rudkowska, I., *Functional foods for health: Focus on diabetes.* Maturitas, 2009. **62**(3): 263-269.
26. Vanschoonbeek, K., *et al.*, *Cinnamon supplementation does not improve glycemic control in postmenopausal type 2 diabetes patients.* The Journal of Nutrition, 2006. **136**(4): 977-980.
27. Ardestani, A. and R. Yazdanparast, *Antioxidant and free radical scavenging potential of Achillea santolina extracts.* Food Chemistry, 2007. **104**(1): 21-29.

16

Functional Food in Reducing The Risk of Obesity

Introduction

The search for weight controlling foods began in the mid 19th century when industrialization led to widespread obesity. Over the years, people have tried to lose weight by avoiding particular food groups like carbohydrates, eating specific foods such as cabbage soup, or adding a certain ingredient, like vinegar or grapefruit to each meal. Recently scientists have demonstrated that green tea, dairy products, and cinnamon have the potential to improve weight control by altering fat and/or carbohydrate metabolism at the sub-cellular level. Considerable research is needed, however, before these food items can be used to develop effective weight management products.

Diet's Role in Prevention of Obesity

Although obesity treatments are described in ancient texts, the search for corpulence curing functional foods did not take-off until the mid 19th century industrialization made corpulence a prevalent problem [1]. Vinegar became one of the earliest widely touted obesity cures. Like its successors, grapefruit, pineapple and cabbage soup, it has fallen in and out of favour with each new generation of dieters. The common rationale for the use of each of these foods is that their acidic makeup and enzyme content literally chew up fat. In actuality, their transient success as weight management agents has more to do with the psychological phenomena of stimuli narrowing than any unique chemical properties of the individual foods. Conversely, many popular 20th century obesity treatments advocate avoidance of dysfunctional (adipose promoting) foods. Dietary carbohydrates and fats have each been accused of causing obesity. The research supporting these theories often neglects to access the role that dietary protein and total calorie intake played in the observed weight loss. Nonetheless, the data suggest that low carbohydrate or low fat diets may improve weight control in individuals with selected metabolic conditions, like Syndrome X, noninsulin dependant diabetes mellitus, and various dislipidemias, respectively [2]. Increased intake of lean proteins may help formerly over weight people maintain a new lower weight. New functional foods that improve weight control may be on the horizon. Recently, unrelated studies concluded that fat catabolism was stimulated in rodents fed with a low-calorie, high-calcium diet, in men given a catechin rich green tea extract, and in people consuming conjugated linoleic acid (CLA). A second CLA study suggested it suppressed fat synthesis while promoting muscle growth. More research is needed before any of these data can be used to develop weight management recommendations or products for the general population [3]. The high calcium effect must be corroborated in humans, the 24 h green tea study, must be carried out long enough to confirm that the modest (4%) increase in energy production would actually result in weight loss, and the precise role of CLA in fat metabolism needs to be reevaluated [4].

Pathologies associated with obesity

Over the past 30 years the incidence of obesity has increased dramatically amongst children and adults living in industrialized nations, causing significant adverse physical and economic consequences [5]. Obesity is a complex condition resulting from a mixture of psychological, environmental and physiological factors. It is a proven causative agent in the so called 'diseases of affluence' such as, cardiovascular disease and adult onset diabetes. The obesity epidemic and its associated health problems are not restricted to the United States. According to the World Health Organization (WHO) the prevalence of obesity is rising to epidemic proportions at an alarming rate in both developed 'Westernized' and less developed countries around the world. For example, the prevalence of obesity has increased from 10 to 50% in the majority of European countries since 1990. As Europeans turn away from traditional healthy diets and active lifestyles, their incidence of obesity is catching up with that of their American counterparts. WHO estimates nearly one third of all Europeans are obese and 3,20,000 die each year from obesity related diseases [6].

The increasing incidence of obesity coincides with two other important trends, an aging population and a growing interest in alternative therapies. Residents of industrialized nations are growing heavier and older. As people age the health risk factors associated with the lifestyle they adopted during their younger years begins to manifest as chronic disease. Given this fact, and the fact the greatest rates of obesity are typically seen during late middle age and in senior citizens, there is considerable concern that the combination of increasing obesity and advancing age will strain health care budgets to the breaking point [7].

Functional Food in Prevention of Obesity

Nutritional anthropologists contend our hunter gatherer ancestors created the concept of functional foods by deeming particular foods capable of enhancing physical prowess, increasing fertility, fighting specific illnesses or inducing a supernatural vision [8]. Since a scarcity, rather than a surplus of food was the typical condition for most of human history, evidence of reducing diets is rare. A few ancient texts, such as the following quote attributed to Hippocrates, mention diet prescriptions but, the search for obesity curing functional foods did not really take off until the mid 19th century when industrialization made corpulence a wide spread problem. The first widely touted obesity treatments were acidic foods like vinegar, cabbage, cabbage water, and the juice of lemons, grape fruits and pineapple. Supposedly the acids and enzymes they contained were capable of chewing up body fat. Although, these foods failed to produce lasting weight loss, the myth that they melt fat has never really disappeared.

While no one functional weight loss food has been identified, there has been some progress in determining the most functional combination of foods for weight maintenance. It appears a diet lower in fat and higher in protein than that recommended by the USDA Food Guide Pyramid assists weight maintenance. According to data presented by the USDA, during the diet summit held at the 2000 American Dietetic Association Meeting, people who self-select at lower fat, higher protein diet seemed better able to control their weight than those who opted to follow a high carbohydrate plan [9].

Low fat dairy products

It is reported that a high calcium, low calorie diet helped obese mice lose weight at rates double than that of mice fed with a low calcium, low calorie diet. The effect was even more dramatic in mice fed with diets containing dairy calcium [6]. These animals lost over five times as much weight and 16% more body fat than did the mice fed with a low calcium diet. Subsequently, the researchers observed a 51% inhibition of fatty acid synthase expression and activity and a 300-500% increase in lypolysis in adipocytes harvested from mice fed with the high calcium diets.

This study lead them to hypothesize that high levels of dietary calcium markedly stimulate lypolyisis while inhibiting lipogenesis, resulting in significant weight loss. Why dairy calcium was more effective than supplemental calcium carbonate, which is the form found in most antacid tablets, remains to be elucidated. The explanation may relate to the amino acid composition of the protein in the daïry calcium supplement (dried powder milk). It remains to be seen if the enhanced weight loss observed with dairy calcium is related to dairy proteins, or other biologically active agents. If amino acids are responsible for the enhanced effect of dairy calcium on weight loss, combining other animal proteins with calcium supplements may yield a similar result. Researchers also need to determine if a low-calorie, high-calcium diet provides adequate physiological (hunger) and psychological (appetite) satiety to achieve long term weight management. Another concern is that the relatively high levels of calcium in these diets may interfere with absorption of other divalent minerals. The low calorie diets used in many of the studies suggest the intake of minerals other than calcium may be relatively low. Despite these gaps in knowledge, the role of calcium as a weight loss stimulant has already made its way into popular american culture.

Conjugated linoleic acid (CLA)

Other researchers are exploring the value of conjugated linoleic acid (CLA), a fatty acid found in dairy products, as a weight control aid. CLA, a by-product of digestive process in the rumen of cattle, is plentiful in milk, dairy products and beef; it is also found in human breast milk leading to speculation that it may be an unrecognized essential nutrient. Extensive animal studies suggest that CLA could also help overweight people to lose fat and increase muscle mass. Although, CLA supplements are being sold as weight management aids in health food stores and over the Internet, well designed studies of CLA activity in humans are sparse.

Only in the past 15 years has it been widely recognized that individual fatty acids can have powerful and unpredictable effects on metabolism through modulation of the inflammatory response and other metabolic pathways.

Tea and spice

Two other promising lines of functional food research concern commonly consumed herbal products, green tea and cinnamon. In late 1999, Swiss researchers reported that green tea extract increased thermogenesis by 4% over a 24 h period in healthy male volunteers [9]. It is important to recognize that many questions must be answered before green tea can be widely touted as a weight loss therapy. The study was too brief to indicate if the body would eventually compensate for the increase in thermogenesis.

Even if the increase in thermogenesis can be maintained for a long term, which is currently under investigation, weight loss would be modest at best. Thermogenesis typically accounts for 30-40 % of total calorie usage. If you assume a 2000 kcal/day intake, a 4 % increase in thermogenesis would only burn an additional 24-32 kcal/day. At that rate it would take over 100 days to loose a pound. It would be easy for people to make up this increased calorie loss by eating an extra cracker or two each day.

How green tea increases thermogenesis has not been investigated directly. Scientists involved in these studies have proposed that green tea extract stimulates brown adipose tissue thermogenesis to an extent that is much greater than can be accounted for solely by its caffeine content. According to these researchers, the thermogenic properties of green tea may reside primarily in an interaction between its high content of catechin-polyphenols and caffeine on sympathetically released noradrenaline. Catechin-polyphenols are known to be capable of inhibiting catechol-O-methyl-transferase (the enzyme that degrades noradrenaline), and caffeine has been shown to inhibit transcellular phosphodiesterases (enzymes that break down NA-induced cyclic AMP). Therefore, it has been proposed that green tea extract, via its catechinpolyphenol and caffeine content, is effective in stimulating thermogenesis by relieving inhibition at different control points along the NA–cAMP axis [19, 26].

Scientist at the USDA's Beltsville, Maryland Human Nutrition Center, report that a water soluble compound in cinnamon may aid glucose metabolism. The compound, identified as methylhydroxy chalcone polymer (MHCP), makes fat cells more responsive to insulin by activating the enzyme that causes insulin to bind to cells, and inhibiting the enzyme that blocks this process. In cultured cells and animal studies, MHCP increased glucose metabolism by 20%. Human studies are currently underway. The preliminary results have been promising enough for the project's lead scientist to suggest that people with insulin resistance or NIDDM can safely consume 1/4 to 1 teaspoon of ground cinnamon each day. Although no direct weight management benefit would be achieved and cinnamon extract may not be effective in controlling blood glucose in all people with diabetes, it could provide a low-cost, low-risk method of improving glucose utilization in many overweight individuals.

Phenolic compounds and plant extracts

Obesity consists in an excess of body fat resulting from an imbalance between energy intake and expenditure. It is a growing worldwide health problem considered as a risk factor for chronic diseases including e.g. cardiovascular diseases, type 2 diabetes, hypertension, stroke, mental health disorders and some forms of cancer. Insufficient physical activity, coupled with high calorie, high fat food consumption, is an environmental factor regarded as the main cause of obesity, beside genetic predisposition [10, 11]. Although prevention with lifestyle modifications is obviously the most appropriate approach, therapeutic strategies comprise antiobesity agents and surgery. Only two drugs are currently approved for long term obesity treatment, even though both of them have undesirable side effects. They either regulate food intake by acting on neural circuits (sibutramine™) or reduce nutrient absorption from gut (orlistat™). Consequently, there is a need for weight loss effective and safe compounds. At present, the potential of natural products is largely unexplored and might be an alternative strategy for the development of antiobesity approaches. Among phytochemicals, "generally recognized as safe", with no side effects at plausible concentrations, the evidence that phenolic compounds (PCs) have beneficial effects in fighting or preventing obesity is increasingly reported, underlying their potential interest for a nutritional approach [12]. The pancreatic lipase (PL) is the major enzyme responsible for the digestion of 50-70% of dietary triglycerides into monoacylglycerides and free fatty acids, both absorbable by enterocytes. The inhibition of PL could, therefore, result in a reduced fat absorption, and thereby energy uptake, which is one of the key targets thought to mediate obesity [13]. Orlistat™ was shown to act by this mechanism [13]. Polyphenolic rich extracts from teas, herbal and fruit sources, such as berry, apple, Melinjo or grape seeds, oolong, black or green tea from different varieties, various high fat or high tannin content plants, were reported as PL inhibitors during *in-vitro* experiments. PCs such as proanthocyanidins or ellagitannins, gallate esters (gallocatechingallate, epigallocatechin gallate (EGCG), catechin gallate) or stilbenoids (gnetin), were considered as the main active ingredients. However, the inhibitory mechanism of these substances on PL remains unclear. It was generally attributed to the ability of tannins to bind, complex and precipitate proteins although some studies reported non-competitive or mixed inhibitions of PL.

Natural Products for Obesity Prevention and Intervention

Due to the high prevalence of overweight and obesity, development and marketing of nondrug products for preventing weight gain or for weight loss have been expanding during the last decade. In 2006, around 41% of US adults and 30% of European adults attempted to lose weight. Sales of diet products for weight control were estimated to be US$ 54 billion in Europe and US$ 50 billion in the United States in 2007. It is predicted that diet products for weight loss will increase by 3.4% annually until 2010 [14].

Diet products for weight loss are actually traditional diets that have low fat or low carbohydrates. Unfortunately, the success rate of using diet products for weight loss is quite low, and therefore, the market is changing. Consumers have recently become more interested in weight management products and techniques that target 'how' to suppress appetite, promote satiety, inhibit fat or carbohydrate digestion and absorption, and/or increase fat oxidation. Accordingly, a variety of natural products, mainly functional foods, have been introduced to the market. Based on market survey and analysis, natural products for weight management covers four categories [15]:

- Fat burners
- Fat and carbohydrate blockers
- Appetite suppressants
- Satiety promoters

Between 2004 and 2006, the number of new weight-loss products launched around the world increased by 50%. A large portion of these products are herbal extracts. Despite the negative view of herbs by the mainstream medical community, botanical dietary supplements for weight loss remain extremely popular with the general public. Although the safety and efficacy of available herbal products for weight reduction remain major issues that need to be resolved through research, herbal products for weight management may ultimately prove to be helpful in overcoming clinically significant obesity when combined with healthydiet and lifestyle changes [16].

Fat burners

Fat burners are the most significant part of the new generation of weight control products. Several botanical ingredients have been used either alone or combined together with others, and they are generally delivered in foods or drinks. Japan and the United States are the leading countries in producing weight control products in this category. The following ingredients are marketed as dietary supplements by companies mainly in these two countries.

Citrus aurantium

This plant is commonly named bitter orange, sour orange or Seville orange. Components of the fruit are sometimes used as a food, but the plant is more widely used as a medicinal or dietary supplement [17]. The extract has been used in dietary supplements as an aid to fat loss and also as an appetite suppressant although it is not recommended for use in isolation. In traditional Chinese medicine, *Citrus aurantium* is always prescribed in concert with other support herbs. *C. aurantium* contains a number of phytochemicals, including p-octopamine and synephrine alkaloids, which are adrenergic agonists and often cited as its active ingredients. Synephrine alkaloids may increase energy expenditure through thermogenesis and reduce food intake by decreasing gut motility.

Green tea catechins

Green tea prepared by heating or steaming the leaves of *Camelia sinensis* is widely consumed on a regular basis throughout Asia. Green tea contains both caffeine and catechins, which make up approximately 30% of the dry matter of green tea leaves. Epigallocatechin gallate is one of the most powerful polyphenols and is the most abundant catechin in tea, especially green tea [18]. Green tea extract protects from free radicals and suppresses inflammation in addition to promoting energy expenditure and fat oxidation. Because caffeine occurs naturally in green tea extract, it has been difficult to separate the effects of catechins from caffeine in humans drinking green tea. In fact, it has been shown that green tea extracts containing both catechins and caffeine are more potent in stimulating brown adipose tissue thermogenesis than equimolar concentration of caffeine alone. It is possible that there exists a synergistic interaction between catechins and caffeine in increasing energy expenditure by promoting thermogenesis [19].

Capsaicin

Capsaicin comes from red peppers and chili peppers. Capsaicin stimulates fat oxidation and thermogenesis and reduces food intake. Capsaicin may increase thermogenesis by enhancing catecholamine secretion from the adrenal medulla involving capsaicin sensitive neurons [18]. It is also implied that capsaicin induces thermogenesis through adrenergic stimulation. The long term use of capsaicin may be limited by its strong pungency. A possible solution for this may be using the fruit of a non pungent cultivar of pepper, such as CH-19 Sweet.

Garcinia cambogia extract

Garcinia cambogia is a subtropical species of Garcinia native to Indonesia. Hydroxycitric acid is one of 16 isomers of citric acid in the extract of *G. cambogia* and is the only one that inhibits citrate lyase, the enzyme that catalyzes the first step in fatty acid synthesis outside the mitochondrion. It is suggested that hydroxycitric acid suppresses de novo fatty acid synthesis and food intake [20]. Overall, the evidence of this extract for weight control is encouraging.

Few and mild adverse effects were reported for both *G. cambogia* and its active component, hydroxycitric acid. However, further studies are needed to determine their efficacy and safety on weight loss [16].

Fucoxanthin

Fucoxanthin is a characteristic carotenoid of brown seaweeds and has been shown to have antiobesity effects. Fucoxanthin promotes thermogenic fat burning in white adipose tissue through the expression of uncoupling protein-1 at both m-RNA and protein levels [21]. Uncoupling protein-1 is usually expressed in brown adipose tissue and functions to release chemical energy and induce heat production. However, it has been found that feeding of fucoxanthin increases uncoupling protein-1 expression in white adipose tissue, which in turn leads to fatty acid oxidation and energy expenditure, and ultimately reduces fat mass [21, 22]. Xanthigen is an extract from deep sea water cultivated marine vegetables standardized to fucoxanthin, neoxanthin and violaxanthin with stronger thermogenic and weight control properties.

Yohimbe

Yohimbe (*Pausinystalia yohimbe*) is a tall ever green tree, which is native to Central Africa. Yohimbine, an -2 receptor antagonist, is the main active constituent of the ground bark of yohimbe. It is often promoted as a weight loss supplement. The adverse effects associated with the use of yohimbine include headache, hypertension, anxiety and agitation.

Fat and Carbohydrate Blockers

As fat and carbohydrates are the main energy supply for the body and the cause of over energy intake and obesity in most cases, inhibition of fat and carbohydrate absorption is beneficial to weight management. However, few fat blocker products have been made available to consumers. The most dynamic area is in carbohydrate blockers, in particular the best known ones *Phaseolamin vulgaris* and the branded ingredient Svetol.

Phaseolamin vulgaris

Phaseolamin vulgaris is an extract of white kidney beans (*Phaseolus vulgaris*) and is an inhibitor of amylase (which is an enzyme that helps the breakdown of starches). Therefore, phaseolamin may reduce the rate at which starches are broken down into sugars in the digestive tract [23]. The extracts are potential ingredients in foods for increasing carbohydrate tolerance in diabetics, decreasing starch digestion and energy intake for weight control.

Chitosan

Chitosan is a mucopolysaccharide component of the shells of crab, lobster, shrimp and other marine organisms, and is one of the more popular weight-loss supplements. It is reported that chitosan binds to fat in the gastrointestinal tract and thus decreases fat digestion and absorption. It is also reported that chitosan oligosaccharides exert antiobesity effects by inhibiting adipocyte differentiation through down regulating the expression of adipogenic transcription factors and other specific genes.

Green Coffee Bean Extract

A possible mechanism for the efficiency of green coffee bean extract against weight gain and fat accumulation is by the inhibition of fat absorption in the intestine and the activation of fat metabolism in the liver. Caffeine is a known suppressor of fat absorption, but chlorogenic acid has also been found to be partially involved in the suppressive effect of green coffee bean extract on fat absorption, which results in the reduction of hepatic triglyceride level. Phenolic compounds such as neochlorogenic acid and feruloylquinic acid mixture, except chlorogenic acid, can enhance hepatic carnitine palmitoyl transferase activity, which is important in fatty acid transport from cytosol into mitochondria for oxidation.

Appetite suppressants

Appetite suppressants are an exciting area in the new generation of weight management products. Current market growth is mainly focusing on appetite suppression products. Several products have been marketed in the United States, Europe and South Africa.

Oat and palm oils

Oat and palm oil mixture is one of the new categories of weight loss products that have been developed recently. The mechanism is based on appetite control and feelings of fullness. Oat and palm oils in combination have an important effect on satiety. Polar lipids from the oat oil extract enclose palm oil droplets, preventing digestion of the palm oil until it reaches the small intestine. When the palm oil reaches the ileum, it is interpreted as undigested fat, which triggers the brain to release satiety hormones and suppresses the hunger signals that would normally be sent. The branded ingredient, Fabuless (once called Olibra and also offered in a consumer product called Slimthru), is taking advantage of functional oils for weight loss, and is currently the market leader with its patent protected combination of palm and oat oils (both naturally occurring dietary lipids). Fabuless has been clinically shown to create and maintain a feeling of satiety by slowing absorption [24].

Pinolenic acid

Pinolenic acid is an appetite suppressant extracted from the Korean pine nut (*Pinus koraiensis*). Korean pine nut oil consists of more than 92% of poly and monounsaturated fatty acids such as linolenic acid (C18:3), linoleic acid (C18:2) and oleic acid (C18:1). It is unique in that it contains approximately 15% linolenic acid (C18:3) [25]. Korean pine nut may work as an appetite suppressant through an increasing effect on satiety hormones such as cholecystokinin and glucagon like peptide-1 and a reduced prospective food intake [25].

Yerba mate

Yerba mate is prepared from *Iles paraguariensis*, an evergreen tree native to subtropical South America. Yerba mate tea, prepared by steeping the dry leaves and twigs of yerba mate in hot water, is widely consumed as a non alcoholic beverage in South America. Yerba mate has been rapidly introduced into the world market, either as tea itself or as ingredient in formulated foods or dietary supplements. The indigenous people have used it for centuries as a social and medicinal beverage. Yerba mate has been shown to be hypocholesterolemic, hepatoprotective, central nervous system stimulant, diuretic, and to benefit the cardiovascular system. Yerba mate has been suggested to induce weight loss through suppressing appetite [26] and enhancing the expression of uncoupling proteins and elevating AMPK phosphorylation. Increases of AMPK phosphorylation lead to upregulation of fatty acid oxidation and down regulation of fatty acid synthesis. Enhanced expression of uncoupling proteins promotes heat production and/or energy expenditure.

Satiety Promoters

Satiety promoter is a subset of the appetite suppressants and refers to the feeling of fullness and disappearance of appetite after a meal. The feeling is controlled by the hypothalamus and is triggered by various hormones. Satiety is a weight management weapon that reduces caloric intake by managing hunger. Apart from the products mentioned above for appetite suppression which also work through satiety promotion, dietary fiber is a typical ingredient that functions as a satiety promoter to reduce food intake.

Dietary Fiber

There are two different types of fibers viz. soluble and insoluble fibers which are commonly used in diet and weight management. Fibers are beneficial to weight control through promoting satiety, decreasing absorption of macronutrients such as fat and glucose, and changing the secretion of gut hormones. A variety of foods and drinks containing fiber have been marketed by different companies. Evidence has demonstrated that dietary fiber increases fullness and decreases hunger, leading to lower food intake, especially in over weight or obese subjects.

Fiber intake is associated with weight loss in both children and adults, and males and females. The inclusion of fiber in low calorie diets helps to reduce the feeling of hunger. Soluble fiber appears to be more efficient than insoluble fiber. The most commonly used soluble dietary fibers include guar gum and glucomannan, while insoluble fibers include mainly bran fiber. Although weight loss associated with dietary fiber is less than 5% of the initial body weight, it is believed that this small loss of body weight confers clinically significant health benefits.

Summary

More functional diet food candidates will be identified as the search for phytochemicals useful in human nutrition broadens. Creating a functional food to treat obesity is much more challenging than creating one to treat a condition like osteoporosis. Bone health can be promoted by adding calcium and other nutrients involved in bone formation to commonly consumed foods. Whereas a functional weight loss food needs to somehow influence the other foods eaten in order to control total calorie intake. Some fibers, for example, reduce absorption of cholesterol and small fat molecules. There is concern, however, that a functional food which limits absorption or utilization of a particular class of nutrients could cause vitamin and mineral deficiencies just as the weight management drug Orlistat can impair fat soluble vitamin absorption. The ideal functional weight loss food would control calorie intake by producing physiological (hunger) and psychological (appetite) satiety, stimulate calorie usage by stimulating metabolic pathways that burn calories, and be available in disease specific versions. It would also have no, or at most minimal side effects. The side effects of any healthcare treatment need to be less severe than the consequences of not providing therapy. Given the many different factors that can cause obesity, as well as the many different consequences it can have on human health, it is doubtful that a single food, or herbal extract will be found which is an effective treatment for all overweight individuals. It is more likely that an array of functional foods addressing different weight management needs will be developed. Or perhaps, a unique product that contains several functional food ingredients, each of which address a different weight management need, will be formulated. Whatever type of functional weight loss foods evolve it is essential to remember that the level of calorie burning activity a person engages in, as well as the calories he or she consumes influences body weight.

Focusing on developing a functional food as the sole treatment for obesity overlooks this important aspect of the energy balance equation. It also neglects to address the emotional issues underlying many people's eating behaviors. To maximize their effectiveness, functional weight loss foods may need to be used in conjunction with behavioral therapy to ensure adequate dietary variety, sufficient activity and development of coping skills for limiting appetite driven eating. While no magic bullet functional weight loss food is currently on the horizon, there are many intriguing lines of research to pursue. To date, the most feasible approach for combating the increased chance of developing overweight and obesity is to control food/energy intake while in the meantime increasing physical activity. Due to an increase of sedentary environments in both the workplace and home, self initiated exercise is becoming increasingly important. Although there are natural products available that may be able to help maintain or control a healthy body weight, the effect is minimal. Furthermore, medical treatment or surgical removal of excessive fat mass should always be the last solution rather than the best option due to the risk associated with these procedures and to the difficulty in maintaining the weight during the postsurgery period. Instead of resorting to these measures, the most efficient way to prevent obesity for individuals is to change their lifestyle by eating healthier and being more physically active.

References

1. Stice, E., *et al.*, *Relation of reward from food intake and anticipated food intake to obesity: a functional magnetic resonance imaging study.* Journal of Abnormal Psychology, 2008. **117**(4): 924.
2. Seiders, K. and R.D. Petty, *Obesity and the role of food marketing: a policy analysis of issues and remedies.* Journal of Public Policy and Marketing, 2004. **23**(2): 153-169.
3. Silveira, M.-B., *et al.*, *Conjugated linoleic acid (CLA) and obesity.* Public Health Nutrition, 2007. **10**(10A): 1181-1186.
4. Lai, C.S., J.C. Wu, and M.H. Pan, *Molecular mechanism on functional food bioactives for anti-obesity.* Current Opinion in Food Science, 2014.
5. Galuska, D.A., *et al.*, *Are health care professionals advising obese patients to lose weight?* JAMA, 1999. **282**(16): 1576-1578.
6. Kos, K. and J.P. Wilding, *SPARC: a key player in the pathologies associated with obesity and diabetes.* Nature Reviews Endocrinology, 2010. **6**(4): 225-235.
7. Rankinen, T., *et al.*, *The human obesity gene map: the 2005 update.* Obesity, 2006. **14**(4): 529-644.
8. Maguelonne, T.S., *A History of Food*, 1993, Cambridge: Blackwell Publishers.
9. Mermel, V.L., *Old paths new directions: the use of functional foods in the treatment of obesity.* Trends in Food Science & Technology, 2004. **15**(11): p. 532-540.
10. Noel. E. Obesity, *The practical guide: identification, evaluation, and treatment of overweight and obesity in adults* 2011: National Heart, Lung and Blood Institute.
11. Puhl, R.M. and C.A. Heuer, *The stigma of obesity: a review and update.* Obesity, 2009. **17**(5): 941-964.

12. Hsu, C.L. and G.C. Yen, *Phenolic compounds: evidence for inhibitory effects against obesity and their underlying molecular signaling mechanisms.* Molecular Nutrition & Food Research, 2008. **52**(1): 53-61.
13. Birari, R.B. and K.K. Bhutani, *Pancreatic lipase inhibitors from natural sources: unexplored potential.* Drug Discovery Today, 2007. **12**(19): 879-889.
14. Smyth, S. and A. Heron, *Diabetes and obesity: the twin epidemics.* Nature Medicine, 2006. **12**(1): 75-80.
15. Graham, I., *et al.*, *European guidelines on cardiovascular disease prevention in clinical practice: executive summary Fourth Joint Task Force of the European Society of Cardiology and Other Societies on Cardiovascular Disease Prevention in Clinical Practice (Constituted by representatives of nine societies and by invited experts).* European Heart Journal, 2007. **28**(19): 2375-2414.
16. Jordan, M.A. and T. Haywood, *Evaluation of internet websites marketing herbal weight-loss supplements to consumers.* The Journal of Alternative and Complementary Medicine, 2007. **13**(9): 1035-1043.
17. Stohs, S. and M. Shara, *A review of the safety and efficacy of Citrus aurantium in weight management.* Obesity: Epidemiology, pathophysiology, and prevention, 2007: 371-382.
18. Diepvens, K., K.R. Westerterp, and M.S. Westerterp-Plantenga, *Obesity and thermogenesis related to the consumption of caffeine, ephedrine, capsaicin, and green tea.* American Journal of Physiology-Regulatory, Integrative and Comparative Physiology, 2007. **292**(1): R77-R85.
19. Potenza, M.A., *et al.*, *EGCG, a green tea polyphenol, improves endothelial function and insulin sensitivity, reduces blood pressure, and protects against myocardial I/R injury in SHR.* American Journal of Physiology-Endocrinology and Metabolism, 2007. **292**(5): E1378-E1387.
20. Oluyemi, K.A., *et al.*, *Erythropoietic and antiobesity effects of Garcinia cambogia (bitter kola) in Wistar rats.* Biotechnology and Applied Biochemistry, 2007. **46**(1): 69-72.
21. Maeda, H., *et al.*, *Effect of medium-chain triacylglycerols on anti-obesity effect of fucoxanthin.* Journal of Oleo Science, 2007. **56**(12): 615-621.
22. Bhandari, K. and S. Srivastava, *Obesity: Overview & Management.* Current Bioactive Compounds, 2007. **3**(4): 278-290.
23. Celleno, L., *et al.*, *A dietary supplement containing standardized Phaseolus vulgaris extract influences body composition of overweight men and women.* International Journal of Medical Sciences, 2007. **4**(1): 45.
24. Diepvens, K., *et al.*, *Long-term effects of consumption of a novel fat emulsion in relation to body-weight management.* International Journal of Obesity, 2007. **31**(6): 942-949.
25. Hughes, G.M., *et al.*, *The effect of Korean pine nut oil (PinnoThin) on food intake, feeding behaviour and appetite: a double-blind placebo-controlled trial.* Lipids Health Dis, 2008. **7**(6): 1-10.
26. Heck, C. and E.G. De Mejia, *Yerba Mate Tea (Ilex paraguariensis): a comprehensive review on chemistry, health implications, and technological considerations.* Journal of Food Science, 2007. **72**(9): 138-151.

Section VI: Future of Functional Foods

17

Future Aspects of Functional Food

Introduction

The future of functional foods is bright and there are several reasons for the growing interest in "functional" foods, such as awareness of personal health deterioration, led by busy lifestyles with poor choices of convenience foods and insufficient exercise; increased incidence of self-medication; increased level of information from health authorities and media on nutrition and the link between diet and health; scientific developments in nutrition research; and a crowded and competitive food market, characterized by pressurized margins. Now, the consumer is well aware of the relationship between diet and health. Consumers understand that the risk for chronic diseases such as heart disease, stroke, colon cancer, and type II diabetes is possibly preventable by modification in their diet pattern. This scientific progress in understanding the relationship of diet to disease, along with increasing health care costs and the consumer's desire to initiate healthy eating and lifestyle habits, makes this a potentially fertile ground for the development of food products that meet these demands.

Extensive research is currently directed towards increasing our understanding of "functional foods." Academic, government and private research institutes around the globe are devoting substantial efforts to identifying how functional foods and food ingredients might help prevent chronic diseases or optimize health, thereby reducing healthcare costs and improving the quality of life for many consumers. An emerging discipline that will have a profound effect on future functional foods research and development efforts is nutrigenomics, which investigates the interaction between diet and development of diseases based on an individual's genetic profile. Nutrigenomics will have a profound effect on future disease prevention efforts including the future of the functional foods industry.

Another technology that will greatly influence the future of functional foods is biotechnology. Recent examples of biotechnology derived crops which have tremendous potential to improve the health of millions worldwide include golden rice and iron-enriched rice. These grains are genetically engineered to provide enhanced levels of iron and β-carotene which could, in turn, help prevent iron deficiency anemia and vitamin A deficiency related blindness worldwide.

Nutrigenomics

As scientific and technological advances develop in the field of health and nutrition, more focus has been directed towards the emerging field of nutrigenomics, or "personalized nutrition." The science of nutrigenomics involves the application of the human genome to nutrition and personal health to provide individual dietary recommendations. By using an individual's unique genetic make-up and nutritional requirements to tailor recommendations, consumers may one day have a greater ability to reduce their risk of disease and optimize their health. Personalizing nutrition to an individual's unique genetic makeup has the potential for positive health outcomes. Choosing an individualized approach, over a more traditional or general approach, to health and nutrition recommendations can provide consumers with the most appropriate and beneficial information for their specific nutritional needs. While personalized nutrition seems promising, research is still in the preliminary stages, and years may pass before accurate and effective recommendations can be made for individuals. Nutrition and genetics both play an important role in human health as well as the development of chronic diseases such as cancer, osteoporosis, diabetes and cardiovascular disease. Nutrigenomics describes the scientific approach that integrates nutritional sciences and genomics and includes the application of other high throughput 'omics' technologies such as transcriptomics, proteomics and metabolomics to investigate the effects of nutrition on health. Nutrigenomics is both the examination of how nutrients affect genes (i.e. influence gene expression and function) and how genes affect diet (i.e. what an individual eats and how an individual responds to nutrients), with the latter sometimes being referred to as nutrigenetics. Nutrigenomics can include the full spectrum of research strategies from basic cellular and molecular biology to, clinical trials, epidemiology and population health [1].

Dietary intake of a nutrient does not necessarily result in the same concentrations in the blood or tissue because substantial individual variability in the absorption, distribution, metabolism and elimination can exist. The mechanisms responsible for the between person differences in dietary response are very complex and have been poorly understood. Research to date has indicated that diet gene interactions play a significant role in this between-person variability, and has clarified some of these genetic differences. The interaction between genetic and dietary influences can result in a higher risk of disease in certain individuals and populations. Currently, diet-gene association studies are revealing evidence on which to base gene-specific dietary intervention trials to confirm results.

Although the term nutrigenomics is relatively new, the concept has been around for some time. Perhaps the most familiar example is lactose intolerance, which is a condition resulting from an inadequate production of lactase in the small intestine due to genetic variation in the lactase gene [2]. Individuals with lactose intolerance are unable to efficiently break down the primary milk sugar (lactose) from dairy products. Consequently, the dietary recommendation is to limit lactose containing foods or to use lactase supplements or lactose-free dairy products to prevent gastrointestinal discomfort [3]. Phenylketonuria (PKU) is an in-born error of metabolism, which represents another classic example of nutrigenomics. PKU can result from a genetic variation in phenylalanine hydroxylase (the enzyme needed to convert phenylalanine to tyrosine), which leads to a decrease in phenylalanine hydroxylase activity [4]. Individuals with PKU can develop neurological damage (severe mental retardation and seizures) from excess phenylalanine unless they follow the recommended low-phenylalanine diet [5].

Lactose intolerance and PKU are examples that involve a single genetic defect along with a single dietary exposure. A major challenge of nutrigenomics, however, is to identify gene-nutrient interactions that affect complex polygenic disorders such as obesity, diabetes, cancer and cardiovascular disease that take several years or even decades to develop and have multifactorial etiologies. As such, it is often difficult to determine the role of specific dietary factors and gene variants in the development of these diseases. The concept of diet-gene interactions involves a genetic variant modulating the effect of a dietary factor on a specific phenotype or health outcome measure such as serum lipid concentrations, high blood glucose or obesity. Conversely, diet-gene interactions can refer to the dietary modification of the effect of a genetic variant on the phenotype or health outcome measure [6]. A major goal of nutrigenomics is the prevention of the onset and progression of chronic disease. Research strategies currently contribute to this goal by building the body of evidence linking nutrients to metabolic pathways that affect disease outcomes. The incorporation of genetics into nutritional epidemiologic studies aims to improve their consistency. Current research could lead to the development of personalized nutrition guidelines for individuals and specific sub-populations, which could decrease the risk of chronic diseases.With the application of genomics our knowledge about nutrient-gene interaction will increase substantially and this will provide novel entries for the prevention and therapy of complex nutrition related disorders for instance by the knowledge based discovery or design of functional foods.

Some companies which already supply a number of functional foods are not only using nutrigenomics to search for new targets, but also to broaden the health claim of products that are already on the market. For instance, a product which has a beneficial influence on the serum lipid profile, can be studied by DNA microarray analysis for its effect on the expression of genes in well selected target tissues. Suppose it turns out that genes involved in free radical scavenging or in immune function are up-regulated. This new information could then be employed to enlarge the health claim and thereby broaden the market for the product.

Biotechnology

Biotechnology in the food processing sector makes use of microorganisms for the preservation of food and for the production of a range of value added products such as enzymes, flavour compounds, vitamins, microbial cultures and food ingredients. Biotechnology applications in the food processing sector, therefore, target the selection and manipulation of microorganisms with the objective of improving process control, product quality, safety, consistency and yield, while increasing process efficiency. Crops are now being enhanced through biotechnology to increase levels of important biologically active substances for improved nutrition, to increase body's resistance to illnesses, and to remove undesirable food components. Phytosterols and phytostanols are cholesterol like molecules found in all plant foods, but the highest concentrations occur in unrefined plant oils, including vegetable, nut and olive oils. Nuts, seeds, whole grains and legumes are also good dietary sources of phytosterols [7]. Studies have shown that these compounds can lower the risk of cardiovascular diseases and the levels of "bad cholesterol" [8, 9]. As phytostanols are more stable than phytosterols during food processing, genetic engineering has been applied for the development of rapeseed and soybean oils with modified ratios of phytosterols to phytostanols. Plants were transformed with a gene from yeast encoding the enzyme 3-hydroxysteroid oxidase, which converts phytosterols to phytostanols [10].

Another example of biotechnological application for production of functional foods is the alteration of glycemic index of foods by modifying starch. Root and tuber crops such as cassava and yams are typically high in amylopectin with lower ratios of amylose to amylopectin. This yields highly digestible, high glycemic index foods. These high-glycemic index starches have been associated with diet-related conditions such as type 2 diabetes and insulin resistance.

Modification of starch to increase the amylose to amylopectin ratios would be greatly useful in improving glycemic index and physiological responses to the starchy foods that are prevalent in the diet in many developing countries. Modifications in the starch synthesis pathway have been effected to modify ratios of amylose to amylopectin in potato and cassava starch. Improving the glycemic indices of staple starchy food crops such as cassava, yams, potatoes and grains will improve their nutritional quality, and subsequently have an effect on the prevalence of attendant disease conditions [11].

Active compounds such as carotenoids from tomatoes, glucosinolates from *Brassica* vegetables, phytoestrogens from soybean and phenolics and antioxidants from various plants have been shown to protect and prevent against numerous diseases such as cancers and cardiovascular disease. These antioxidants are also associated with slowing the aging process and improving overall health. Most of these bioactive components are plant secondary metabolites. Their levels and expression can be improved and optimized through genetic modification of their synthetic pathways. The vitamin A content has been increased in rice (Golden rice) using genetic engineering [12, 13]. In another study, overexpression of genes encoding biosynthetic enzymes or transcription factors have resulted in tomatoes with improved carotenoid content [14].

Fermentation markedly improves nutritional and health quality in food products. Protein content of certain grain foods is increased by fermentation. A study evaluated the effect of fermentation on the nutritional quality of food grade soybeans and feed grade soybean meals. Soybeans and soybean meals were fermented by *Aspergillus oryzae* GB-107 in a bed-packed solid fermentor for 48 hours. Fermentation increased protein content, eliminated trypsin inhibitors, and reduced peptide size in soybeans and soybean meals making soy foods more useful in human diets as a functional food and benefit livestock as a novel feed ingredient [15]. Furthermore, fermentation and staling results in modification of starches to form resistant starches which are associated with protection against colon cancer and gastrointestinal disease [16].

A probiotic is a viable microbial dietary supplement that beneficially affects the host through its effects in the intestinal tract. Probiotics are widely used to prepare fermented dairy products such as yogurt or freeze dried cultures. The use of probiotics has been demonstrated to be effective for the treatment or control of several diseases. With few exceptions, most probiotic products currently available contain lactic-acid-producing bacteria, which mainly belong to the genera *Lactobacillus and Bifidobacterium*. Evidence suggests the beneficial effects of probiotics such as normalization of the intestinal microflora, ability to block the invasion of potential pathogens in the gut, prophylactic or therapeutic treatment for several types of diarrhea, relief of symptoms of irritable bowel syndrome and inflammatory bowel disease, amelioration of lactose intolerance, prevention of colon cancer, modulation of immune function, inhibition of *Helicobacter pylori*, and possible enhancement of calcium absorption and reduction of blood cholesterol levels [17, 18]. Most important, probiotic bacteria such as Bifidobacteria and various *Lactobacilli* improve immune status of the colon, and are involved in preventing and depleting pathogenic and infectious bacteria in the gastrointestinal tract such as *Helicobacter pylori* [19]. Furthermore, probiotic bacteria are effective against lactose intolerance and diarrhea [20]. Most commercial probiotics are produced by fermentation. Mechanisms for the above benefits have been proposed, but none has been proven. An adequate level of viable bacteria in a probiotic product and an appropriate daily dose are critical to achieve a health benefit. Because probiotics are not known to be pathogenic and their upper tolerable level is high, they could be promoted as a beneficial food supplement.

Market Scenario and Innovation in Functional Foods

Global market

Worldwide market for functional foods and nutraceutical products is growing at a tremendous pace and outpacing the traditional processed food market. There is growing demand for functional foods, especially in developed economies due to increasing awareness towards health benefits of functional foods and increase in disposable incomes. In 2010, the global nutraceuticals market was estimated to be worth \$128.6 billion, after increasing at a compound annual growth rate (CAGR) of 4.4% during 2002-2010. Furthermore, the market is forecast to reach \$180.1 billion by 2017, after growing at a CAGR of 4.9% throughout 2010-2017 [21]. Another report by Industry ARC forecasts even higher sales for functional foods. It also says that with an annual average growth rate of about 15% the global market for functional food is forecast to exceed \$230 billion by 2015 [22]. The primary reason for the growth rate is an increase in the elderly population, the affluence of the working population, and awareness of, and preference for, preventive medicine. Fig. 17.1 shows the revenue generation by different categories of functional foods in the year 2013 [22].

The US is the largest market for functional foods and is expected to witness 21% growth in the coming years owing to the growing demand for functional foods. This growth is mainly driven by the continuously growing demand for energy drinks and fortified dairy products. The boom is on in Asia too, with India's \$ 1 billion to \$ 2 billion nutraceutical market projected to grow to \$ 5 billion by 2015. Indeed, the Asia Pacific market, with its culture of non-Western healing remedies, is second only to the United States in global sales of all functional foods. The global functional foods market is categorized as: bakery and cereals, confectionery, fortified dairy products, frozen and chilled food and functional drinks.

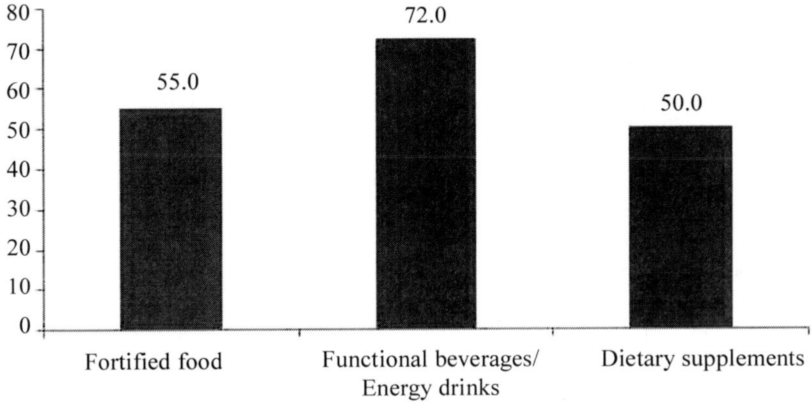

Fig. 17.1: Functional food revenue, by segment – 2013

Some of the major players in the global functional foods market include Abbott Laboratories, Arla Foods, BNL Food Group, The Coca Cola Company, Dean Foods, Dr Pepper Snapple Group, General Mills Inc., Glanbia Plc, GlaxoSmithKline Company, Groupe Danone, Kellogg Company, Kirin Holdings, Kraft Foods Inc., Mars Inc., Meiji Group, Murray Goulburn, Nestlé, Ocean Spray Cranberries Inc., PepsiCo Inc., Raisio Group, Red Bull GmbH, Royal Friesl and Campina, Sanitarium Health & Wellbeing Company, Unilever, and others. The various ingredients added to the foods so as to provide the functionality to the product are on rise. Earlier consumers were interested only in vitamins and minerals but now they are focussing on other things as well like antioxidant, omega-3-fatty acids, etc. The Table 17.1 shows the percentage of increased interest in various functional ingredients which means that there are various opportunities of growth in this market [22, 24].

Table 17.1: Increased interest of consumers in functional ingredients

Functional ingredients	Increased interest, %
Antioxidants	57.3
Dietary fibre	55.8
Organic ingredients	44.2
Calcium	43.5
Soy protein	43.4
Omega fatty acids	41.8
Vitamins	37.5
Lycopene	32.9
Whey proteins	30.9
High oleic fats/Oils	30.6
Beta carotene	30.2
Vitamin E	29.7
Lutein	26.6
Probiotics	26.6
Fruits (whole-not extract)	25.9
Prebiotics	25.8
Minerals	25.5
Isoflavones	23.8
Phytosterols/stanols	23.1
Resistant starch	23.1
Botanicals	22.9
Folic acid	22.5
Amino acids	22.0
Fruit extracts	20.8
Oligosaccharides	19.2
L-carnitine	12.5

The health and wellness trend is becoming one of the key growth areas and marketing strategies within the food industry. Due to the saturation in many categories of food and the emergence of private labels which create an increasingly competitive market, the health and wellness trend has provided food manufacturers with a point of differentiation to help their products stand out from competitors. Fig. 17.2 shows the market share of different health and wellness foods, with fortified and functional foods having a big share of it [22].

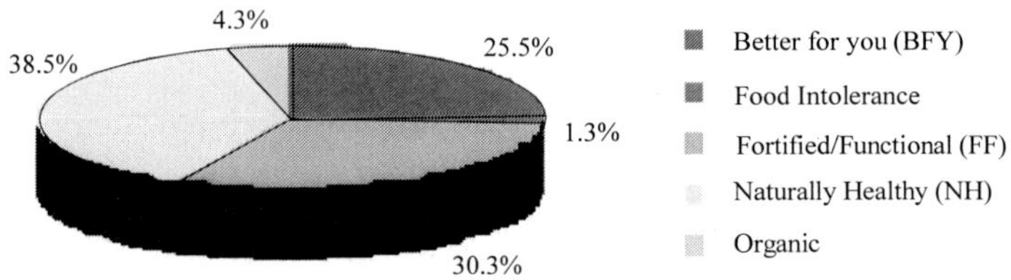

Fig. 17.2: Market share of health and wellness products in year 2010

The overall outlook for incorporating functional ingredients in packaged foods and beverages is very positive, as consumers are becoming more educated about the health benefits of fortified/functional foods. According to industry estimates, the largest increases in nutraceutical ingredient demand during the next few years are expected to be for glucosamine, probiotics, sterol esters, whey protein, and ω-3 fatty acids [23]. The Fig. 17.3 shows the global sales of health and wellness products by region. It shows that America, Europe and Asia Pacific are the largest markets for health and wellness foods [22].

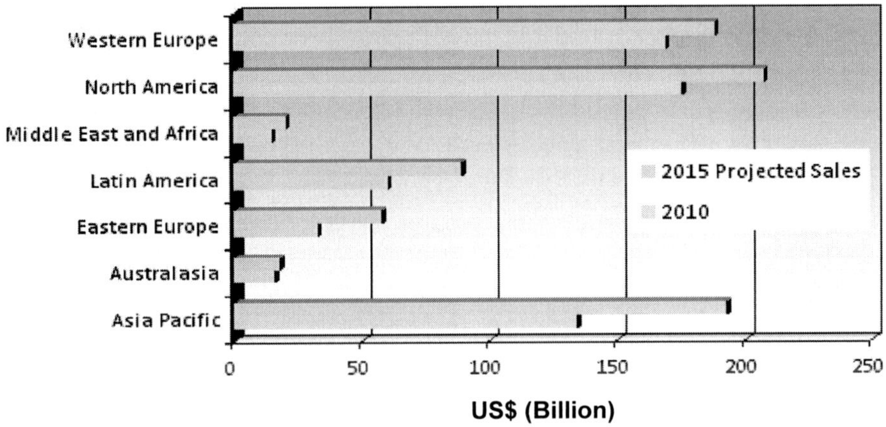

Fig. 17.3: Sale of health and wellness products by region

Functional Food Market in USA

The functional food industry in the United States has also been growing and, along with Japan and Europe, is one of the largest in the world. Growth in the US functional food sector has been attributed to the realization of the link between diet and health by consumers, the increasing cost of health care, the aging US population and the challenge of consumers meeting their daily nutritional needs through conventional food [24].

The United States is the world's largest functional food market with sales of \$43.9 billion in 2012, +6.9% over 2011 [24]. Six in 10 U.S. adults consume specially formulated functional foods/beverages at least occasionally. Yogurt for digestive health and cereal for heart health are the most consumed items, followed by cholesterol lowering butter/margarine, cholesterol lowering orange juice, shakes/bars to reduce hunger, orange juice for joint health, immune-boosting dairy beverages, and medicinal teas. Beverages remain the largest U.S. functional product sector with sales of \$27.8 billion in 2012, +9.8% vs 2011. Functional bread/grain posted sales of \$6.4 billion, down 3.3%; packaged foods \$3.6 billion, +1.4%; snack foods \$3.5 billion, +11.7%; dairy \$2.2 billion, +6.9%; condiments \$172 million, +7.0%; and fruit/vegetables \$45 million, +12.2% [24].

One in five adults cut back on their use of supplements because they are eating so many fortified foods; 88% regularly consume vitamin or mineral fortified foods [24]. Firms in the health food sector can be categorized into three groups. First are the mainstream foods and drink companies such as Danone, PepsiCo and Quaker, which make up the greatest proportion of firms in the industry. The second group is comprised of pharmaceutical companies such as Johnson & Johnson, while the last group consists of single-product specific companies [22]. The US functional food market is dominated by multinational corporations. In 2005, PepsiCo had approximately 16% of the US market for functional food; Kellogg Co., Coca-Cola Co. and General Mills followed with approximately 14%, 11%, and 11% respectively (Table 17.2).

Table 17.2: Manufacturers of functional foods in US market

Manufacturer	Sales (million US \$)	Share (%)
PepsiCo	2397	16.3
Kellogg Co.	2100	14.3
Coca Cola Co.	1591	10.8
General Mills	1558	10.6
Kraft Foods, Inc.	847	5.8
Group Danone	643	4.4
Nestle USA, Inc.	545	3.7
Ocean Spray	427	2.9
Welch Foods, Inc	330	2.2
Dean Foods	292	2
Ferolito Vultaggio & Sons	241	1.6
Red Bull North America, Inc	226	1.5

Functional foods in the US exist across a broad range of categories, including dairy products, bakery products, cereal products, soy products, beverage products, and meat, fish and egg products. The US market for functional food is mostly domestically driven. Thus, most of the functional food products are domestically produced and consumed in the US with a small proportion of the total US production exported. The rapid development of the US functional food market has been associated with the approval of health claims by the Food and Drug Administration (FDA). The FDA has sixteen approved health claims and over twenty qualified health claims.

Functional Food Market in Europe

The European functional food market is very much concentrated in northern Europe, with the United Kingdom leading in terms of per capita consumption/spending and absolute levels. In 2006, the UK value for functional food was estimated to be US$ 1998 million, with a per capita spending on functional food equal to $33. In 2006, the UK and Germany represented almost half of the total European functional food markets [31]. Between 2001 and 2006, the EU market grew by 36%, while Italy and Sweden had the highest growth rate among the country members with 54%, followed by Spain (49%), Germany (38%), the UK (28%) and France (28%) [31].

A majority of European consumers are believers in positive nutrition, with 56% using food/drink to improve health, according to a 2014 Datamonitor Consumer Survey presented by the firm's head of food and drink Mark Whalley at Vitafoods Europe in Geneva, Switzerland. Women (60%) are slightly bigger believers than men (51%). The survey also showed 19% of Europeans consume functional foods at least a few times a week, which is more than potato chips or ice cream. Additionally, 16% consume functional drinks at least a few times a week. Nearly three quarters of consumers said they are not loyal to specific brands (69% for functional foods; 72% for functional drinks). "There is an absolute wealth of opportunity to build up your brand within the functional food/drink space," said Mr. Whalley. About 10% of consumers claim to purchase at the "premium" end of the market for functional foods (9% for functional drinks).

An estimated 168 companies market at least one functional food in the EU, however, the European functional food industry is dominated by a relatively small number of large multinational companies [31]. Examples include PepsiCo, GSK, Danone, Nestle, and Unilever. The market for functional food in Europe is dominated by soft drinks (energy drinks, juices, enriched milk).

A study identified 385 functional food products in the European market in 2004, containing 503 different ingredients, the most common of which were probiotics, prebiotics and plant extracts. Almost 30% of the functional foods contained more than one functional ingredient. Bacteria cultures, which are primarily probiotics, were used in about 45% of all functional food products. Saccharides (mostly prebiotics) were the next most commonly (42%) used ingredient in the European market in about 20% of all products, followed by plant extracts, Terpenes, Fibres, Phenols, Peptides and lipids [32].

Functional Food Market in Japan

Japan is often credited with creating the term "functional foods" in the late 1980's. Japan is the only nation that has legally defined functional foods, and the Japanese functional food market is now one of the most advanced in the world. For those reasons, developments in Japan are often cited as indicative of possible developments in Europe and the United States. Japan has a large, mature and sophisticated functional food and beverages market. Japanese consumers are some of the most health conscious in Asia, and are willing to spend money on high quality, safe, natural and tasty products.

The Japanese functional foods market is the most developed in the world as illustrated by the cutting edge innovations that continue to deliver more sophisticated and niche health benefits. In terms of size on an international scale, Japan's functional products market is second only to the United States, and is valued at US $19.6 billion annually.

The government has had a strong involvement in this industry and continues to encourage the development of foods and beverages with proven benefits, as a way to help maintain the health of the general population, while potentially lessening the increases in health care costs that are expected in the future. Consequently, despite the shrinking population, as Japan's older demographic continues to expand, too much will be the demand for functional foods. Japan represents one of the most sophisticated markets for these products, and foods that are perceived or labelled to have some sort of health benefit as shown in Fig. 17.4 [24].

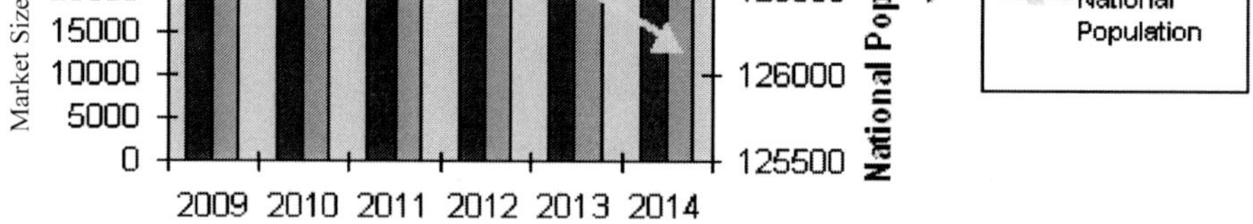

Fig. 17.4: Market size for functional food

There were about 755 FOSHU approved products in 2007 with an estimated retail value of US$ 6.5 billion. FOSHU products can either take the form of capsules, tablets or conventional food, however, the majority are in the form of conventional food [25]. Targeted health categories include digestive health, weight management, dental (teeth) health and metabolic syndrome risk factors. The most commonly used ingredient in FOSHU products is dietary fibre followed by probiotics [26]. Currently, the ingredients gaining popularity in

Japan include soy peptides, green tea catechin and gamma amino butyric acid [27].

Many science based functional food products have been developed and approved by the Japanese government as "Food for specified health use (FOSHU)" in the past two decades. However, the functions already approved are limited, and new categories of functional food are being expected by the society. Expected future FOSHU categories may include immune-modulating food, anti-fatigue food, skin health-promoting food, food preventing joint ache, and, so on. More studies on the molecular mechanisms for food functions, establishment of biomarkers to assess objective evaluation of food functions, and development of novel technology for functional food processing will be necessary. Social agreement on the evidence based functional food educating consumers as how to evaluate and how to accept functional food is also important. A huge amount of scientific data on functional food substances, in terms of their potential in disease prevention and health promotion, may be useful for many purposes. Recently, new crops or agricultural products fortified with specific functional substances have been created in Japan, and they are expected to supply functional substances to consumers in everyday meal, not via packaged food products.

Functional Food Market in India

Exceptional growth in the Indian economy and an increasing awareness of healthy functional ingredients is driving the Indian functional foods and beverages market, according to a report by Frost and Sullivan [28]. A strong desire among Indian consumers to maintain a healthy lifestyle is an additional factor driving this market, according to the analysts, who found that the market earned revenues of over US$ 185 million in 2007 and estimates this to reach US$ 1161 million in 2012. While the market is still in its infancy, along with sports and energy drinks, the demand for functional foods is predicted to witness an expanding consumer base due to their specific health benefits. "It is difficult to convince Indian fast moving consumer goods (FMCG) companies to fortify their products with innovative ingredients," said Aditi Paul, 2008. "The vague regulatory and industry approval protocols for functional foods further dampen the speed of market growth." Survival in the functional foods business requires cutting edge R & D and high technical expertise, and it seems most Indian companies do not invest enough in research, yet this may change in the near future.

Challenges in Functional Foods

Formulation challenges

When the functional foods were launched, consumers were willing to compromise on taste in exchange for health benefits offered by these products but now functional foods manufacturers have begun dealing with a new consumer for whom taste is the top priority.

Identifying a valuable, sought-after ingredient is one thing; getting it into a food or beverage matrix successfully is another. The latter can be complicated, but challenges are overcome through inventiveness, creativity, and the rigorous application of science. The ingredient as such can have functional properties but when incorporated in food which undergoes various stages of processing may result in loss of its functional properties. For example, high heat, humidity, and pressure sometimes involved in extruding food products may effect the functionality of added ingredient. These sensitive nutrients can still be added to products, but the conditions of processing must be considered when selecting the amount of nutrients to add.

Incorporating an insoluble fiber can prove difficult to be incorporated into functional foods as it effects mouthfeel of the product. Most insoluble fibers contribute some chalkiness or grittiness to end products, and a serving of food must contain at least 2.5g of fiber to make a fiber claim. This limits the scope of food products that can be targeted for insoluble fiber fortification. For drinks, solubility is the key. Agglomeration can come to the rescue and increase the solubility of an ingredient when that ingredient is mixed with liquids. Agglomeration can also enhance "flowability" and assist with ingredient dosing and dispersion when working with a formula requiring expensive ingredients. Solubility of functional ingredient in foods is sometimes an issue of concern. Many functional foods and beverages require miscibility and water solubility to properly mix in the ingredient.

Most functional ingredients will contribute off flavors to foods and beverages and present challenges to the product developer. Two areas where there is a lot of (flavor masking) activity are vegetable protein enhancement and natural, high-intensity sweeteners, such as stevia. The proteins can contribute beany, musty off flavors that require masking, and the sweetness profile of stevia can be made to more closely resemble that of sucrose via masking as well.

Fortification challenges

Fortifying foods with calcium and vitamin D is very much there in the market because of their potential to reduce the risk of osteoporosis (FDA). However, it becomes a problem particularly in beverages with a low pH (<4.4), as increasing calcium level above 30% results in tooth etching and a chalky mouth feel. Zinc and magnesium have not much impact on flavour, so, they can be used in any amount. However, iron and potassium leads to bitterness and metallic aftertaste if used above a certain concentration. So, flavourist needs to work with taste modulators so as to overcome these challenges.

Herbals are the newest flavour challenge. They are getting more and more of attention because of their calorie burning capabilities. However, weight-loss ingredients such as hoodia, hydroxyl citric acid and epigallocatechin gallate also cause bitterness or astringent notes in the mouth. To overcome these off-notes, flavourist modulate the taste receptors by masking the bitter receptors and increasing the sweetness perception through receptor binding or through steric blocking of the receptor. Through manipulation of the flavour chemicals, the flavourist can make the product taste like whatever the beverage manufacturer desires, whether it is mango flavour or herbal notes.

Addition of vitamins such as B_1 which has a meaty flavour is creating a lot of problem. This is because a meaty note would make sense in a beverage with a base such as tomato juice, but no one wants to drink a light orange juice or lemonade with a meaty aftertaste. So, research is going on in this direction as to how to mask the undesirable flavor [33] .

Protein challenges

The various protein enriched functional drinks use soy protein or whey protein, often with casein to round out the amino acid profile and to add viscosity. Pea, rice, and egg proteins (albumin and yolk) are used more often in drinks fortified specifically for body building or protein recovery. The challenge faced by manufacturers when formulating a protein drink having high pH (>4.5), is the harsher processing conditions such as retorting, high temperature short time processing, or UHT processing. These processes often leads to off flavours such as burnt, caramelized, nutty, beany, bitter, etc. Earlier, flavourist used to counter these off-notes with high-sweetness flavors such as chocolate, vanilla, and strawberry. But now as flavor is becoming more and more important, there is need for vast array of masking agents, and modulating flavors. For example, to make the protein in soy milk (delivering greater than 6.25g of protein/serving) palatable, the raw soy concentrate must be processed to minimize the inherent green/beany/nutty notes present in bean. So, the soy protein must be processed to eliminate strong base notes and be stabilized, sweetened, and modulated to reduce these off-notes and flavored to taste like milk because it is taken as a milk substitute [33].

In ready-to-drink, low pH (<4.4) protein beverages, pectins, alginates and other stabilizers are used to protect and suspend the protein in the beverage. Use of partially hydrolyzed whey protein isolates has enabled manufacturers to make low pH, clear, fruit flavored beverages with very high levels of protein (≥30g/8 fl oz), ideal for body builders. However, these high levels of protein create salty and bitter flavor which must be removed or masked off, hence, base modifications are done.

Processing and pH

When a functional beverage is made, processing conditions to get better shelf life needs should be determined first. Determining this will allow off-notes to be masked at the forefront rather than after the complete product is manufactured. To determine the processing requirements, functional drinks producers need to establish the pH of the system they intend to produce.

For low pH (<4.4) drinks, preservation methods used are hot filling (at 82 °F) and chemical preservation (using potassium sorbate, sodium benzoate, and sodium hexametaphosphate) so as to control and inhibit microbial growth. Products preserved with chemical preservatives must start "clean" (low bacterial counts for bacteria, yeast and mold) and are generally cold filled. But these preservatives create kind of off notes in the product (produced by the chemical burn). Consumers consider addition of preservatives as "negative". This coupled with the off-notes restricts the application of this method.

For products with pH>4.5, more harsh measures needs be taken for preservation, since spoilage microorganisms like *Clostridium botulinum* could grow in this pH. Aseptic filling and UHT processing followed by refrigeration are two ways to produce product with neutral pH. These treatments ensure killing of any spoilage microbes. Other factors should be kept in mind while adjusting the pH of the beverage. For example, some components change their taste with pH. EGCG found in teas tends to become increasingly bitter as the pH is lowered. The use of stronger flavours such as heavier peach notes and vanillas may aid the masking of the bitter off-notes from the product.

Technological challenges

The main areas for technological challenge that have been identified are as follows.

- The ***creation*** of new food components in traditional and novel raw materials that add or increase functionality. Examples of such challenges are genetic modifications, use of underutilised or unconventional natural sources (e.g. algae, seaweeds), development of bioreactors based on immobilised enzymes or live microorganisms.
- The ***optimisation*** of functional components in raw material and in foods to ensure maximal preservation of the component(s), to modify function, to increase their bioavailability. Examples of such challenges are development of membrane processing techniques, use of controlled and modified atmospheres, use of high hydrostatic pressure, high intensity electric field pulse technology or ultrasound treatments.
- The ***effective monitoring***, throughout the entire food chain, of the amount and functionality of the component(s) in raw materials and foods. Examples of such challenges are monitoring of microbial viability and productivity for probiotic functions, development of sensitive markers to record changes in speciation and interactions with food components during processing, especially fermentation.

Consumer Awareness

Even though it is estimated that growth rates of functional foods are rising and more new products are launched, the functional food market is experiencing several difficulties. Consumers have many grounds for criticism. From the consumer perspective, there is undoubtedly more information available than ever before, but it is quite fragmented and specific (not to mention contradictory), and seems to be getting even more so. Specialised terminology that is not easy for consumers to comprehend such as peptide and plantstanol-adds to the confusion and complexity. In addition, consumers will not buy products that they think are of a poor quality or unsafe, and this applies in particular to product categories such as functional foods that are consumed by eating and directly affect health and life. Consumers can only be expected to choose functional over conventional foods if they are aware of and know enough about both functional and conventional products and believe that the differences justify the higher prices and that the health claims will benefit them without risk.

The relatively low degree of familiarity with and acceptance of functional foods among consumers poses a distinct market challenge. In a global survey polling over 21,100 respondents, regular internet users in 38 markets across Europe,

Asia Pacific, North America, Latin America, and in South Africa, the consumer's awareness of different health foods including functional foods was remarkably low. Other studies have reported equivalent findings [34, 35]. For example, a US survey conducted by the American Dietetic Association found that only 21% of Americans had ever heard of functional foods, and even fewer could relate them to health issues. On average, about a third of consumers in all countries covered in the survey did not believe that functional foods really offer additional health benefits. It seems that extensive comparative studies have not been made lately, but earlier [36] showed significant differences in consumer attitudes to food, nutrition and health between fifteen pan-European countries. For example, 13% of Finns consider knowing enough about healthy food as a criterion to healthy eating compared to 24%, 22% and 20% of Danes, Swedes, and Brits, respectively. Finnish consumers have been found to be comparatively positive towards functional foods, more so than Americans and other Europeans [37].

However, this scenario is now changing and more and more consumers are becoming aware of the health benefits of various food products present in market. The 2013 'Functional Foods Consumer Survey' focused on American consumers shows that 90% of consumers agree that certain foods have health benefits beyond basic nutrition. This represents a significant increase compared to 85% in 2002. When asked an open ended question about a food or food component that has benefits beyond basic nutrition, most consumers (89%) can name at least one functional food and its associated health benefit. Specifically, fruits and vegetables are by far the most commonly mentioned functional food. When asked to name the health benefit that the food or food component is good for, 27% of consumers say it provides a specific nutrient or food component such as vitamins or protein. Cardiovascular and digestive health are also benefits named by consumers. Consumer interest in learning more about functional foods remains high. Almost 9 in 10 Americans (86%) are interested in learning more about foods that have health benefits beyond basic nutrition. Similar to 2011, almost half of all consumers (45%) are "very interested" while another 42% are "somewhat interested." Approximately 3 out of 4 consumers report concerned that they are not getting all of the nutrients or food components necessary for good health. Nearly 63% of consumers believe that functional foods will provide health benefits and have confidence they could make the necessary changes to integrate functional foods into their diet. About 50% of consumers believe they get at least most of the nutrients/food components they need for good health from food. Cost is the most common barrier, with over half of consumers (55%) identifying this as a major reason. Other commonly identified barriers included taste and a preference for pure, basic foods.

Summary

As functional food alternatives are being launched at a faster rate and commonly displayed with more conventional food products, it becomes increasingly difficult for consumers to choose an appropriate product. Although functional foods have received attention in the marketing literature, research has not focused specifically on one important component of functional foods i.e. consumer knowledge. Knowledge of health claims is considered one driver for consumers to engage in health related behaviour. With specific reference to on 606 American and Canadian consumers showed that the level of nutritional knowledge about soy did not necessarily influence how much people liked soy, but was related to how much people consumed soy. This would indicate that buying tendency might also be affected by knowledge. It is valuable for companies selling and marketing functional foods to gain insights specifically into consumer's knowledge in order to segment their target markets as well as plan and direct marketing efforts. Considering that marketing costs for functional foods have been claimed to be high and that the awareness and buying propensity of consumers varies, new customer understanding could be useful.

References

1. Mariman, E., *Nutrigenomics and nutrigenetics: the 'omics' revolution in nutritional science.* Biotechnology and applied Biochemistry, 2006. **44**(3): 119-128.
2. Swallow, D.M., *Genetics of lactase persistence and lactose intolerance.* Annual Review of Genetics, 2003. **37**(1): 197-219.
3. Swagerty Jr, D.L., A.D. Walling, and R.M. Klein, *Lactose intolerance.* American Family Physician, 2002. **65**(9): 1845-1850.
4. DiLella, A.G., *et al., Molecular structure and polymorphic map of the human phenylalanine hydroxylase gene.* Biochemistry, 1986. **25**(4): 743-749.
5. Surtees, R. and N. Blau, *The neurochemistry of phenylketonuria.* European Journal of Pediatrics, 2000. **159**(2): 109-113.
6. Ordovas, J.M., *Genotype–phenotype associations: modulation by diet and obesity.* Obesity, 2008. **16**(S3): 40-46.
7. Ostlund Jr, R.E., *Phytosterols in human nutrition.* Annual Review of Nutrition, 2002. **22**(1): 533-549.
8. Jones, P.J. and S.S. AbuMweis, *Phytosterols as functional food ingredients: linkages to cardiovascular disease and cancer.* Current Opinion in Clinical Nutrition & Metabolic Care, 2009. **12**(2): 147-151.
9. Moruisi, K.G., W. Oosthuizen, and A.M. Opperman, *Phytosterols/stanols lower cholesterol concentrations in familial hypercholesterolemic subjects: a systematic review with meta-analysis.* Journal of the American College of Nutrition, 2006. **25**(1): 41-48.
10. Venkatramesh, M., *et al., Expression of a Streptomyces 3-hydroxysteroid oxidase gene in oilseeds for converting phytosterols to phytostanols.* Phytochemistry, 2003. **62**(1):39-46.
11. Niba, L.L., *The relevance of biotechnology in the development of functional foods for improved nutritional and health quality in developing countries.* African Journal of Biotechnology, 2004. **2**(12): 631-635.

12. Beyer, P., *et al.*, *Golden rice: Introducing the â-carotene biosynthesis pathway into rice endosperm by genetic engineering to defeat vitamin A deficiency.* The Journal of Nutrition, 2002. **132**(3): 506S-510S.
13. Paine, J.A., *et al.*, *Improving the nutritional value of Golden Rice through increased pro-vitamin A content.* Nature Biotechnology, 2005. **23**(4): 482-487.
14. Davuluri, G.R., *et al.*, *Fruit-specific RNAi-mediated suppression of DET1 enhances carotenoid and flavonoid content in tomatoes.* Nature Biotechnology, 2005. **23**(7): 890-895.
15. Hong, K.J., C.H. Lee, and S.W. Kim, *Aspergillus oryzae GB-107 fermentation improves nutritional quality of food soybeans and feed soybean meals.* Journal of Medicinal Food, 2004. **7**(4): 430-435.
16. Ahmed, R., I. Segal, and H. Hassan, *Fermentation of dietary starch in humans.* The American Journal of Gastroenterology, 2000. **95**(4): 1017-1020.
17. Lin, D.C., *Probiotics as functional foods.* Nutrition in Clinical Practice, 2003. **18**(6): 497-506.
18. Stanton, C., *et al.*, *Fermented functional foods based on probiotics and their biogenic metabolites.* Current Opinion in Biotechnology, 2005. **16**(2): 198-203.
19. Hamilton-Miller, J., *The role of probiotics in the treatment and prevention of Helicobacter pylori infection.* International Journal of Antimicrobial Agents, 2003. **22**(4): 360-366.
20. Scheinbach, S., *Probiotics: functionality and commercial status.* Biotechnology Advances, 1998. **16**(3): 581-608.
21. Market, R.A. *Nutraceuticals Market to 2017 - Food Additives such as Omega-3 Fatty Acids, Probiotics, Soy and Energy Drinks to Perform Strongly*. 2011.
22. IndustryARC. *Global Functional food and Nutraceuticals Industry to Cross $ 230 Billion Mark in 2015*. 2013 December 28, 2013].
23. Canada. *Health and Wellness Trends for Canada and the World.* 2011; Available from: http://www.ats-sea.agr.gc.ca/inter/4367-eng.htm.
24. Bigliardi, B. and F. Galati, *Innovation trends in the food industry: the case of functional foods.* Trends in Food Science & Technology, 2013. **31**(2): 118-129.
25. Arai, S., *Functional food science in Japan: state of the art.* Biofactors, 2000. **12**(1): 13-16.
26. Saito, M., *Role of FOSHU (Food for Specified Health Uses) for healthier life.* Yakugaku zasshi: Journal of the Pharmaceutical Society of Japan, 2007. **127**(3): 407-416.
27. Arai, S., *et al.*, *Recent trends in functional food science and the industry in Japan.* Bioscience, Biotechnology, and Biochemistry, 2002. **66**(10): 2017-2029.
28. Sullivan, F. *Driven by High Penetration Potential, the Indian Nutraceuticals Market Expected to Reach US $4 Billion by 2018, Finds Frost & Sullivan.* 2014; Available from: http://www.frost.com. Future Aspects of Functional Food 309.
29. Aditi Paul, *India: increasing health awareness driving Indian food market.* 2008; Retrieved from http://www.just-food.com/news/increasing-health-awareness-driving-indian-food-market_id101819.aspx.
30. Eckert, Markus, and Paul Riker. "*Overcoming challenges in functional beverages*." Food Technology 2007. **61**(3):25-26.
31. Nelson, *Growing Chinese Health Awareness Could Mean Big Business for Manufacturers,* 2015; Retrived from http://www.nielsen.com/cn/en/press-room/2015/growing-chinese-health-awareness-could-mean-big-business-for-manufacturers-en.html.
32. Menrad, Klaus. "*Market and marketing of functional food in Europe*." Journal of Food engineering 2003. **56** (2): 181-188.
33. Verbeke, Wim. *Consumer acceptance of functional foods: socio-demographic, cognitive and attitudinal determinants*. Food Quality and Preference 2005. **16**(1): 45-57.

34. Killackey-Jones, Brenda, *et al. An Effective One-Hour Consumer-Education Program on Knowledge, Attitude, and Behavior Toward Functional Foods*. Journal of Extension 2004. **1**(1):1-12.

35. Bech-Larsen, Tino, and Klaus G. Grunert. *The perceived healthiness of functional foods: A conjoint study of Danish, Finnish and American consumers' perception of functional foods*. Appetite 2003 **40**(1): 9-14.

36. Urala, Nina, and Liisa Lähteenmäki. *Reasons behind consumers' functional food choices*. Nutrition & Food Science 2003 **33**(4): 148-158.

37. Canada. *Health and Wellness Trends for Canada and the World*. 2011; Available from: http://www.ats-sea.agr.gc.ca/inter/4367-eng.htm.

18

Functional Food Regulation Around The World

In most countries there is no legislative definition of the term drawing a border line between conventional and functional foods is challenging even for nutrition and food experts. Functional foods do not have a specific place in food law; consequently there is a great deal of confusion about what they are, could or should be. A common perception is that functional foods are foodstuffs either with one or more specific added ingredients (components), or certain substances of common ingredients (components) removed and which are claimed to have a positive influence on human health or well being. The term 'functional foods' for this group of products is considered to be an unfortunate one since all foodstuffs are or should be functional and a more appropriate term might be helpful.

In a regulatory sense, functional foods and nutraceuticals are not clearly classified and defined. Other than in Japan, the term 'functional food' is neither defined nor recognised as a distinct product category by any regulatory authority [1]. Even the Japanese definition fails to encompass all products that could be considered functional foods or nutraceuticals. The lack of a formal definition essentially requires regulatory authorities to place functional food products into one of their already established and regulated categories [2]. Functional foods and nutraceuticals have properties inherent to both foods and drugs; they are at border between conventional food and drugs. In general, a discussion on the regulation of functional food and nutraceuticals in a particular jurisdiction involves discussion of that jurisdiction's framework to allow claims for products that are or resemble foods. Although these products need not make an explicit claim, the definitions for functional foods and nutraceuticals declare these products as exerting some sort of effect on health and are viewed by regulatory authorities to be within the realm of drugs, medicines and/or therapeutic products.

This chapter discusses how functional foods and nutraceuticals are addressed by regulatory authorities of different countries and provides a basic overview of the frameworks used and a general sense of how functional foods and nutraceuticals are treated in each authority.

United States Regulatory Framework

The United States Food and Drug Administration (FDA) controls food and drug regulations in the United States. Functional foods and nutraceuticals are not defined in the United States regulatory framework. Although a legal definition in the USA does not exist, the Institute of Medicine of the US National Academy of Sciences defined functional foods as those that, to encompass potential healthful products, "include any modified food or food ingredient that may provide a health benefit beyond the traditional nutrients it contains" [3]. This is consistent with the definitions of Japan and ILSI in that they should have a health benefit beyond traditional nutritional values [4].

In the USA, a number of health claims have been already permitted under the rules of the Nutrition Labelling and Education Act 1990 [5]. Functional foods can be sold with a health claim on the label once permission from the USFDA has been obtained. In 1994, the US Congress passed the Dietary Supplements Health and Education Act (DSHEA). The DSHEA was designed to regulate products that supplemented the diet. This act amended the US Food Drug and Cosmetic Act to add dietary supplements as a subcategory of food. As a consequence, those generally considered to be functional food and nutraceutical products are likely to fall within food law in the United States.

Dietary supplements and the dietary supplements health and education act

The Dietary Supplement Health and Education Act of 1994 ("DSHEA"), is a 1994 statute of United States Federal legislation which defines and regulates dietary supplements. Under the act, supplements are mainly unregulated, without proof of effectiveness or safety needed to market a supplement, as well as dietary supplements being classified as foods instead of as drugs [5, 6].

DSHEA defines the term "dietary supplement" to mean a product (other than tobacco) intended to supplement the diet that bears or contains one or more of the dietary ingredients such as a vitamin, a mineral, herb or other botanical, an amino acid, a dietary substance for use by man to supplement the diet by increasing the total dietary intake, or concentrate, metabolite, constituent, extract, or combination of any of the afore mentioned ingredients.

Furthermore, a dietary supplement must be labeled as a dietary supplement and be intended for ingestion and must not be represented for use as conventional food or as a sole item of a meal or of the diet. In addition, a dietary supplement can not be approved or authorized for investigation as a new drug, antibiotic, or biologic, unless it was marketed as a food or a dietary supplement before such approval or authorization. Under DSHEA, dietary supplements are deemed to be food, except for purposes of the drug definition [7].

Dietary supplements do not require premarket approval from the FDA [3]. FDA reviews neither the safety nor the efficacy of dietary supplements prior to their introduction to the market. The FDA may take enforcement action on products in the market place if the products pose a health risk or do not meet the requirements set out in the legislation. Also, DSHEA does not set a limit on the amount of vitamins or minerals which can be put into a pill. The manufacturers of the products are responsible for ensuring that the requirements are met.

Health claims

Health claims describe a relationship between a food substance (a food, food component, or dietary supplement ingredient), and reduced risk of a disease or health related condition. There are three ways in which FDA exercises its oversight in determining which health claims may be used on a label or in labeling for a conventional food or dietary supplement viz. (i) the 1990 Nutrition Labeling and Education Act (NLEA) provides for FDA to issue regulations authorizing health claims for foods and dietary supplements after reviewing and evaluating the scientific evidence, either in response to a health claim petition or on its own initiative; (ii) the 1997 Food and Drug Administration Modernization Act (FDAMA) provides for health claims based on an authoritative statement of the National Academy of Sciences or a scientific body of the U.S. government with responsibility for public health protection or nutrition research; such claims may be used 120 days after a health claim notification has been submitted to FDA, unless the agency has informed the notifier that the notification does not include all the required information; and (iii) as described in FDA's guidance entitled "Interim Procedures for Qualified Health Claims in the Labeling of Conventional Human Food and Human Dietary Supplements", the agency reviews petitions for qualified health claims where the quality and strength of the scientific evidence falls below that required for FDA to issue an authorizing regulation. If FDA finds that the evidence supporting the proposed claim is credible and the claim can be qualified to prevent it from misleading consumers, the agency issues a letter of enforcement discretion specifying the qualifying language that should accompany the claim and describing the circumstances under which it intends to exercise enforcement discretion for use of the claim in food labeling.

A "health claim" by definition has two essential components viz. (i) a substance (whether a food, food component, or dietary ingredient), and (ii) a disease or health related condition. A statement lacking either one of these components does not meet the regulatory definition of a health claim. For example, statements that address a role of dietary patterns or of general categories of foods (e.g., fruits and vegetables) in maintaining good health are considered to be dietary guidance rather than health claims. Dietary guidance statements used on food labels must be truthful and non-misleading. Statements that address a role of a specific substance in maintaining normal healthy structures or functions of the body are considered to be structure/function claims. Unlike health claims, dietary guidance statements and structure/function claims are not subject to pre-market review and authorization by FDA.

Summary of the United States regulatory framework

It is very difficult to obtain health claim in USFDA and manufacturers need to be prepared to invest significant resources in order to accomplish this task. As far as dietary supplements are concerned, manufacturers have broad freedom in making claims on their products without any pre-market authority which is not seen in other jurisdictions. Although this freedom is limited to products meant to supplement the diet and not on food products per se, this distinction has not prevented products appearing as conventional foods from being sold as dietary supplements. The lack of pre-market approval means the FDA can not perform enforcement action on these products until after they have reached the market. Even then, the burden of proof is upon the FDA to demonstrate the illegality of a product.

The European Union's Regulatory Framework

As a result of increasing interest in the concept of "Functional Foods" and "Health Claims", the European Union set up a European Commission Concerted Action on Functional Food Science in Europe (FUFOSE). The programme was coordinated by the International Life Sciences Institute (ILSI) Europe and the aim was to develop and establish a science based approach to the evidence needed to support the development of food products that can have a beneficial effect on an identified physiological function in the body, that can improve an individual's state of health and well being and/or reduce the risk of disease [2, 6]. The FUFOSE project looked at six areas of science and health, which are growth, development and differentiation, substrate metabolism, defence against reactive oxidative species, functional foods and the cardiovascular system, gastro intestinal physiology and function, and the effects of foods or behaviour and psychological performance. The final document was published in the British Journal of Nutrition [8].

The report takes the position that functional foods should be in the form of normal foods and they must demonstrate their effects in amounts that can normally be expected to be consumed in the diet. A functional food can be a natural whole food, a food to which a component has been added, or a food from which a component has been removed by technological or biotechnological means. It can also be a food in which the nature of one or more components has been modified, or a food in which the bioavailability of one or more components has been modified, or any combination of these possibilities. A functional food may be targeted at the whole population or for particular groups, which may be defined, for example, by age or by genetic constitution.

The EU Commission Concerted Action supports the development of two types of health claims relevant to functional foods, which must always be valid in the context of the whole diet and must relate to the amounts of foods normally consumed [9]. These are:

(i) Type A: "Enhanced function"claims that refer to specific physiological, psychological functions and biological activities beyond their established role in growth, development and other normal functions of the body. This type of claim makes no reference to a disease or a pathological state, e.g. certain non-digestible oligosaccharides improve the growth of a specific bacterial flora in the gut; caffeine can improve cognitive performance.

(ii) Type B: "Reduction of disease-risk"claims that relate to the consumption of a food or food component that might help reduce the risk of a specific disease or condition because of specific nutrients or non-nutrients contained within it (e.g. folate can reduce a woman's risk of having a child with neural tube defects, and sufficient calcium intake may help to reduce the risk of osteoporosis in later life).

The FUFOSE conclusions and principles need to be implemented. Therefore, a new EU Commission Concerted Action programme, the Process for the Assessment of Scientific Support for Claims on Foods (PASSCLAIM) project is aiming to resolve some of the ongoing issues of validation, scientific substantiation of claims and communication to the consumer.

The project started with and will build upon the principle that "enhanced function" and "reduced risk of disease" claims should be based on well designed studies using appropriately identified, characterised and validated biomarkers. PASSCLAIM aims to establish common criteria to assess the scientific substantiation of health claims, providing the framework to prepare scientific dossiers supporting claims. The PASSCLAIM consensus document will assist those making claims, those who regulate claims and it will also improve the credibility of claims for consumers. This integrated strategy will generate more consumer confidence in the science base related to claims on foods and will better address the concerns of consumers.

Although there is no European legislation regarding safety of functional foods as such, the food safety aspects are already covered by existing EU regulations. However, foods with health claims must consider the overall dietary significance, including the amount and frequency of consumption, any potential interactions with other dietary constituents, any impact on metabolic pathways and potential for adverse effects, including allergy and intolerance factors.

Regulation EC No. 258/97 (Regulation Concerning Novel Foods and Novel Food Ingredients)

Regulation (EC) No. 258/97 of the European Parliament and of the Council of 27 January 1997 concerning novel foods and novel food ingredients lays down an authorization procedure for a number of foods or food ingredients that have not been consumed to a significant degree in the European Union. In order to protect public health, it was deemed necessary to ensure that such novel foods and novel food ingredients are subject to a safety assessment through a Community procedure before they are placed on the market. This procedure is, in essence, a procedure managed by the member states. But in cases where concern is raised by Member States because of public health or food safety concern (which is rather the rule than the exception), the community procedure of risk analysis comes into play. Products defined as novel foods may be:

- Foods and food ingredients with a new or intentionally modified primary molecular structure;
- Foods and food ingredients consisting of or isolated from microorganisms, fungi or algae;
- Foods and food ingredients consisting of or isolated from plants and food ingredients isolated from animals, except for foods and food ingredients obtained by traditional propagating or breeding practices and having a history of safe food use;

- Foods and food ingredients to which has been applied a production process not currently used, where that process gives rise to significant changes in the composition or structure of the foods or food ingredients which affect their nutritional value, metabolism or level of undesirable substances.

The regulation sought to protect public health by ensuring that new or novel foods or food ingredients in Europe would be subjected to a safety assessment before being placed in the market. The deciding factor on whether a food would be considered novel is whether it has already been sold under food law in the EU market prior to 1997. To show that a food or food ingredient is not novel, the manufacturer needs to prove significant sale of the product in the EU as food prior to May 1997. Otherwise, the manufacturer is required to submit the product for authorization prior to introducing it to the market.

The Novel Foods Regulation provides two different procedures for authorization viz. a simplified procedure for substantially equivalent food or food ingredients and a full authorization procedure for all other foods and food ingredients. In order to use the simplified procedure, the applicant needs to provide evidence showing the novel food or food ingredient is substantially equivalent to an existing food ingredient in regards to composition, nutritional value, metabolism, intended use and level of undesirable substances (European Union Publications Office, 1997). Foods can show substantial equivalence through generally recognized scientific evidence or from an opinion of a competent food assessment body in the member states. The regulation does not clearly define what is meant by generally recognized scientific evidence, so all the applications submitted thus far have relied on opinions of food assessment bodies.

Regulation EC No 1924/2006 (Regulation on Health and Nutrition Claims)

EC 1924/2006, also known as the Health and Nutrition Claims Regulation, was published in the official journal of the EU at the end of December 2006 (European Union Publications Office, 2007a). Prior to passage, each state had their own framework allowing, or in some cases preventing, health and nutrition claims on foods. Like the previously mentioned pieces of legislation, the Health and Nutrition Claims Regulation strives to harmonize the rules and procedures and allow for consistent application of them throughout all the member states. The goals of free movement of goods across all member states and the assurance of safety and reliance of scientific evidence are very prominent themes in this piece of legislation. All previous frameworks used by individual states were required to conform to this framework. Upon passage of the regulation, claims previously approved by the individual states were no longer valid and had to be reassessed under the new principles set out by this regulation. Under this regulation, nutrition and health claims are only permitted if the following conditions are fulfilled:

- The presence, absence or reduced content in a food or category of food of a nutrient or other substance in respect of which the claim is made has been shown to have a beneficial nutritional or physiological effect, as established by generally accepted scientific evidence.
- The nutrient or other for which the claim is made is contained in the final product in a significant quantity as defined in community legislation or, where such rules do not exist, in a quantity that will produce the nutritional or physiological effect claimed as established by generally accepted scientific data; or is not present or is present in a reduced quantity that will produce the nutritional or physiological effect claimed as established by generally accepted scientific data.
- Where applicable, the nutrient or other substance for which the claim is made is in a form that is available to be used by the body; the quantity of the product that can reasonably be expected to be consumed provides a significant quantity of the nutrient or other substance to which the claim relates, as defined in community legislation or, where such rules do not exist, a significant quantity that will produce the nutritional or physiological effect claimed as established by generally accepted scientific evidence.

Summary of the EU's regulatory framework

In the consensus document on "Scientific Concepts of Functional Foods in Europe" of the European Commission Concerted Action on Functional Food Science in Europe (FUFOSE) the following working definition was used: "A food can be regarded as 'functional' if it is satisfactorily demonstrated to affect beneficially one or more target functions in the body, beyond adequate nutritional effects, in a way that is relevant to either an improved state of health and well being and/or reduction of risk of disease. Functional foods must remain foods and they must demonstrate their effects in amounts that can normally be expected to be consumed in the diet; they are not pills or capsules, but part of a normal food pattern." or "A functional food can be a natural food, a food to which a component has been added, or a food from which a component has been removed by technological or biotechnological means. It can also be a food where the nature of one or more components has been modified, or a food in which the bioavailability of one or more components has been modified, or any combination of these possibilities."

A functional food might be functional for all members of a population or for particular groups of the population, which might be defined, for example, by age or by genetic constitution. The amount of intake and form of the functional food should be as it is normally expected for dietary purposes. Therefore, it could not be in the form of pill or capsule just as normal food form [2].

There is not, as such, a regulatory framework for 'functional foods' or 'nutraceuticals' in EU Food Law. The rules to be applied are numerous and depend on the nature of the foodstuff. The rules of the General Food Law Regulation, including responsibility for food safety, traceability, recall and notification, definitely are applicable to all foods. In the EU, rather than regulating the product group per se, legislative efforts currently being developed are directed towards restricting the use of health claims on packages and in marketing. According to the EU regulation on nutrition and health claims made on foods (EC No. 1924/2006), a list of authorized claims has to be published for all member states, and nutrient profiles also has to be established for foods containing health claims. Health claims can be "function claims" and "reduction of disease risk claims". There is no single framework for functional food regulation in the EU. Decisions within the EU rely heavily on scientific evidence. The safety, efficacy and quality of products are all accessed through review of scientific evidence. The regulatory framework in relation to functional food is still in the process of evolving [2].

Japanese Regulatory Framework

The concept of functional foods first evolved in Japan in the mid 1980s. The regulation of foods and drugs in Japan is the responsibility of the Ministry of Health Labor and Welfare (MHLW) (Ministry of Health, Labor and Welfare 2006). The Japanese Ministry of Health, Labor, and Welfare (MHLW) set up 'Foods for Specified Health Use' (FOSHU) in 1991 as a regulatory system to approve the statements made on food labels concerning the effect of the food on the human body. Food products applying for approval by FOSHU are scientifically evaluated in terms of their effectiveness and safety by the Council of Pharmaceutical Affairs and Food Hygiene under the MHLW [10]. The regulatory range of FOSHU was broadened in 2001 to accept the forms of capsules and tablets in addition to those of conventional foods. FOSHU increased the total to about 330 items in January 2003. Functional food and nutraceutical products are classified as either drugs or non-drugs. Non drugs are further classified as foods with health claims (FHCs) or general foods. FHCs are further classified based on the type of claim used, either foods with nutrient function claims (FNFC) or foods for specified health uses (FOSHU) [11].

Foods for specified health uses

FOSHU refers to foods containing ingredient with functions for health and officially approved to claim its physiological effects on the human body [12, 13]. FOSHU is intended to be consumed for the maintenance/promotion of health or special health uses by people who wish to control health conditions, including blood pressure or blood cholesterol. In order to sell a food as FOSHU, the assessment for the safety of the food and effectiveness of the functions for health is required, and the claim must be approved by the MHLW [14, 15]. To gain FOSHU approval, the safety and efficacy of the product must be demonstrated and the claim must be approved by the MHLW. The requirements listed by the MHLW are as follows:

- Effectiveness on the human body is clearly proven.
- Absence of any safety issues (animal toxicity tests, confirmation of effects in cases of excess intake, etc.
- Use of nutritionally appropriate ingredients (e.g. no excessive use of salt, etc.).
- Guarantee of compatibility with product specifications by the time of consumption.
- Established quality control methods, such as specifications of products and ingredients, processes and methods of analysis.

The flow chart below shows the steps in FOSHU approval process (Fig. 18.1). Demonstration of efficacy requires evidence from both animal and human studies. The evaluation of the efficacy evidence is performed by the Council on Pharmaceutical Affairs and Food Sanitation division of the MHLW. The safety of the product is evaluated by the Food Safety Commission. Safety is typically demonstrated through toxicity studies and presentation of historical dietary use of the product and ingredient in Japan and foreign countries [4, 10].

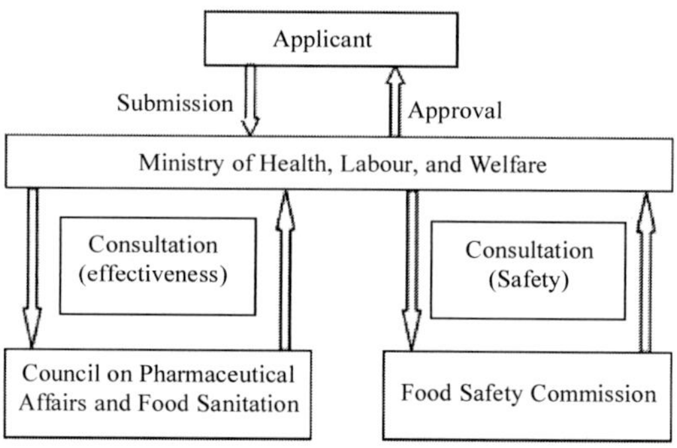

Fig. 18.1: Flow chart of the applicant

Qualified FOSHU

Food with health function which is not substantiated on scientific evidence that meets the level of FOSHU, or the food with certain effectiveness but without established mechanism of the effective element for the function will be approved as qualified FOSHU.

Standardized FOSHU

Standards and specifications are established for foods with sufficient FOSHU approvals and accumulation of scientific evidence. Standardized FOSHU are approved when it meets the standards and specifications.

Reduction of disease risk FOSHU

Reduction of disease risk claim is permitted when reduction of disease risk is clinically and nutritionally established in an ingredient.

There are currently nine categories of approved FOSHU products, which are as follows.

- Foods to modify gastrointestinal conditions
- Foods related to blood cholesterol level
- Foods related to blood pressure
- Foods related to dental hygiene
- Cholesterol plus gastrointestinal conditions, triacylglycerol plus cholesterol
- Foods related to mineral absorption
- Foods related to osteogenesis
- Foods related to triacyglycerol.

Summary of Japanese regulatory system

The Japanese were the first to have regulatory framework in place for functional foods. The current Japanese system for regulation of health foods, designated FHC, consists of 2 categories viz. FNFC and FOSHU. The label FNFC may be freely used if a product satisfies the standard for the minimum and maximum levels per daily portion consumed. In terms of approval, the Food Safety Commission examines the safety of the product, and the Pharmaceutical Affairs and Food Sanitation Council evaluates its effectiveness. Subsequently, the MHLW individually approves claims, which allows the manufacturer to officially carry the claim and special FOSHU logotype on their product. Getting a product with a health claim into the Japanese market is not a simple thing. Safety and efficacy needs to be demonstrated both on animal and humans. Any manufacturer who applies to the government for approval under the FOSHU code for its product must tabulate both published literature and internal reports on the effectiveness of the product and/or its ingredients and provide a summary of each available publication or report. The tabulation must include *in-vitro* metabolic and biochemical studies, *in-vivo* studies, and randomized controlled trials on Japanese people. However, the MHLW has recognized the concern regarding generation of this data and is looking at ways to make claims easier to get.

Australian and New-Zealand Regulatory Framework

In Australia, health claims are split into two groups viz. general level claims and high level claims. General level claims are equivalent to structure/ function claims, in that they describe the relationship between a nutrient and a health function. High level health claims are disease risk reduction claims. They elucidate a relationship between the consumption of a nutrient and the risk of a disease [16].

Food Standards Australia New Zealand (FSANZ) has evaluated seven high level health claims and approved five of those seven. The proposed approved health claims are:

- Sodium (with or without potassium) and heart disease
- Fruits and vegetables and heart disease
- Saturated and /or trans fat and cholesterol or heart disease
- Calcium (with or without vitamin D) and osteoporosis
- Folic acid and neural tube defects.

Rejected health claims sought to establish a relationship between whole grains and heart disease and between ω-3 fatty acids and heart disease [17]. Before using a new high level health claim that is not currently approved, the firm must obtain pre-market approval from FSANZ who will evaluate whether or not there is "convincing" scientific evidence for the claim. If the evidence is less convincing, (for example, "probable" or "insufficient") then the claim will not be approved. General level health claims do not require pre-market approval but they do require scientific substantiation (similar to the US situation). FSANZ has a list of approved general level claims that count as scientific substantiation for firms wishing to use those claims. Examples of general level health claims are:

- Vitamin D and calcium for phosphorus utilization and absorption
- Vitamin E as antioxidant
- Vitamin K for proper coagulation
- Thiamine and normal metabolism of carbohydrates
- Riboflavin and metabolism
- Niacin and metabolism
- Pantothenic acid and metabolism of fat
- β-glucan and reduced blood cholesterol.

Nutrient content claims are also permitted. New nutrient content claims have come into force with the implementation of the new standard 1.2.7 in January 2013. No specific wordings were provided for claims about ω fatty acids (any), however, "good source" and "increased" are used for ω-3, while "increased" is used for ω-6, ω-9 and polyunsaturated fatty acids. There is a minimum quantity required to make a claim on ω fatty acids. Australia and New Zealand are the only countries to regulate and authorize the use of gluten content claims. The "free" claims in Australia and New Zealand are also much stricter than other countries, requiring no detectable levels of nutrients for those claims. There are also regulations for nutrition labelling. The Australia New Zealand Food Standards Code sets out regulations for "nutrition information panels" in Standard 1.2.8. Novel foods must also undergo a registration process. Novel foods are foods without a history of safe use. These foods undergo a safety assessment to determine if they are safe to eat. A unique factor in the Australia and New Zealand case is that the firm applying for permission for a novel food can receive exclusive access to the Australia and New Zealand markets for that food for 15 months.

Approval is granted on a product-by-product basis. This approval is contingent upon evidence of both safety and efficacy of the claim because complementary medicines must possess claims. Claims are split into two types with different levels of evidence for the two types. One type is traditional use evidence and another is scientific evidence. In Australia, nutraceuticals are regulated like a drug and involve product specific claims. In New Zealand, nutraceuticals are called dietary supplements and are treated more or less like food. Moreover, there is no pre-market approval necessary for dietary supplements in New Zealand [3].

Indian Regulatory Framework

The act that regulate the functional foods and nutraceuticals in India is known as Food Safety and Standards Act, 2006; this act consolidates 8 laws governing the food sector.The Food Safety and Standards Authority of India (FSSAI) is the implimenting authority. The FSSAI would consist of a Chairperson and 22 members. FSSAI regulates manufacturing, distribution and sale of nutraceuticals, functional foods and dietary supplements in India [18]. However, due to lack of clarity of specific regulations for registration of nutraceuticals, permitted additives, etc, entrepreneur intending to launch nutraceutical in India still face challenges. Though the structure function claims are permitted, there is no clarity as to the permitted structure-function claims for nutraceuticals and dietary supplements.

New regulations have come into effect in January 2013, requiring the labelling of products containing biotechnology ingredients. Also, India has embarked on a process to harmonize Indian food standards with the Codex standards and other international best practices. This process was initiated in January 2013 with scheduled completion and implementation by December 2014 [19]. Any claims that imply medicinal and therapeutic functions are not permitted by law. However, nutrition structure-function claims that are not misleading can generally be used on these products. There has to be enough scientific evidence to support the claim. While there is no pre-market approval for these products, the responsibility lies with the manufacturer to ensure that all claims satisfy the stipulated regulations [19].

Labelling of these products is mandatory and it is an offence to sell a product without a proper label. All information on the label should be truthful and must not mislead the consumer. Information about ingredients, nutrients and quantity of the product on the label must not be different from the actual product in the package. Labels that falsely represent the product to promote its sale and give the public any guarantee of effectiveness that is not scientifically justified are not permitted. In order to enter Indian nutraceutical market, some of the very important areas to focus include product evaluation, actual product analysis, procuring licenses and developing India specific health and label claims.

Product evaluation

In order to perform product assessment as per Indian regulatory definition, it is of utmost importance to examine each active ingredient and additive in the context of permissibility, standards and dosage of vitamins/minerals allowed as per the therapeutic, prophylactic or recommended daily allowance for Indians.

There is confusion amongst manufacturers whether their products will be classified as food or food supplement or drug in the context of the Prevention of Food Adulteration Act, 1954 and Rules, 1955, Food Safety and Standards Act, 2006 and Drugs and Cosmetics Act, 1940 and Rules, 1945.The Food Safety and Standards Rules, 2011 highlights regulatory enforcement structure and procedures which Central Government proposes to make .

Licenses

Though new FSSAI promises to simplify licensing and registration processes for nutraceuticals, the actual process vary depending on number of parameter. To get the product registered in India, number of licenses (almost 4-5) will be required depending on the actual product status. Number of documents will have to be furnished by the food importer to the government authority along with registration application dossiers. Interlink through its regulatory product offerings provides regulatory support for following licensing procedures which need to be taken care of before launching these products in India.

Health and label claim

Developing health and label claims specific to Indian regulatory guidelines is the major element to be focused while entering Indian market. Based on the results of regulatory assessment of the product, India specific label content and claims are developed.

Codex Alimentarius

The Codex Alimentarius has defined two types of claims viz. (i) nutrition claim, and (ii) health claim.

Nutrition claim

Guidelines for the use of nutrition claim by the Codex Committee on Food Labeling proposed that 'nutrient claim means any representation which states, suggests or implies that a food has particular nutritional properties including but not limited to the energy value and to the content of protein, fat and carbohydrate, as well as the content of vitamins and minerals'.

It includes (i) nutrient content claim that describes the level of a nutrient contained in a food, and (ii) nutrient comparative claim that compares the nutrient levels and/or energy value of two or more foods.

Health claim

Guidelines for the use of nutrition claims by the Codex Committee in Food Labelling proposed that "health claim means any representation which states, suggests or implies that a relationship exits between a food or a constituent of that food and health". It includes (i) nutrient function claim that describes the physiological role of the nutrient in growth, development, and the normal function of the body, (ii) enhanced function claim that concerns specific beneficial effects of the consumption of foods and their constituents in the context of the total diet and relate to a positive contribution to health or to improvement of a function or to modifying or preserving health, and (iii) disease risk reduction claim which relates to the consumption of a food or food constituent in the total diet, to the reduced risk of developing a disease or a health-related condition. Risk reduction means significantly altering a major risk factor(s) for a disease or a health related condition. Diseases have multiple factors and altering one of these risk factors may or may not have a beneficial effect [2,7].

Summary

Functional foods are found virtually in all food categories, however products are not homogeneously scattered over all segments of the growing market. The development and commerce of these products is rather complex, expensive and risky, as special requirements should be answered. Besides potential technological obstacles, legislative aspects, as well as consumer demands need to be taken into consideration when developing functional food. In particular, consumer acceptance has been recognized as a key factor to successfully negotiate market opportunities. Consumer acceptance of the concept of functional foods has been widely recognized as key success factors for market orientation, consumer led product development and successfully negotiating market opportunities. The development and commerce of functional food products is rather complex, expensive and risky, as special requirements should be answered. In the case of a successful product development attention should be paid both to consumer demands and technical conditions. Further more, the legislation background should not be neglected; the companies should meet the special requirements occurring during the development and marketing of functional foods. They should possess adequate R&D activities, proper know-how and economic potential to introduce branded products to the market.

References

1. Siro, I., *et al.*, *Functional food. Product development, marketing and consumer acceptance—A review.* Appetite, 2008. **51**(3): 456-467.
2. Roberfroid, M., *Global view on functional foods: European perspectives.* British Journal of Nutrition, 2002. **88**(S2): 133-138.
3. Wildman, R.E.C., *Handbook of nutraceuticals and functional foods* 2006: CRC Press.
4. Bailey, R., *Functional Foods in Japan: FOSHU (Foods for Specified Health Uses) and Foods with Nutrient Function Claims.* Regulation of functional foods and nutraceuticals: A global perspective, 2008: 247.
5. Milner, J.A., *Functional foods: the US perspective.* The American Journal of Clinical Nutrition, 2000. **71**(6): 1654-1659.
6. Hasler, C.M., *Functional foods: Benefits, concerns and challenges—A position paper from the American Council on Science and Health.* The Journal of Nutrition, 2002. **132**(12): 3772-3781.
7. Bagchi, D., *Nutraceutical and functional food regulations in the United States and around the world* 2008: Academic Press.
8. Bech-Larsen, T. and J. Scholderer, *Functional foods in Europe: consumer research, market experiences and regulatory aspects.* Trends in Food Science & Technology, 2007. **18**(4): 231-234.
9. Hasler, C.M., *The changing face of functional foods.* Journal of the American College of Nutrition, 2000. **19**: 499-506.
10. Shimizu, T., *Health claims on functional foods: the Japanese regulations and an international comparison.* Nutrition Research Reviews, 2003. **16**(02): 241-252.
11. Arai, S., *et al.*, *A mainstay of functional food science in Japan—history, present status, and future outlook.* Bioscience, Biotechnology, and Biochemistry, 2001. **65**(1): 1-13.
12. Saito, M., *Role of FOSHU (food for specified health uses) for healthier life.* Yakugaku zasshi: Journal of the Pharmaceutical Society of Japan, 2007. **127**(3): 407-416.
13. Bailey, R., *15 FunctionalFoodsinJapan: FOSHU ("Foods for Specified Health Uses") and "Foods with Nutrient Function Claims".* Regulation of functional foods and nutraceuticals: A Global Perspective, 2008: 247.
14. Shimizu, T., *Newly established regulation in Japan: foods with health claims.* Asia Pacific Journal of Clinical Nutrition, 2002. **11**(2): S94-S96.
15. Tokunaga, T., *[A present situation of Foods for Specified Health Use (FOSHU) in Japan].* Nihon yakurigaku zasshi. Folia Pharmacologica Japonica, 1997. **110**: 17P-22P.
16. Tapsell, L.C., *Evidence for health claims: A perspective from the Australia–New Zealand region.* The Journal of Nutrition, 2008. **138**(6): 1206S-1209S.
17. Maubach, N. and J. Hoek, *A qualitative study of New Zealand parents' views on front of pack nutrition labels.* Nutrition & Dietetics, 2010. **67**(2): 90-96.
18. Umali Deininger, D. and M. Sur, *Food safety in a globalizing world: opportunities and challenges for India.* Agricultural Economics, 2007. **37**(s1): 135-147.
19. FSSAI, *Food Safety and Standards Act 2006, Food Safety and Standards Authority, Government of India, New Delhi,* 2013.